Lipids and Tumors

Progress in Biochemical Pharmacology

Vol. 10

Series Editor: R. Paoletti, Milan

S. Karger · Basel · München · Paris · London · New York · Sydney

Lipids and Tumors

Edited by K. K. Carroll, Department of Biochemistry, University of Western Ontario, London, Ont.

List of Contributors: F. M. Archibald, New York, N.Y.; M. Barclay, New York, N.Y.; K. K. Carroll, London, Ont.; S. Hakomori, Seattle, Wash.; B. V. Howard, W. J. Howard, Philadelphia, Pa.; J. Hradec, Prague; H. T. Khor, London, Ont.; J. R. Sabine, Adelaide, S.A.; V. P. Skipski, New York, N.Y.; C. Snyder, F. Snyder, Oak Ridge, Tenn.; A. A. Spector, Iowa; C. C. Stock, New York, N.Y.; J. F. Weiss, New York, N.Y.

32 figures and 60 tables, 1975

S. Karger · Basel · München · Paris · London · New York · Sydney

Progress in Biochemical Pharmacology

Vol. 4: Recent Advances in Atherosclerosis. International Symposium, Athens 1966. Editors: MIRAS, C. J. (Athens); HOWARD, A. N. (Cambridge) and PAOLETTI, R. (Milan). X + 633 p., 236 fig., 151 tab., 2 cpl., 1968. ISBN 3-8055-0383-0
Vol. 5: Synthesis and Use of Labelled Lipids and Sterols. Symposium, Milan 1968. Editor: GROSSI-PAOLETTI, E. (Milan). VIII + 176 p., 93 fig., 40 tab., 1969. ISBN 3-8055-0384-9
Vol. 6: Biochemistry and Pharmacology of Free Fatty Acids. Editors: HOLMES, W. L. and BORTZ, W. M. (Philadelphia, Pa.). XII + 395 p., 55 fig., 41 tab., 1971. ISBN 3-8055-1211-2
Vol. 7: Drugs Affecting Kidney Function and Metabolism. Editor: EDWARDS, K. D. G. (Sydney). XIV + 538 p., 73 fig., 20 tab., 1972. ISBN 3-8055-1386-0
Vol. 8: Effect of Carbohydrates on Lipid Metabolism. Editor: MACDONALD, I. (London). X + 314 p., 24 fig., 38 tab., 1973. ISBN 3-8055-1600-2
Vol. 9: Drugs and the Kidney. Editor: EDWARDS K. D. G. (Sydney). XI + 274 p., 64 fig., 43 tab., 1974. ISBN 3-8055-1693-2

S. Karger · Basel · München · Paris · London · New York · Sydney
Arnold-Böcklin-Strasse 25, CH-4011 Basel (Switzerland)

All rights, including that of translation into other languages, reserved.
Photomechanic reproduction (photocopy, microcopy) of this book or parts thereof without special permission of the publishers is prohibited.

© Copyright 1975 by S. Karger AG, Verlag für Medizin und Naturwissenschaften, Basel
Printed in Switzerland by Buchdruckerei Graphische Anstalt Schüler AG, Biel
ISBN 3-8055-1708-4

Contents

Glycerolipids and Cancer
F. SNYDER and C. SNYDER, Oak Ridge, Tenn.

I. Introduction	2
II. Occurrence of Ester and Ether Glycerolipids in Neoplastic Cells	3
A. Nonphosphorus Lipids	3
1. General	3
2. Triacylglycerols	4
3. Alkyl Analogs of Triacylglycerols	5
4. Alk-1-Enyl Analogs of Triacylglycerols	7
5. Diacylglycerols, Monoacylglycerols, and their Ether Analogs	8
B. Phosphorus-Containing Lipids	10
1. General	10
2. Subfractions of Ethanolamine and Choline Phospholipid Classes	14
III. Pathways for Ester and Ether Glycerolipids and their Significance in Neoplastic Cells	17
A. General	17
B. Biosynthesis	17
1. The *sn*-1-Position	17
a) Acyl Moieties	17
b) Alkyl Moieties	18
c) Alk-1-Enyl Moieties	19
2. The *sn*-2-Position	20
3. The *sn*-3-Position	23
C. Catabolic Enzymes	23
1. The *sn*-1-Position	23
a) Acyl Moieties	23
b) Alkyl Moieties	25
c) Alk-1-Enyl Moieties	25
2. The *sn*-2-Position	26
3. The *sn*-3-Position	26

IV. Regulation of Glycerolipids and their Precursors in Cancer Cells 27
   A. The Glycerol Portion of Lipids 27
   B. Biosynthesis of Alkyldihydroxyacetone-P 28
   C. Biosynthesis of Fatty Alcohols 29
   D. Enzymic Cleavage of Ether Lipids 29
   E. Factors Affecting the Formation of Ester- and Ether-Linked Lipids in Intact Cells 30
V. Conclusion 31
VI. References 31

Fatty Acid Metabolism in Tumors
A. A. SPECTOR, Iowa City, Iowa

I. Introduction 43
II. Sources of Tumor Fatty Acids 44
   A. Fatty Acid Biosynthesis in Tumors 44
      1. *De novo* Biosynthesis 45
      2. Desaturation and Chain Elongation 46
         a) Essential Fatty Acids 46
      3. Regulation of Fatty Acid Biosynthesis 48
      4. Glucose Incorporation into Lipids 49
      5. Relative Contributions of Biosynthesis and Host Lipids 50
   B. Lipids Supplied by the Host 51
      1. Blood Plasma Lipids 51
         a) Role of Dietary Lipids 52
      2. Ascites Plasma Lipids 52
         a) Comparison of Cell and Extracellular Fluid Fatty Acid Composition 54
III. Free Fatty Acid Utilization by Tumors 54
   A. Free Fatty Acid Utilization *in vivo* 55
   B. Mechanism of Free Fatty Acid Utilization in Tumors 56
      1. Uptake 57
      2. Oxidation 58
      3. Esterification 60
IV. Esterified Fatty Acid Utilization by Tumors 62
   A. Lipoprotein Triglycerides 62
V. Turnover of Esterified Fatty Acids in Tumors 64
   A. Free Fatty Acid Release from Tumor Cells 65
VI. Summary and Comment 66
   A. Future Directions 67
VII. Acknowledgments 69
VIII. References 69

Lipoproteins in Relation to Cancer
M. BARCLAY and V. P. SKIPSKI, New York, N.Y.

I. Introduction	76
II. Lipoproteins in Serum or Plasma, Determinations and Quantitation	77
III. Selection and Description of Normal Subjects and Patients with Cancer	79
IV. Lipoproteins in Serum from Normal Subjects and Patients with Cancer	81
A. Classes of Lipoproteins	81
1. Very Low-Density Lipoproteins (VLDL)-Chylomicra or Particles	81
2. Low-Density Lipoproteins (LDL or β-Lipoproteins)	86
3. High-Density Lipoproteins (HDL_2 and HDL_3) or α-Lipoproteins	89
4. Albumin-Bound Free Fatty Acid Complexes ('Ultracentrifugal Residue'; Very High-Density Lipoproteins)	96
5. Commentary	97
V. Lipoproteins in Serum or Plasma from Animals with Experimental Tumors	100
A. Classes of Lipoproteins	100
1. Chylomicra or Particles, and the VLDL	100
2. Low-Density Lipoproteins (LDL or β-Lipoproteins)	101
3. High-Density Lipoproteins (HDL_2 and HDL_3) or α-Lipoproteins	102
VI. Summary	106
VII. Acknowledgments	107
VIII. References	107

Tumor Proteolipids
V. P. SKIPSKI, M. BARCLAY, F. M. ARCHIBALD and C. C. STOCK, New York, N.Y.

I. Introduction	112
II. Presence of Neoproteolipids in Tumors	113
III. Neoproteolipid-W	115
A. Chemical Composition	115
B. Occurrence of Neoproteolipid-W in Tissues of Normal and Cancer-Bearing Rats	118
IV. Neoproteolipid-S	123
V. Neoproteolipids in Serum of Cancer Patients	124
VI. Discussion	126
VII. Summary	130
VIII. Acknowledgments	131
IX. References	131

Lipids in Normal and Tumor Cells in Culture
B. V. HOWARD and W. J. HOWARD, Philadelphia, Pa.

I. Introduction	135
II. Nutrition	138

III.	Lipid Composition	140
	A. Major Lipid Classes	140
	B. Fatty Acids	141
	C. Lipid Ethers	144
IV.	Lipid Metabolism	144
	A. Lipid Biosynthesis	144
	B. Regulation of Lipid Metabolism	147
V.	Membranes	151
	A. Evidence for Alteration after Transformation	151
	B. Lipid Composition of Membranes	152
	C. Phospholipid Metabolism	152
	D. Glycolipid Metabolism	153
	E. Lipid Transport	154
VI.	Summary	156
VII.	References	157

Fucolipids and Blood Group Glycolipids in Normal and Tumor Tissue
S. HAKOMORI, Seattle, Wash.

I.	Introduction	167
II.	Fucolipids as a New Glycolipid Class and Cell Surface Marker	169
III.	Fucolipids as a Blood Group Hapten of Human Erythrocyte Membrane	170
IV.	Fucolipids of Gastrointestinal Tract and Glandular Tissues	175
V.	Fucolipids of Tumor Tissue	177
	A. Blood Group A and B Fucolipids and Glycoproteins in Tumors	180
	1. Deletion of A and B Determinants	180
	2. Blocked Synthesis of A and B Determinants in Tumors	182
	B. Lewis Fucolipids in Tumors	185
	C. Presence of Incompatible Blood Group Antigens in Human Tumors	186
	D. Fucolipid Changes in Transformants *in vitro*	187
	E. Relation of 'Carcinoembryonic' Antigen and Blood Group Substances	188
VI.	Epilogue	188
VII.	References	190

The Role of Cholesteryl 14-Methylhexadecanoate in Gene Expression and its Significance for Cancer
J. HRADEC, Prague

I.	Introduction	197
II.	Cholesteryl 14-Methylhexadecanoate and Protein Synthesis	199
	A. Formation of the Aminoacyl-tRNA Complex	200
	B. Transcription of the Genetic Message	208
	C. Translation of the Genetic Message	208
III.	Cholesteryl 14-Methylhexadecanoate and Malignant Growth	217

Contents

IV. The Significance of Cholesteryl 14-Methylhexadecanoate for Malignant Growth 219
V. References 222

Sterols and Other Lipids in Tumors of the Nervous System
J. F. WEISS, New York, N.Y.

I. Introduction 228
II. Developmental Changes in Neural Lipids 230
   A. Characteristic Neural Lipids 230
   B. Sterol Synthesis and Composition 231
   C. Desmosterol 234
   D. Sterol Esters 236
   E. Effect of Hypocholesteremic Agents on the Developing Nervous System 237
III. Human Tumors of the Nervous System 238
   A. Classification of Tumors 238
   B. Changes in Lipid Classes, Neutral Lipids, and Fatty Acid Composition 239
   C. Sterol Composition and Synthesis and the Effect of Drugs 240
   D. Phospholipids 241
   E. Glycolipids 242
   F. Cerebrospinal Fluid 243
      1. Comparison of Blood and Cerebrospinal Fluid Lipids 245
      2. Cerebrospinal Fluid Lipids in Neoplasia 245
      3. Use of Triparanol to Augment CSF Desmosterol 246
IV. Models for Brain Tumor Research 248
   A. Primary Tumors 248
      1. Induction of Tumors in Animals 248
      2. Sterol Metabolism in Nitrosourea-Induced Tumors 250
   B. Transplanted Tumors 252
      1. Lipid Composition and Synthesis 252
      2. Sterol Metabolism and the Effect of Drugs 253
   C. Tissue Culture 254
      1. Lipid Metabolism 254
      2. Sterols and the Effect of Drugs 255
V. Summary 260
VI. Acknowledgments 260
VII. References 261

Defective Control of Lipid Biosynthesis in Cancerous and Precancerous Liver
J. R. SABINE, Adelaide, S.A.

I. Introduction and Significance 269

II. Defective Controls 270
   A. Hepatoma and Cholesterol Biosynthesis 270
   B. Hepatoma and Fatty Acid Biosynthesis 272
   C. Lipid Biosynthesis in Other Tumors 275
   D. Lipid Biosynthesis in Pretumorous Tissue 275
III. Mechanisms of Control 277
   A. Hepatic Cholesterol Synthesis................................... 278
   B. Hepatic Fatty Acid Synthesis 283
   C. Defective Controls in Hepatomas 286
IV. Defective Control and Carcinogenesis 290
   A. Possible Correlative (Causative?) Mechanisms 290
   B. Correlation with Other Precancerous Changes 293
   C. Some Doubts 294
V. Summary 295
VI. References 296

Dietary Fat in Relation to Tumorigenesis
K. K. CARROLL and H. T. KHOR, London, Ont.

I. Introduction.. 308
II. Studies with Experimental Animals 309
   A. Tumors Enhanced by Dietary Fat 309
      1. Skin Tumors in Mice 309
      2. Mammary Tumors in Mice and Rats............................ 313
      3. Hepatomas in Mice and Rats 318
      4. Other Tumors Enhanced by Dietary Fat 319
   B. Tumors not Enhanced by Dietary Fat 319
   C. Mode of Action of Dietary Fat 322
      1. Cocarcinogenesis and the Two-Stage Theory of Tumor Induction ... 322
      2. Effects on the Initiating Agent 323
      3. Promoting Action of Dietary Fat 325
III. Epidemiological Data on Humans 329
IV. Discussion 339
V. Summary 344
VI. Acknowledgments 345
VII. References 345
   Author Index 354
   Subject Index 373

Glycerolipids and Cancer[1]

FRED SNYDER and CATHY SNYDER

Medical Division, Oak Ridge Associated Universities, Oak Ridge, Tenn.

Contents

I. Introduction	2
II. Occurrence of Ester and Ether Glycerolipids in Neoplastic Cells	3
A. Nonphosphorus Lipids	3
1. General	3
2. Triacylglycerols	4
3. Alkyl Analogs of Triacylglycerols	5
4. Alk-1-Enyl Analogs of Triacylglycerols	7
5. Diacylglycerols, Monoacylglycerols, and their Ether Analogs	8
B. Phosphorus-Containing Lipids	10
1. General	10
2. Subfractions of Ethanolamine and Choline Phospholipid Classes	14
III. Pathways for Ester and Ether Glycerolipids and their Significance in Neoplastic Cells	17
A. General	17
B. Biosynthesis	17
1. The *sn*-1-Position	17
a) Acyl Moieties	17
b) Alkyl Moieties	18
c) Alk-1-Enyl Moieties	19
2. The *sn*-2-Position	20
3. The *sn*-3-Position	23
C. Catabolic Enzymes	23
1. The *sn*-1-Position	23
a) Acyl Moieties	23
b) Alkyl Moieties	25

[1] From the Medical Division, Oak Ridge Associated Universities, Oak Ridge, Tenn., under contract with the United States Atomic Energy Commission. This work was supported in part by the American Cancer Society, Grant No. BC-70E.

      c) Alk-1-Enyl Moieties 25
    2. The *sn*-2-Position 26
    3. The *sn*-3-Position 26
IV. Regulation of Glycerolipids and their Precursors in Cancer Cells 27
  A. The Glycerol Portion of Lipids 27
  B. Biosynthesis of Alkyldihydroxyacetone-P 28
  C. Biosynthesis of Fatty Alcohols 29
  D. Enzymic Cleavage of Ether Lipids 29
  E. Factors Affecting the Formation of Ester- and Ether-Linked Lipids
     in Intact Cells 30
V. Conclusion 31
VI. References 31

I. Introduction

Naturally occurring glycerolipids, those with phosphorylbase groups as well as so-called 'neutral' lipids, contain aliphatic moieties that are linked to the glycerol portion by ester and ether bonds. The latter linkage can be either alkyl[2] or alk-1-enyl[2] *(cis)* and is always in the *sn*-1-position of glycerol (D configuration) except for the 2,3-dialkyl-*sn*-glycerophosphatides that occur in halophilic bacteria. The pertinent literature in this field was discussed in a series of chapters on the chemistry and biology of the ether lipids [101].

Considerable work has been published on the glycerolipids of cancer cells. Much of it relates to composition, but in recent years, many metabolic studies have centered on glycerolipids containing ether linkages, since it is this group of lipids that is characteristically elevated in most tumors. The purpose here is to discuss data available on specific glycerolipid classes in cancer, and to compare these findings to those for normal tissues. Unfortunately, interpretation of much of the work in the field is complicated by the inadequate methods available at the time of the investigations or by the fact that the existence of ether and ester subfractions of the specific lipid classes studied was ignored. In addition, any comparison of lipid features of tumors and normal cells is difficult to make, because normal tissue is in all likelihood not homogeneously made

---

[2] The R groupings of the alkyl ($-OCH_2CH_2R$) and alk-1-enyl ($OCH=CHR$) ether-linked moieties can be saturated or unsaturated; see SNYDER [102] for a discussion of the nomenclature used for these lipids. Abbreviations: GPE = glycerophosphorylethanolamine; GPC = glycerophosphorylcholine.

up of the particular cell type considered to be the origin of the tumor, and the tumor cells may have dedifferentiated to such an extent that they are no longer validly comparable to 'normal' cells of origin. The large number of dying and necrotic cells in a rapidly proliferating neoplasm further confuses the issue. In the subsequent sections we discuss papers that have investigated individual molecular species of glycerolipids and attempt to generalize only when warranted. When discussing the phospholipids in tumors, we have specifically concentrated on those that contain choline and ethanolamine, since limited information is available on the other glycerolipids that contain phosphorus, e. g., polyglycerolphosphatides, phosphoinositides, and serine phosphatides. The metabolic studies included are primarily those dealing with specific enzymes responsible for the biosynthesis and degradation of the three substituted portions of glycerol, but we have tried to relate pertinent *in vivo* findings to the enzyme results. A bibliography including general publications on glycerolipids and cancer is available [98]. A review by HAVEN and BLOOR [49] in 1956 and a more recent review by SNYDER [99] in 1971 discuss most of the relevant literature available at that time on the subject of lipids in cancerous tissues.

II. Occurrence of Ester and Ether Glycerolipids in Neoplastic Cells

A. Nonphosphorus Lipids

1. General

Thin-layer chromatographic resolution of acyl, alkyl, and alk-1-enyl types of nonphosphorus glycerolipids is excellent, even permitting separation of the disubstituted and monosubstituted isomers; see KATES [57] and SNYDER [100, 105, 106] for recent reviews. The order of migration (R_f) is alk-1-enyl type > alkyl type > acyl type. Later in this section, we summarize detailed investigations where *specific* molecular species of the various glycerolipids found only in tumors were isolated and analyzed.

It should be pointed out at the outset that both alkyl and alk-1-enyl glycerolipids are prominent intracellular components of human [124] and animal [123] cancer cells, although found in extremely minor amounts in most normal tissues. The values reported in most studies are based on the quantity of alkylglycerols or alk-1-enylglycerols liberated from intact glycerolipids by chemical reduction with metal halides. Approximately

95% of the ether-linked moieties in the reduction products are 16:0, 18:0, and 18:1 carbon chains in both normal and neoplastic tissues from mammals, although significant amounts of longer chains have been detected in lipids from the gastrointestinal tract of rats [74], harderian gland tumors in mice [56], and in human leukemic cells [115].

2. Triacylglycerols

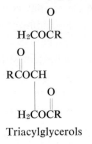

Triacylglycerols

The level of triacylglycerols is elevated from the normal in tumors of the brain [95]. According to SMITH and WHITE [95], triacylglycerols are a nonspecific component of brain tumors, especially high in the less differentiated primary sarcoma and metastatic carcinoma. They postulate that these high levels may be associated with Wallerian degeneration. Their findings are supported by the work of STEIN et al. [129], GOPAL et al. [41], and SLAGEL et al. [91], who found increased levels of total glycerides in brain tumors they studied: gliomas, meningiomas, neurinomas, carcinomas, and glioblastomas. STEIN et al. [130, 131] reported greatly decreased levels of triacylglycerols in the liver of glioma-bearing mice, and suggested that since there is a concomitant fivefold increase in diacylglycerols, the tumor does not just act as a sponge for triacylglycerols, but actually influences liver metabolism to support its own growth.

In the Yoshida ascites hepatoma investigated by RUGGIERI and FALLANI [84], triacylglycerols were also elevated, especially in the mitochondrial and supernatant fractions. The fatty acids in these triacyl lipids contained less palmitic and more stearic than those from normal rat liver. KAWANAMI et al. [59] also found definitely increased amounts of triacylglycerols in the Shionogi carcinoma 115 and Nakahara-Fukuoka sarcoma, both of which also had higher proportions of stearic acid and slightly lower ratios of palmitic acid than those found in normal liver.

In mammary carcinomas, triacylglycerols are very much decreased from those in normal mammary tissue [50, 55, 81]. This is not surprising, since mammary tissue is a depot for neutral fats. REES et al. [80] reported that there is a more marked variation in the proportion of C_{20} and C_{22} fatty acids in the triacylglycerols from mammary carcinoma than in those from mammary glands; they suggest this is related to the fact that the more necrotic neoplasms have a high proportion of C_{22} fatty acids. The triacylglycerols of mammary carcinoma had an increase of C_8, C_{10}, C_{12}, and C_{14} and a decrease of $C_{18:1}$ and $C_{18:2}$ fatty acids when compared to the normal gland; they contained fairly constant proportions of 16:0, 16:1, and 18:0 fatty acids. A decrease in triacylglycerols has been reported for only one other malignant tissue, namely human malignant uterus [79]. This decrease, a fourfold one, was reflected in the mitochondrial fraction, but not in the microsomes.

Ehrlich ascites cells and Landschutz ascites carcinoma cells contain about 7 % neutral lipids on the basis of dry weight, and GRAY [43] reports that these are 56 % triacylglycerols. YAMAKAWA et al. [158] found that only 10.6 % of the neutral lipid fraction was triacylglycerols, but since free fatty acids constituted over 42 % of the neutral lipid values reported by this group, one may suspect that their values reflect a breakdown of ester bonds during isolation and analysis. DIPAOLO et al. [26] isolated lipid granules from Ehrlich ascites cells and found that they contain large amounts of triacylglycerols and cholesterol.

MAHAILESCU et al. [71] separated lipid classes of human lymph node malignant disease by thin-layer chromatography, but found no appreciable amounts of variation in the triacylglycerols. Rauscher leukemia virus also contained glycerides [53] in amounts too small to be investigated quantitatively, being 75–80 % phospholipid in nature. Some tumors, on the other hand, arise from fatty tissue, and contain >98 % triacylglycerols. These human lipomas and hibernomas were investigated by ANGERVALL et al. [3], but the fatty acid composition of the triacylglycerols was not determined.

3. Alkyl Analogs of Triacylglycerols

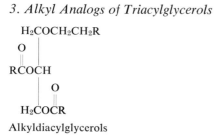

Alkyldiacylglycerols

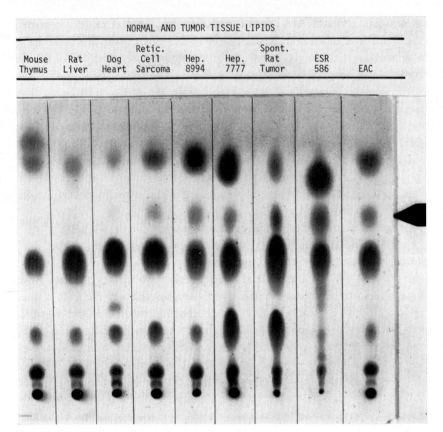

Fig. 1. The occurrence of alkyldiacylglycerols in cancer cells. Thin-layer chromatogram of total lipid extracts obtained from several normal tissues and tumors. The abbreviated notations above each lane designate reticulum-cell sarcoma (human), 8994 Morris hepatoma (rat), 7777 Morris hepatoma (rat), spontaneous rat tumor, ESR-586 preputial gland tumor (mouse), and Ehrlich ascites cells (mouse). The black arrow indicates the location of the alkyldiacylglycerols; the spots directly below the alkyldiacylglycerols are triacylglycerols. A solvent mixture of hexane:diethyl ether:acetic acid (90:10:1, v/v) was used for development, and Silica Gel G was the adsorbent. Reproduced by permission of Academic Press [103].

The alkyl ether analogs of triacylglycerols, 1-alkyl-2,3-diacyl-*sn*-glycerols, are a distinguishing lipid feature of most neoplastic cells (fig. 1). This class of ether-linked lipids, readily visible on charred thin-layer chromatograms as a distinct spot that migrates just ahead of the triacylglycerols [21, 111] was not recognized until the early sixties when our group at Oak Ridge noted an unknown component at this R_f. It always appeared after chromatography of tumor lipid extracts on Silica Gel G developed with hexane:diethyl ether (95:5 or 90:10). Identification of the component as alkyldiacylglycerols [14, 146] demonstrated that neoplastic growth had in some way altered the metabolism of ester-type glycerolipids.

The alkyl grouping in alkyldiacylglycerols isolated from neoplastic cells has been characterized in Ehrlich ascites cells [146, 147], preputial gland tumors [114], Walker-256 carcinosarcomas [14, 97], taper liver tumors [97], a human lymphosarcoma [14, 97], 7777 Morris hepatomas [109], and in cells grown in tissue culture (L-M fibroblasts [2] and Novikoff hepatomas [128]). In all samples, the 16:0, 18:0, and 18:1 chains accounted for more than 95 % of the alkyl moieties present.

Stereospecific analysis of the fatty acids from alkyldiacylglycerols in Ehrlich ascites cells [147] demonstrated that the 2-position was rich in 20:4 and 22:6 fatty acids ($\backsim$11 and 16 %, respectively) in comparison to that of the triacylglycerols from the same tissue ($\backsim$5 and 1 %, respectively). Palmitic acid in the alkyldiacylglycerols was esterified at the 2- and 3-positions in essentially equal amounts. Other data on the total fatty acid composition of the alkyldiacylglycerols from 7777 Morris hepatomas [109] and from Walker-256 carcinosarcoma, taper liver tumors, and a human lymphosarcoma [97] also showed much higher levels of long-chain polyenoic fatty acids in this lipid class than in the triacylglycerols.

There are a few studies published describing the chromatographic isolation and quantitation of the alkyldiacylglycerols from normal mammalian tissues. Such data clearly show that tumors contain much more alkyldiacylglycerols ($>$2 % of the total lipid weight) [56, 109, 111, 114] than do normal tissues, e.g., 0.3 % in subcutaneous fat [88, 89]. Even in reports showing only thin-layer chromatographic patterns [103, 111, 124] it becomes obvious that normal tissues contain essentially no detectable quantities of alkyldiacylglycerols, whereas tumors do (fig. 1).

4. Alk-1-Enyl Analogs of Triacylglycerols

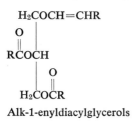

Alk-1-enyldiacylglycerols

1-Alk-1'-enyl-2,3-diacyl-*sn*-glycerols are often referred to as neutral plasmalogens. Data obtained from the *total* neutral lipid fraction show that this lipid class does occur to some extent in a variety of cells [38, 90, 123, 124]. However, the alk-1-enyldiacylglycerols have been separated chromatographically as a class and detected in lipid extracts from tumors [114], adipose tissue [88, 89] and in lower forms [31, 87]. The alk-1-enyl moieties of the alk-1-enyldiacylglycerols are primarily 16:0, 18:0, and 18:1 carbon chains in human adipose tissue [88, 89] and the several tumors analyzed [97].

In adipose tissue the acyl moieties of the alk-1-enyldiacylglycerols were mainly 16:1, 16:0, 18:1, 18:2, and 20:4 chains [88, 89]. Comparable analysis of tumor lipids has not been reported. However, in three tumors where the acyl moieties of both the alkyldiacylglycerols and alk-1-enyl-diacylglycerols were analyzed as a single fraction [97], the major acyl moieties were 16:0, 18:0, 18:1, 18:2, and 20:4.

5. Diacylglycerols, Monoacylglycerols, and their Ether Analogs

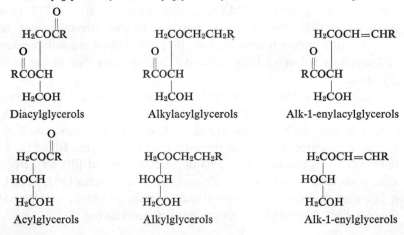

Levels of diacylglycerols and monoacylglycerols are generally very low in most tissues and little effort has been made to quantitate these components. Microsomal membranes isolated from rat liver contained less than 4 % of the total neutral lipids as diacylglycerols and none was found in the plasma membrane fraction [66]. A problem that is difficult to resolve when judging values reported for partial glycerides is whether they represent the effect of lipolytic enzymes after the tissues are removed from the animals or whether they do indeed reflect the actual quantities present under physiological conditions. For example, SMITH and WHITE [95] found essentially no partial glycerides in their analysis of human brain tumors, while GRAY [43] found that Ehrlich ascites cells contained very high levels of monoacylglycerols (17 % of neutral lipid). STEIN *et al.* [130] report that the increased diacylglycerols from the livers of glioma-bearing mice contained a much higher ratio of stearic:palmitic acid than did those from normal liver. The hepatoma diacylglycerol data reported by RUGGIERI and FALLANI [84] reflect the general increase and modifications seen in triacylglycerols in all cell fractions when compared to those of normal liver tissues, except that the diacylglycerols from the hepatoma supernatant fraction had a decreased, rather than an increased, ratio of stearic:palmitic acids. KAWANAMI *et al.* [59], on the other hand, found *decreased* values of partial glycerides along with the increased values of triacylglycerols in the Shionogi carcinoma 115 and Nakahara-Fukuoka sarcoma. They also compared the fatty acid composition of diacylglycerols and monoacylglycerols to that of triacylglycerols found in normal mouse liver and in the two tumors. In normal liver the diacylglycerols and triacylglycerols were similar with respect to the major fatty acids present (16:0, 18:1, and 18:2); in contrast, the monoacylglycerols contained significant quantities of 20:2 and 22:3 acids and about half as much palmitic acid. More variability was seen in the two tumors, but in many ways they also had similar fatty acid patterns for the diacylglycerols and triacylglycerols, whereas the monoacylglycerols again displayed much longer carbon chains. No generalized statement can be made regarding the comparison of these lipid classes in the normal liver and the two tumors. GRAY [43] found that the fatty acid compositions of triacylglycerols, diacylglycerols, and monoacylglycerols of Landschutz ascites carcinoma cells were quite similar except that the diacylglycerols contained a little less 18:0 acid and more 18:2 acid than the other two glycerides analyzed. In the human malignant cervix discussed by RAY and ROY [79], the decreased triacylglycerols were not accompanied by any change in the amount of partial glycerides. However, the microsomes contained more mono- and diacyl-glycerols than those from normal tissue, whereas the mitochondria contained less. The microsomes contained high amounts of monoacylglycerols especially in malignant uteri, when compared to mitochondria. Whether all of the variations discussed in this section are due to differences in cell types or are merely reflections of the amount of lipolytic or artifactual breakdown of lipids after the tissues are removed from the animal is extremely difficult to assess.

The presence of the ether analogs of diacylglycerols and monoacylglycerols in tissues has not yet been unequivocally documented, probably because they too occur in such minute quantities. Enzyme studies have shown that the alkylacylglycerols [12, 107] and alk-1-enylacylglycerols [12, 61] mimic the diacylglycerols as substrates for acyltransferases and phosphorylbase transferases. Lipolytic enzymes that occur in tissues are also capable of utilizing the ether analogs of triacylglycerols [120], so that trace quantities of monoacyl and diacyl ether-linked lipids can be expected in tissues that are rich in alkyl and alk-1-enyl glycerolipids, perhaps at levels similar to those of their ester-linked counterparts.

B. Phosphorus-Containing Lipids

1. General

One of the difficulties in the analysis of phospholipids is that the three subfractions (diacyl, alkylacyl, and alk-1-enylacyl types) of each phospholipid class are generally not completely resolved from each other unless the polar groupings, e.g., phosphorylethanolamine, are masked [82] or hydrolyzed with phospholipase C [40]. Chemical reduction procedures with $LiAlH_4$ [137] or $NaAlH_2(OCH_2CH_2OCH_3)_3$ (Vitride [110]) can also be used to yield information on the relative proportion of diacyl, alkylacyl, and alk-1-enylacyl subfractions in a specific phospholipid class [97, 109, 114], but under these conditions it is not possible to analyze the fatty acids at the *sn*-2-position of each subfraction, since all are removed nonselectively by the metal hydride. VAN GOLDE and VAN DEENEN [39] have described a useful procedure for determining the molecular species of a phospholipid class (phosphatidylcholine) that is based on analysis of lipid fractions obtained after argentation thin-layer chromatography and phospholipase A and C treatment followed by additional chromatography. Positional fatty acid analysis of the diacylglycerol fractions formed after phospholipase C treatment was obtained by pancreatic lipase hydrolysis. Interpretation of data on phospholipid classes not resolved into subfractions requires caution, since values for a specific small subfraction are lost in the overall values reported for the phospholipid classes.

Phospholipid classes in all mammalian tissues generally contain much higher quantities of the diacyl rather than the alkylacyl or alk-1-enylacyl type. However, within a specific phospholipid class the proportion of ether-linked subfractions can be considerable. The plasmalogen (alk-1-enyl) levels in the phospholipids are usually high in neoplastic cells [36, 43, 123, 124, 140], but the qualitative characteristics within the phospholipids are not as strikingly different from those of normal cells as those observed for the alkyldiacylglycerols that we discussed in section II.A. In reading papers in this field, it is helpful to realize that the term 'plasmalogen' usually refers to the ethanolamine plasmalogens, i.e., 1-alk-1′-enyl-2-acyl-*sn*-glycero-3-phosphorylethanolamine, but one must be aware of the fact that the alk-1-enyl grouping has also been found in a number of other phospholipid classes. However, in tumors, it appears that both the alkyl and alk-1-enyl moieties are located mainly in the ethanolamine fraction and to some extent in the choline fraction.

General data on phospholipid classes that have not been subfractionated into acyl, alkyl, or alk-1-enyl types are plentiful and deserve mention, although interpretations are difficult to make. In brain tumors, phospholipid values are generally lower [27, 72, 91] than in the host tissue, an organ unusually rich in phospholipids. SLAGEL *et al*. [91] reported values for a number of brain tumors that included: phosphatidylglycerol, phosphatidylglycerol phosphate, choline and inositol plasmalogens, plasmalogenic acid, and alkylacylphosphorylethanolamine. Only the alkyl ether was more abundant in tumor tissue. The patterns of subclasses within the phospholipids seemed more or less normal in SLAGEL'S observations, but several investigators have reported an increase in the ratio of phosphatidylcholine:phosphatidylethanolamine in brain tumors [22, 27, 41]. In 1963, NAYYAR [72] reported that the major decrease in brain tumor phospholipids was in phosphatidylethanolamine, while GOPAL *et al*. [41] extends NAYYAR'S observations that the ratio of phosphatidylcholine to phosphatidylethanolamine is <1 in normal brain, but >1 in gliomas, meningiomas, and neurinomas. CHRISTENSEN LOU and CLAUSEN [22] observed that this change was most pronounced in more malignant tumors, but DOHR *et al*. [27] did not find any phosphatide difference between malignant and benign tumors.

WEBER and CANTERO [143] observed a loss of phospholipid in Novikoff hepatoma but not in normal regenerating liver, and suggested that this reflects a loss of specialized function. Their observation agrees with that of KAWANAMI *et al*. [59], who observed lower levels of phospholipid in liver tumors, but no difference in the distribution of total classes of phospholipid. These authors also reported the fatty acid composition of the main phospholipid classes, but the classes were not subfractionated. FIGARD and GREENBERG [34] agreed that phosphatide proportions of rat hepatomas were largely normal, but sometimes shifted in the direction of a decreased ratio in choline phosphatides:ethanolamine phosphatides. RUGGIERI and FALLANI [84] studied the Yoshida ascites hepatoma rather extensively, and found that: (a) the mitochondria contained normal amounts of phospholipids, but increased phosphatidylinositol and decreased phosphatidylethanolamine; (b) microsomes were deficient in phospholipid due mainly to drops in phosphatidylcholine and phosphatidylethanolamine levels, and (c) supernatant contained very little phospholipid, but more than that from normal liver. The tumor phosphatidylethanolamine, phosphatidylinositol, and phosphatidylcholine fractions contained less stearic and palmitic, and more oleic and linoleic acids than normal liver phospholipids. There was also a general decrease in polyunsaturated 20–22 carbon fatty acids. The author ascribes these differences to structural and functional changes in tumor membranes. VEERKAMP et al. [140] echo the fact that phospholipids in hepatomas are decreased by a factor of 2 from normal liver, and that there are no differences in the proportions of various phosphatides in the whole cell. The stearic:oleic ratio within the total phospholipid fraction was, however, reversed (decreased) in the whole tumor cell and in the mitochondrial fractions. These authors suggest that the decrease in hepatoma phosphatides may be related to the disappearance of the organized structure of the endoplasmic reticulum of the liver, and that it also correlates with degeneration of cellular and intracellular boundaries in the hepatoma.

Although it is difficult to compare Ehrlich ascites carcinoma to any normal tissue, WALLACH et al. [141] called the level of phopholipids in these cells 'high' and drew a line of agreement with the high values for malignant liver lesions. The correlation between the two tumors failed, however, when YAMAKAWA and co-workers [157, 158] found the predominant fatty acid in the phospholipids of Ehrlich and sarcoma 180 ascites cells to be stearic acid; in fact, it was the only tumor fatty acid not directly influenced by dietary intake. The ratio of choline phosphatides: ethanolamine phosphatides in these cells was 5 : 1. The Landschutz ascites cells examined by GRAY [43] had a choline phospholipid:ethanolamine phospholipid ratio of 2.4 : 1, which the author considers 'normal'. But he makes the observation that the stearic:palmitic acid ratio is high in tumor phospholipids and, in fact, that there is a less selective distribution of fatty acids in tumor phosphatides 'suggesting a loss in specificity in the enzyme systems responsible for the incorporation of fatty acids into phospholipids'.

The mammary tumors examined by Rees et al. [81] did not vary more in phospholipid content than does normal mammary tissue. However, the phospholipid fatty acid composition of a derived transplantable sarcoma differed from that of the mammary carcinoma.

Lindlar and Wagener's data [68] on phospholipid classes and their fatty acid composition in five different animal tumors (Ehrlich ascites carcinoma, solid Ehrlich carcinoma, a transplantable mammary tumor, carcinoma 775, and sarcoma 180) showed a wide range of values for the relative proportions of the various types of lipid classes. These investigators conclude that, 'related neoplasms of the same species may be distinguished by their specific lipid and fatty acid components'. This is also borne out by a number of investigations conducted by Carruthers [18] on the changes in phospholipid and fatty acid composition that occur in malignant epidermis. His 1967 article [18] cites his earlier references on the subject. The main conclusion from these experiments is that malignant epidermis has a higher degree of saturated fatty acids in the choline and ethanolamine phospholipids than the normal epidermis of mice.

Mihailescu et al. [71] reported that lymph node cellular phospholipids in human malignancies rose markedly from normal, even more so in very ill patients. They did not characterize the phospholipids further, but Rapport and Alonzo [78] have observed that the ethanolamine phospholipids of a rat lymphosarcoma are coincident with thromboplastic and antigenic activity. Slot [92] has also reported high phopholipid levels in human malignancies. He studied ovarian tumors, and found that the phospholipid level was directly proportional to the degree of malignancy of the lesion. Again, human cervical tumors contain twice as much phospholipid as normal tissue. This is accounted for by the mitochondrial fraction, and caused by an increase in the phosphatidylcholine and lysophosphatidylcholine classes; the phosphatidylethanolamine is actually lower in the malignant tissue. In human leukemic leukocytes, there is a small but significant increase in the relative content of phosphatidylcholine [42]. In these leukocytes, there was a reciprocal relationship between sphingomyelin and phosphatidylcholine similar to that observed in the erythrocyte membranes from different species. Such a shift was not found in the red cell membranes and plasma from adults with active malignancies unrelated to the circulatory system [23].

Phospholipids were also high in the various solid tumors of the mouse examined by UEZUMI et al. [139] (48.2–70.6 % of total lipids). But in the hibernomas and lipomas studied by ANGERVALL et al. [3], the phospholipids are even lower than in the brown fat to which the tumors were compared (< 4.3 % of total fat); the authors use the similarity in phosphatide composition to prove that the hibernoma arises from brown fat. The phospholipids of the Jensen sarcoma have been studied in Russia: After ascertaining that the phospholipids in the tumor were no different from those of normal tissues [29], this group fractionated the cells of the tumor and found that the phosphatide composition of the microsomes and mitochondria did not differ as much as that from normal cells [30] and that the nuclei, unlike the nuclei from normal liver cells, differed little from whole tissue in phospholipid composition [63]. Although these workers did not subfractionate the phospholipid classes, their results show that the phospholipid patterns in the organelles from solid mouse hepatomas 22 and 27 and Zajdela ascites hepatomas [28] as well as the Jensen sarcoma are not as characteristic as those seen in the organelles from normal liver. Since regenerating liver did not differ from normal liver in the phospholipid composition of the organelles, these investigators felt that rapid cell proliferation by itself does not cause the dedifferentiation of phospholipids in the subcellular fractions of the hepatoma [8].

2. Subfractions of Ethanolamine and Choline Phospholipid Classes

WOOD and SNYDER [147] found that the ethanolamine phospholipids from Ehrlich ascites cells consisted of 56 % diacyl-GPE, 29 % alkylacyl-GPE, and 15 % alk-1-enylacyl-GPE, whereas the choline phospholipids in these cells consisted of 64 % diacyl-GPC, 36 % alkylacyl-GPC, and less than 3 % alk-1-enylacyl-GPC. The 7794 A and 7777 Morris hepatomas, however, contained much lower quantities of the ether-linked lipids in both the ethanolamine and choline phospholipids [109]: 93 and 92 % diacyl-GPE 0.57 and 1.15 % alkylacyl-GPE, 6.3 and 7.1 % alk-1-enylacyl-GPE, 99 and 97 % diacyl-GPC, 0.70 and 2.85 % alkylacyl-GPC, and trace and 0.46 % alk-1-enylacyl-GPC. The same study [109] also documented the minute quantitites of ether lipids found in the ethanolamine and choline phospholipids in normal liver: 0.5 % alkylacyl-GPE, 2.15 % alk-1-enylacyl-GPE, 0.83 % alkylacyl-GPC, and trace quantities of alk-1-enylacyl-GPC were detected. The preputial gland tumor, which has been used in many enzyme studies of the ether lipids, contains 64 % diacyl, 8 % alkylacyl, and 28 % alk-1-enyl species in the GPE fractions,

Phosphatidylethanolamine
(1,2-diacyl-*sn*-glycero-3-phosphorylethanolamine; diacyl-GPE)

$$\begin{array}{l} H_2COCR \\ \quad \parallel \\ \quad O \\ RCOCH \\ \quad \parallel \\ \quad O \\ H_2COP\text{--}OCH_2CH_2NH_2 \\ \quad | \\ \quad OH \end{array}$$

Phosphatidylcholine
(1,2-diacyl-*sn*-glycero-3-phosphorylcholine; diacyl-GPC)

$$\begin{array}{l} H_2COCR \\ \quad \parallel \\ \quad O \\ RCOCH \\ \quad \parallel \\ \quad O \\ H_2COP\text{--}OCH_2CH_2\overset{+}{N}(CH_3)_3 \\ \quad | \\ \quad OH \end{array}$$

1-alkyl-2-acyl-*sn*-glycero-3-phosphorylethanolamine (alkylacyl-GPE)

$$H_2COCH_2CH_2R$$
$$RCOCH$$
$$H_2COP\text{--}OCH_2CH_2NH_2$$

1-alkyl-2-acyl-*sn*-glycero-3-phosphorylcholine (alkylacyl-GPC)

$$H_2COCH_2CH_2R$$
$$RCOCH$$
$$H_2COP\text{--}OCH_2CH_2\overset{+}{N}(CH_3)_3$$

Ethanolamine Plasmalogen
(1-alk-1′-enyl-2-acyl-*sn*-glycero-3-phosphorylethanolamine; alk-1-enylacyl-GPE)

$$H_2COCH=CHR$$
$$RCOCH$$
$$H_2COP\text{--}OCH_2CH_2NH_2$$

Choline Plasmalogen
(1-alk-1′-enyl-2-acyl-*sn*-glycero-3-phosphorylcholine; alk-1-enylacyl-GPC)

$$H_2COCH=CHR$$
$$RCOCH$$
$$H_2COP\text{--}OCH_2CH_2\overset{+}{N}(CH_3)_3$$

and 86 % diacyl, 10.3 % alkylacyl, and 3.4 % alk-1-enylacyl in the GPC fraction. Thus, for the few tumors in which the subfractions of the phospholipid classes were examined in detail, one cannot generalize any relationship on the proportion of the subfractions in the ethanolamine and choline phospholipids; undoubtedly, such proportions can only be validly compared in a tumor and its cells of origin. In this context, the two hepatomas analyzed did show significantly higher quantities of alk-1-enylacyl-GPE than normal liver. One other point that can be made is that generally the choline phospholipids contain only small amounts of alk-1-enyl moieties compared to the ethanolamine fraction, whereas the alkyl moieties in the choline fraction are at least equal to or higher than the alkyl moieties in the ethanolamine fraction for the various tissues analyzed, whether normal or neoplastic.

Early studies by GRAY [43] clearly demonstrated that the alk-1-enyl chains of ethanolamine and choline plasmalogens from the Landschutz ascites carcinoma and the BP8/C3H ascites sarcoma consisted of mainly 16:0, 18:0, and 18:1 aliphatic chains. This simple array for both alkyl and alk-1-enyl ether-linked chains in the ethanolamine and choline phosphatides, like those of the alkyldiacylglycerols and alk-1-enyldiacylglycerols (sect. II. A), has been shown in a wide variety of tumors including Ehrlich ascites cells [147], preputial gland tumors [114], Walker-256 carcinosarcomas, taper liver tumors, a human lymphosarcoma, and 7777 Morris hepatomas [97].

To our knowledge, only one study has been published on the fatty acid composition of the three subfractions of the ethanolamine and choline phospholipids found in a tumor (Ehrlich ascites cells); analyses were carried out on acetate derivatives of the diacylglycerols, alkylacylglycerols, and alk-1-enylglycerols librated from phospholipids by phospholipase C hydrolysis [147]. A significant finding originating from these experiments is that much higher quantities of 20:4 and 22:6 fatty acids occur at the *sn*-2-position of alkylacyl-GPE, alk-1-enyl-GPE, and alkylacyl-GPC in comparison to either diacyl-GPE or diacyl-GPC. The significance of the high degree of unsaturation at the *sn*-2-position of the ether-linked lipids is not known, but enzyme studies done with rat testes preparations [152] have demonstrated that there is a tendency for unsaturated fatty acids to build up in ether-linked lipid classes via deacylation-reacylation reactions.

Data in some of the general papers discussed in section II. B. 1 showed that lipid composition can differ considerably among different tumor types. This finding has been emphasized in recent work done on subfractions of phosphatidylcholine in urethan-induced pulmonary adenomas [96]. In this lung tumor, like normal lung, there are considerable quantities of disaturated phosphatidylcholine ($\simeq 28\%$), of which 60% was 1,2-dipalmitoyl-*sn*-glycero-3-phosphorylcholine, a molecular species probably unique to lung and lung tumors.

III. Pathways for Ester and Ether Glycerolipids and their Significance in Neoplastic Cells

A. General

For some years it was thought that the ether-linked lipids were formed directly from ester linkages. Not only do ether lipids always have ester analogs, but a simple mechanism could be postulated for conversion of an acyl moiety to an alk-1-enyl moiety via a hemiacetal intermediate [5]. However, we now know that the acyl, alkyl, and alk-1-enyl moieties at the *sn*-1-position of glycerol are formed by completely different types of reactions (sect. III. B. 1). But before the enzyme systems described in the next section had been completely worked out, THOMPSON [136] had already reported experiments with a ^{3}H- and ^{14}C-labeled alkylglycerol indicating that an alkyl moiety could be converted to an alk-1-enyl moiety in the terrestial slug. Later, incubations of Ehrlich ascites cells with alkylglycerol [13] or a long-chain fatty alcohol [145, 148] produced data leading to the same conclusion. The properties of the enzyme system responsible for this conversion are discussed along with the other reactions in section III. B. 1 (c). Modifications at the *sn*-2- and *sn*-3-positions of the glycerol portion of any of the ether- and ester-containing glycerolipids, such as acyl transfer, phosphorus transfer, and phosphoryl-base transfer, ocur in a very similar fashion, and they might well be catalyzed by the same enzymes.

Many early investigations were designed to elucidate the pathways for glycerolipids in tumor systems by studying the *in vivo* incorporation of ^{32}P, acetate, fatty acids, glycerol, and related lipid precursors [98]. Because the interpretation of such experiments is complex, and always open to question, we have chosen to exclude most of them from this review.

B. Biosynthesis

1. The sn-1-Position
a) Acyl Moieties

Acyl (ester) bonds are formed by enzyme systems that transfer the entire $R\overset{\text{O}}{\overset{\|}{C}}$O-entity to the glycerol portion of lipids. The acylation of glycerol-3-P [62, 159] or dihydroxyacetone-P [45,48] illustrates the simplest form of this process (fig. 2); such transfers occur with most glycerolipid substrates that possess free hydroxyl moieties. The important aspects of these transferase reactions are that acyl-CoA is required and that either glycerol-3-P, dihydroxyacetone-P, or a lyso-glycerolipid may act as the acceptor molecules. YAMASHITA et al. [159] were able to separate the two enzymes that synthesize phosphatidic acid from glycerol-3-P in rat liver microsomes, i.e., glycerol-3-P acyltransferase and 1-acyl-glycerol-3-P acyltransferase.

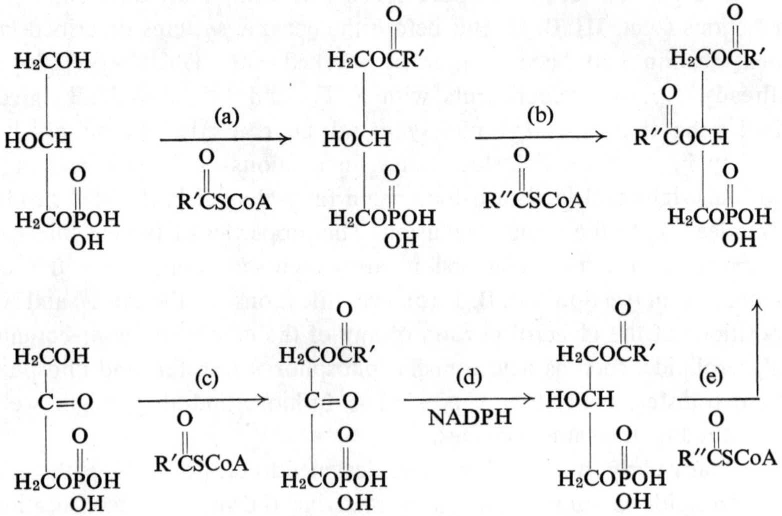

Fig. 2. Biosynthesis of phosphatidic acid from glycerol-3-P or dihydroxy-acetone-P. The reactions are catalyzed by (a) acyl-CoA:*sn*-glycerol-3-P acyltransferase, (b) acyl-CoA:1-acyl-*sn*-glycerol-3-P acyltransferase, (c) acyl-CoA:dihydroxyacetone-P acyltransferase, (d) NADPH$_2$:acyldihydroxyacetone-P oxidoreductase, and (e) acyl-CoA:1-acyl-*sn*-glycerol-3-P acyltransferase.

b) Alkyl Moieties

Alkyl (ether) linkages at the *sn*-1-position of glycerolipids are formed from long-chain fatty alcohols [119, 125] instead of fatty acids, and the glycerol portion originates only from dihydroxyacetone-P [46, 114, 116, 155, 156] and not from glycerol-3-P. The initial reaction that forms the ether linkage substitutes a long-chain fatty alcohol for the acyl moiety of acyldihydroxyacetone-P [47, 153]; the entire chain of the alcohol [114, 156] including the oxygen [122], replaces the acyl group (fig. 3).

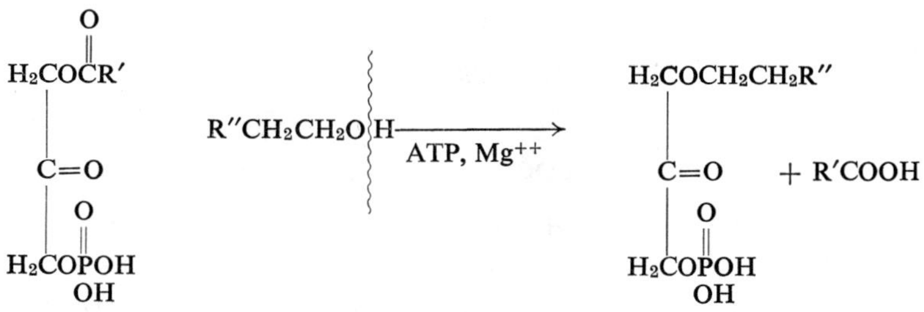

$$\begin{array}{c}\text{H}_2\text{COCR}'\\\|\\\text{O}\end{array}\quad\begin{array}{c}|\\\text{C=O}\\|\\\text{O}\\\|\\\text{H}_2\text{COPOH}\\\text{OH}\end{array}\quad\xrightarrow[\text{ATP, Mg}^{++}]{\text{R}''\text{CH}_2\text{CH}_2\text{O}\{\text{H}}\quad\begin{array}{c}\text{H}_2\text{COCH}_2\text{CH}_2\text{R}''\\|\\\text{C=O}\\|\\\text{O}\\\|\\\text{H}_2\text{COPOH}\\\text{OH}\end{array}\quad+\text{ R}'\text{COOH}$$

Fig. 3. Biosynthesis of alkyldihydroxyacetone-P from long-chain fatty alcohols and acyldihydroxyacetone-P. The reaction, catalyzed by a microsomal enzyme, involves a deacylation-alkylation that substitutes the entire RO- portion of the alcohol for the acyl moiety.

c) Alk-1-Enyl Moieties

Alk-1-enyl (plasmalogen) linkages in glycerolipids originate from the alkyl grouping [110, 151] via a microsomal mixed-function oxidase in tumors [150] and intestinal mucosa [73]. The substrate is an intact alkyl-ethanolamine-containing phosphatide (1-alkyl-2-acyl-*sn*-glycero-3-phosphorylethanolamine) and the requirements for the reaction are molecular oxygen and NADPH; a soluble cytoplasmic factor, ATP, and Mg^{++} also stimulate this conversion (fig. 4). It appears that 1-alkyl-2-acyl-*sn*-glycero-

$$\begin{array}{c}\text{H}_2\text{COCH}_2\text{CH}_2\text{R}\\|\\\text{O}\\\|\\\text{RCOCH}\\|\\\text{O}\\\|\\\text{H}_2\text{COPOCH}_2\text{CH}_2\text{NH}_2\\\text{OH}\end{array}\quad\xrightarrow[\text{O}_2\text{, cyanide-sensitive factor}]{\text{(a)}\;\;\;\text{NADPH + H}^+\text{ or NADH + H}^+}\quad\begin{array}{c}\text{H H}\\\text{H}_2\text{COC=CR}\\|\\\text{O}\\\|\\\text{RCOCH}\\|\\\text{O}\\\|\\\text{H}_2\text{COPOCH}_2\text{CH}_2\text{NH}_2\\\text{OH}\end{array}$$

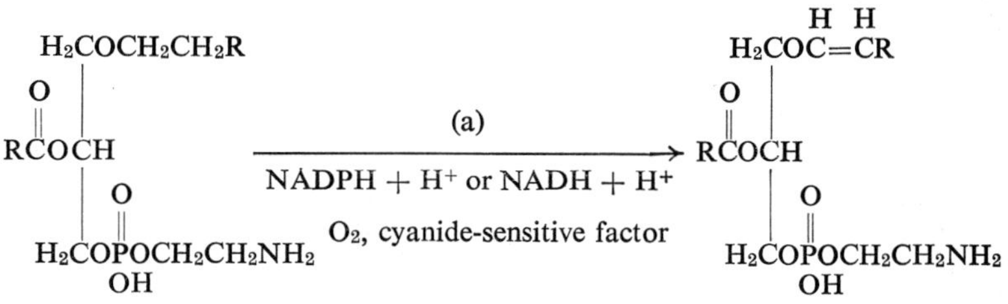

Fig. 4. Biosynthesis of ethanolamine plasmalogens from 1-alkyl-2-acyl-*sn*-glycero-3-phosphorylethanolamine. The reaction is catalyzed by a microsomal desaturase that has the properties of a mixed-function oxidase (a).

3-phosphorylcholine cannot participate in this mixed-function oxidase reaction. However, it is possible for the choline plasmalogens to be synthesized by base exchange reactions [35] or via alk-1-enylacylglycerols and CDP-choline [60], the former being produced from ethanolamine plasmalogens by phospholipase C (fig. 5).

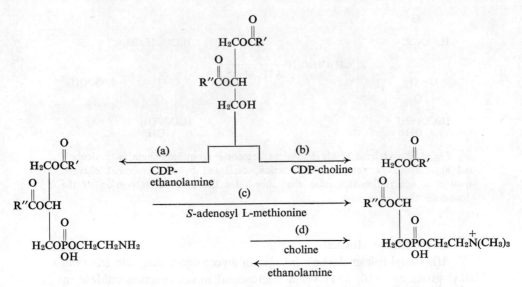

Fig. 5. Biosynthesis and interrelationships of choline and ethanolamine phospholipids. The reactions are catalyzed by (a) CDP-ethanolamine:1,2-diacyl-*sn*-glycerol phosphorylethanolamine transferase, (b) CDP-choline:1,2-diacyl-*sn*-glycerol phosphorylcholine transferase, (c) *S*-adenosylmethionine:1,2-diacyl-*sn*-glycero-3-phosphorylethanolamine methyltransferase, and (d) base exchange enzymes which require **calcium**.

2. The sn-2-Position

The *sn*-2-position of certain glycerolipids also participates in acyltransferase reactions. Perhaps the best known reaction of this type occurs with 1-acyl-*sn*-glycerol-3-P in the formation of phosphatidic acids [62, 159]; its alkyl analog (1-alkyl-*sn*-glycerol-3-P) is acylated in the same fashion [20, 153, 156]. Lyso-phospholipids can also serve as substrates for acyltransferases, as LANDS [65] demonstrated as early as 1960 for 1-acyl-*sn*-glycero-3-phosphorylcholine (fig. 6). In whole animals, the overall result is envisioned as a deacylation-reacylation reaction; the exact mechanism, in terms of whether one enzyme complex or two separate enzymes are required, is unknown. These reactions appear to be widespread and probably account for the diverse molecular species of glycerolipids found in nature.

Glycerolipids and Cancer

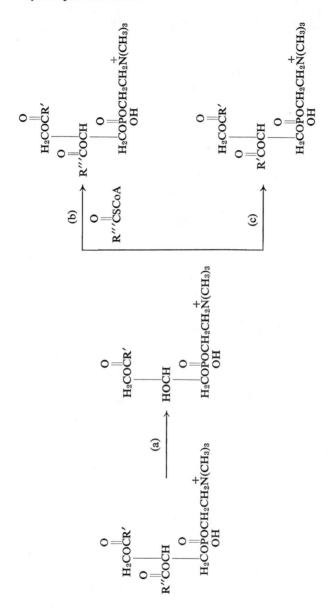

Fig. 6. Acylation of lyso-phospholipids. The reactions are catalyzed by (a) phospholipase A_2, (b) acyl-CoA:1-acyl-*sn*-glycero-3-phosphorylcholine acyltransferase, and (c) 1-acyl-*sn*-glycero-3-phosphorylcholine:1-acyl-*sn*-glycero-3-phosphorylcholine acyltransferase. In deacylation-reacylation reactions, (a) and (b) could be a single enzyme complex.

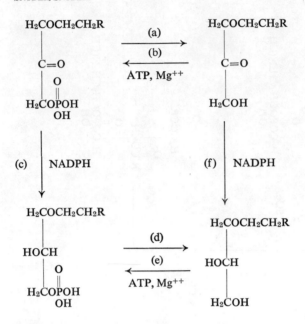

Fig. 7. Phosphotransferase, phosphohydrolase, and reductase activities involved in the metabolism of alkyldihydroxyacetone-P and alkyldihydroxyacetone. The reactions are catalyzed by (a) alkyldihydroxyacetone-P phosphohydrolase, (b) ATP:alkyldihydroxyacetone phosphotransferase, (c) NAD(P)H:alkyldihydroxyacetone-P oxidoreductase, (d) 1-alkyl-*sn*-glycerol-3-P phosphohydrolase, (e) ATP:alkylglycerol phosphotransferase, and (f) NAD(P)H:alkyldihydroxyacetone oxidoreductase.

A second acylation reaction that involves the *sn*-2-position is the one that occurs when two lyso-phospholipids (e.g., 1-acyl-*sn*-glycero-3-phosphorylcholine) react with each other (fig. 6). It is generally referred to as the 'Erbland-Marinetti' pathway [32], and the enzyme catalyzing this reaction transfers the acyl group from one lyso-phosphatidylcholine to the other lyso-phosphatidylcholine to form phosphatidylcholine. Its significance in tumors has not been investigated. However, data obtained by STEIN and STEIN [132] for the Landschutz ascites tumor indicate that lyso-phosphatidylcholine is acylated via acyl-CoA to form phosphatidylcholine instead of participating in the Erbland-Marinetti reaction.

Ketone-containing glycerolipids can be substrates for a third type of reaction that occurs at the *sn*-2-position; it is catalyzed by NADPH and

NADH reductases (fig. 7). Acyldihydroxyacetone-P [64] is reduced by NADPH-linked reductases that are located in microsomes. Alkyldihydroxyacetone-P also is reduced preferentially by an NADPH-linked reductase [20, 64, 153]; however, high levels of NADH can substitute for NADPH with this substrate [20, 153]. After reduction of the ketone moiety, acyltransferases then form 1,2-diacyl-*sn*-glycerol-3-P or 1-alkyl-2-acyl-*sn*-glycerol-3-P. Tumors are particularly rich in the NADPH-linked reductases that utilize the ketone-containing lipids as substrates.

3. The sn*-3-Position*

Biosynthetic reactions that can occur at the *sn*-3-position of glycerolipids most commonly involve phosphotransferases for acyl [51] and alkyl [19, 20] lipids, acyltransferases for acyl [144], alkyl [12, 121], and alk-1-enyl [12] lipids, and phosphorylbase transferases for acyl [60], alkyl [107], and alk-1-enyl [61] lipids, and base exchange enzymes (fig. 5, 6, and 7) for acyl [35, 76] alkyl [35] and alk-1-enyl [35] lipids. At the present time it is unknown whether identical enzymes are involved in the utilization of the ester-linked and ether-linked substrates.

In some tissues enzymes are present that can catalyze the conversion of phosphatidylethanolamine to phosphatidylcholine via methylation (fig. 5) [17]. FIGARD and GREENBERG [34] found that the methylation pathway is largely lacking in a number of mouse ascites tumors: hepatoma 134, Ehrlich carcinoma, Gardner lymphosarcoma 6C3HED, sarcoma 180, lymphatic leukemia 4946, and Novikoff. Morris and Dunning hepatomas grown subcutaneously in rats were also assayed for the methylating enzyme. Of all these tumors, only the Morris hepatoma had a methylating enzyme activity greater than 10 % of that found in normal rat liver.

C. Catabolic Enzymes

1. The sn*-1-Position*
a) Acyl Moieties

Pancreatic lipase, which has been studied extensively, specifically deacylates the *sn*-1- and *sn*-3-positions of triacylglycerols [25]. DE HAAS *et al.* [44] have shown that even after it has been highly purified it can also catalyze the hydrolysis of fatty acids at the *sn*-1-position of phosphoglycerides.

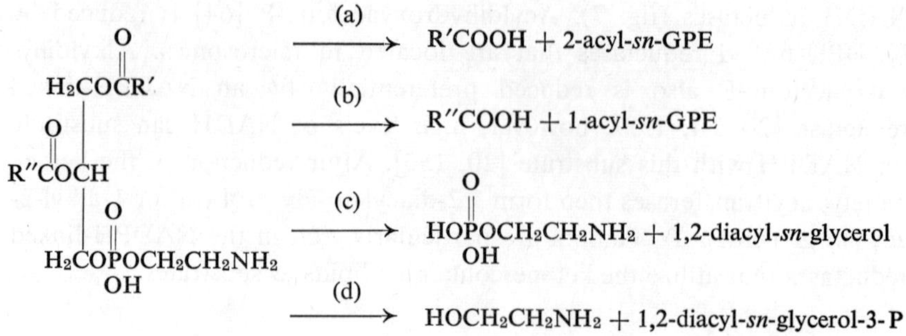

$$
\begin{array}{c}
\text{O} \\
\parallel \\
\text{H}_2\text{COCR}' \\
\text{O} \\
\parallel \\
\text{R}''\text{COCH} \\
\mid \\
\text{O} \\
\parallel \\
\text{H}_2\text{COPOCH}_2\text{CH}_2\text{NH}_2 \\
\text{OH}
\end{array}
\begin{array}{l}
\xrightarrow{(a)} \text{R}'\text{COOH} + \text{2-acyl-}sn\text{-GPE} \\
\xrightarrow{(b)} \text{R}''\text{COOH} + \text{1-acyl-}sn\text{-GPE} \\
\xrightarrow{(c)} \text{HOPOCH}_2\text{CH}_2\text{NH}_2 + \text{1,2-diacyl-}sn\text{-glycerol} \\
\qquad\quad \text{OH} \\
\xrightarrow{(d)} \text{HOCH}_2\text{CH}_2\text{NH}_2 + \text{1,2-diacyl-}sn\text{-glycerol-3-P}
\end{array}
$$

Fig. 8. Phospholipases. The reactions shown for phosphatidylethanolamine are catalyzed by (a) phospholipase A_1, (b) phospholipase A_2, (c) phospholipase C, and (d) phospholipase D.

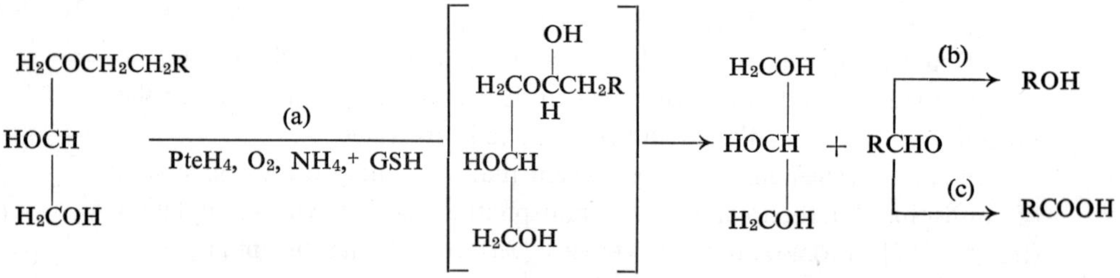

$$
\begin{array}{c}
\text{H}_2\text{COCH}_2\text{CH}_2\text{R} \\
\mid \\
\text{HOCH} \\
\mid \\
\text{H}_2\text{COH}
\end{array}
\xrightarrow[\text{PteH}_4,\ \text{O}_2,\ \text{NH}_4^+,\ \text{GSH}]{(a)}
\left[
\begin{array}{c}
\text{OH} \\
\mid \\
\text{H}_2\text{COCCH}_2\text{R} \\
\mid \\
\text{H} \\
\mid \\
\text{HOCH} \\
\mid \\
\text{H}_2\text{COH}
\end{array}
\right]
\longrightarrow
\begin{array}{c}
\text{H}_2\text{COH} \\
\mid \\
\text{HOCH} \\
\mid \\
\text{H}_2\text{COH}
\end{array}
+ \text{RCHO}
\begin{array}{l}
\xrightarrow{(b)} \text{ROH} \\
\xrightarrow{(c)} \text{RCOOH}
\end{array}
$$

Fig. 9. Alkylglycerol cleavage enzyme. This reaction is catalyzed by a tetrahydropteridine-dependent hydroxylase (a); the hemiacetal is thought to break down spontaneously and the fatty aldehyde can be reduced (b) or oxidized (c) by oxidoreductases.

Figure 8 depicts the mode of action of various phospholipase activities that can occur in mammalian cells. Enzymes that deacylate only the *sn*-1-position of phospholipids are called phospholipase A_1 and they have been reported in many tissues [15]. Lyso-phospholipases, widely distributed in nature, are responsible for completing the deacylation of lyso-phospholipids; these enzymes are not stereospecific like other phospholipases [15]. Little is known about the activity of these enzymes or other phospholipases in cancer cells, but on the basis of *in vitro* data obtained with lyso-phosphatidylcholine in the Landschutz ascites tumor, STEIN and STEIN [132] proposed that lyso-phospholipases are localized at the cell surface of this tumor.

b) Alkyl Moieties

Liver microsomes are a particularly good source of a tetrahydropteridine-dependent cleavage enzyme for alkylglycerols (fig. 9) [75, 117, 118, 127, 138]. It appears to exist to some extent in nearly all normal tissues [75] but not in most tumors [126]. Maximum activity of this system requires a soluble cytoplasmic protein fraction [127, 138], reduced glutathione, and ammonium ions [127]. Products of the reaction are long-chain fatty aldehydes [75, 127, 138] and glycerol [138]; under specified conditions the aldehydes are further converted to either fatty acids [75, 127, 138] or fatty alcohols [75, 127].

c) Alk-1-Enyl Moieties

Cleavage of the alk-1-enyl portion of plasmalogens (fig. 10) by mammalian enzymes has been investigated in liver [142] and brain [4, 160, 161], but not so far in tumor cells. WARNER and LANDS [142] found that rat liver microsomes contained a hydrolytic enzyme highly

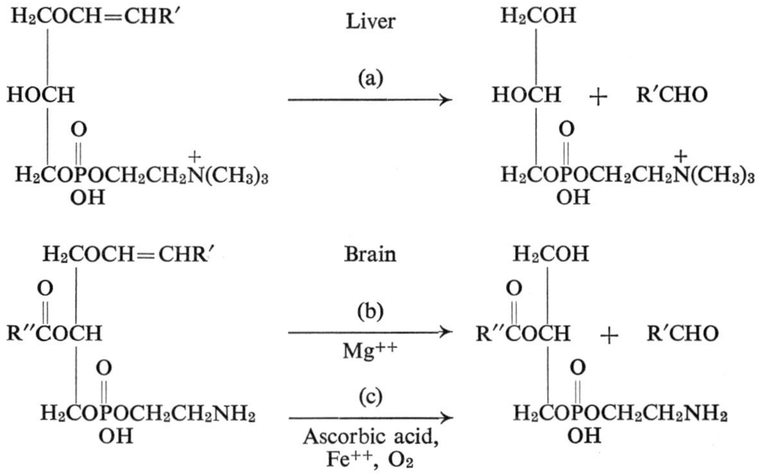

$$\begin{array}{c}H_2COCH=CHR' \\ | \\ HOCH \\ | \\ O \\ \| \\ H_2COPOCH_2CH_2\overset{+}{N}(CH_3)_3 \\ OH\end{array} \xrightarrow[\text{(a)}]{\text{Liver}} \begin{array}{c}H_2COH \\ | \\ HOCH \quad + \quad R'CHO \\ | \\ O \\ \| \\ H_2COPOCH_2CH_2\overset{+}{N}(CH_3)_3 \\ OH\end{array}$$

$$\begin{array}{c}H_2COCH=CHR' \\ | \\ O \\ \| \\ R''COCH \\ | \\ O \\ \| \\ H_2COPOCH_2CH_2NH_2 \\ OH\end{array} \xrightarrow[\substack{\text{Mg}^{++} \\ \text{(c)} \\ \text{Ascorbic acid,} \\ \text{Fe}^{++}, O_2}]{\text{Brain} \atop \text{(b)}} \begin{array}{c}H_2COH \\ | \\ O \\ \| \\ R''COCH \quad + \quad R'CHO \\ | \\ O \\ \| \\ H_2COPOCH_2CH_2NH_2 \\ OH\end{array}$$

Fig. 10. Cleavage of the alk-1-enyl moiety of plasmalogens. These hydrolytic reactions are catalyzed by (a) liver microsomes, (b) brain particles (acetone powder), and (c) a soluble fraction from brain that appears to be a ferrous-ascorbate complex. Phosphatidylcholine also serves as a substrate for (b) and (c).

specific for the alk-1-enyl moiety of choline-containing lyso-plasmalogens (1-alk-1'-enyl-*sn*-glycerol-3-phosphorylcholine). No cofactors were required and the products of the reaction were long-chain fatty aldehydes and *sn*-glycero-3-phosphorylcholine. ANSELL and SPANNER [4] discovered a similar enzymic activity in acetone powders prepared from particulate matter of rat brain. With the brain preparation, Mg^{++} was required and the substrate specificity was highest for 1-alk-1'-enyl-2-acyl-*sn*-glycero-3-phosphorylethanolamine, although its lyso form also served as a substrate.

More recently, YAVIN and GATT [160, 161] reported the oxygen-dependent cleavage of the alk-1-enyl group of plasmalogens by a supernatant fraction of rat brain (fig. 10). Isolation of the active component indicates that it is identical or very similar to ascorbic acid. Both choline and ethanolamine plasmalogens served as substrates, but the latter was only cleaved in the presence of a detergent. Products of the reaction were long-chain fatty aldehydes and either lyso-phosphatidylcholine or lyso-phosphatidylethanolamine. One possible mechanism envisioned by the authors [161] is that a ferrous-ascorbate complex forms an active oxygen (OH^+) that attacks the alk-1-enyl linkage to produce a diol. The diol is then hydrolyzed to an α-hydroxy fatty aldehyde, which is then oxidatively cleaved to the fatty aldehyde having one less carbon atom.

Since plasmalogens are a significant component of tumors, it will be necessary to know something about the enzymes that degrade alk-1-enyl lipids before regulatory controls in cancer cells or the lack of them can be understood.

2. The sn*-2-Position*

Lipases that attack the *sn*-2-position of nonphosphorus lipids have been described [25], but their importance in catabolic processes is not known. Phospholipase A_2 deacylates the *sn*-2-position of glycerol-containing phospholipids (fig. 8) and this activity has been demonstrated in many mammalian tissues [15]. LUMB [69] studied microsomes from normal rat liver and Novikoff hepatoma and recently obtained interesting kinetic data indicating that the properties of phospholipase A_2 in this tumor are different from those in liver.

3. The sn*-3-Position*

Fatty acid groups at the *sn*-3-position of glycerolipids can be hydrolyzed by pancreatic lipase. As with other lipolytic enzymes, there appears to be little distinction made between the ester and ether lipid analogs, since triacylglycerols [25], alkyldiacylglycerols [120], and alk-1-enyl-diacylglycerols [37, 93] all serve as substrates.

Phospholipase C hydrolyzes phosphorylbase groupings from phospholipids and phospholipase D hydrolyzes the base groupings only (i.e., choline, ethanolamine, or serine) from phospholipids (fig. 8). Little is known about these enzyme activities in mammalian tissues [15]; the evidence for their existence is based primarily on turnover studies and phospholipase D-type exchange reactions (see VAN DEN BOSCH et al. [15] for a recent work on this subject). WYKLE and SCHREMMER [154] have described a lyso-phospholipase D pathway in rat brain microsomes that utilizes lyso ether-linked phospholipids. A similar enzymic reaction has been observed in normal rat liver and Fischer rat sarcomas, but the activity of the enzyme in the tumor is much lower than that in brain or liver [149].

Phosphohydrolases are also important in the metabolism of phospholipid intermediates such as phosphatidic acid [94] and its ether analogs [20, 108, 125, 156], alkyldihydroxyacetone-P [20, 153], and acyldihydroxyacetone-P [153]. These enzymes hydrolyze the free phosphate moiety; however, because of the wide range of specificity of phosphatases, it is not clear whether any of the enzyme activities observed are actually catalyzed by separate enzymes. Acid phosphatase and alkaline phosphatase from nonmammalian sources can also hydrolyze the phosphate moieties of the lipids mentioned above; however, when an acyl grouping is substituted at the *sn*-2-position of the glycerol moiety, the glycerolipid is no longer a suitable substrate for the alkaline phosphatase [10].

IV. Regulation of Glycerolipids and their Precursors in Cancer Cells

A. The Glycerol Portion of Lipids

HOWARD et al. [52] found a correlation between increased ether-linked glycerolipids and decreased glycerol-3-P dehydrogenase activities in rapidly growing Morris hepatomas (3924 A, 7288 CTC, 9618 A, 7288 C, 7777, 3683 F). The slowest growing tumors, which are more differentiated, contained the highest glycerol-3-P dehydrogenase activities and the lowest levels of ether lipids. In the same investigation, five cell cultures (L- fibroblast, Ehrlich ascites cells, normal WI 38 cells, transformed WI-38VA13A cells, and KB carcinoma cells) had higher levels of ether lipids than normal tissues and the glycerol-3-P dehydrogenase activities were decreased to levels found in the fastest growing hepatomas.

Ehrlich ascites cells also have low glycerol-3-P dehydrogenase activity [1, 16]. However, LETNANSKY and KLC's work [67] demonstrated that the glycerol-3-P dehydrogenase activity varies considerably according to the strain of Ehrlich ascites cells; their data suggested that this is not the only factor that controls the activity of the glycerophosphate cycle in Ehrlich ascites cells. In this connection it is interesting that RAO and ABRAHAM [77] also found a low glycerol-3-P dehydrogenase activity in slow-growing mammary tumors in comparison to that found in normal mammary glands of lactating mice. Although the implication from these investigations is that dihydroxyacetone-P levels would be low, RAO and ABRAHAM [77] found just the opposite, i.e., the glycerol-3-P content was 12 times higher than that of dihydroxyacetone-P. The latter data indicate that the absolute cellular level of dihydroxyacetone-P is probably not the most important factor in explaining the higher levels of ether lipids in tumors.

AGRANOFF and HAJRA [1], using tritiated NADH and NADPH to quantitate the relative importance of glycerol-3-P and acyldihydroxyacetone-P in the biosynthesis of glycerolipids, obtained results showing that the acyldihydroxyacetone-P pathway is the dominant one in Ehrlich ascites cells. WYKLE et al. [153] also found that significant quantities of acyldihydroxyacetone-P are synthesized by the Ehrlich ascites cells. These results emphasize the important role of acyldihydroxyacetone-P in the biosynthesis of alkyl glycerolipids.

B. Biosynthesis of Alkyldihydroxyacetone-P

On the basis of available data, it does not seem that the enzyme catalyzing the formation of alkyldihydroxyacetone-P from fatty alcohols and acyldihydroxyacetone-P is responsible for the simple array of ether-linked chains found in neoplasms. The substitution of a fatty alcohol for the acyl moiety of acyldihydroxyacetone-P in microsomal preparations of preputial gland tumors and Ehrlich ascites cells occurs with a wide variety of substituted alcohols [112]. Data obtained by BANDI et al. [6, 7] also show that when animals are fed long-chain unsaturated alcohols that do not occur naturally, they too are incorporated into ether lipids. However, the substitution of a hydroxyl group at either end of the fatty alcohol chain precludes its utilization by the enzyme [112].

C. Biosynthesis of Fatty Alcohols

Free and bound fatty alcohols are found in a variety of mammalian cells [9, 135] and their levels appear to be higher in tumors [9]. Our laboratory has shown that fatty acids are converted to fatty alcohols by microsomal preparations from preputial gland tumors [113] and Ehrlich ascites cells [110] by NADPH-linked reductases, whereas the enzymes responsible for the reverse action, the oxidation of the alcohol to the acid, require NAD as a cofactor. The interconversion of fatty acids, fatty aldehydes, and fatty alcohols has been demonstrated in a number of preparations from mammals [33, 54, 58, 133, 134] and lower animals [104]. The specificity of the reductases has not been clearly documented in enzymic systems. However, in fish, the saturated acids appear to be reduced to the alcohol more readily than unsaturated fatty acids [85]. Furthermore, the lack of selectivity in the utilization of fatty alcohols in the biosyntheses of alkyldihydroxyacetone-P [112] indirectly supports the idea that the acyl-CoA reductase is responsible for the simplicity of the aliphatic moieties found in alkyl- and alk-1-enylglycerolipids.

D. Enzymic Cleavage of Ether Lipids

The O-alkyl cleavage enzyme is essentially absent from tumors, although some activities are found in highly differentiated slowly growing tumors such as 7794 A Morris hepatomas that possess only normal trace quantities of alkyl lipids [126]. In contrast, microsomes from 7777 Morris hepatomas, rapidly growing undifferentiated neoplasms, with the higher levels of alkyl lipids characteristic of malignant tumors, possess very low O-alkyl cleavage activities [126]. These data suggest that the O-alkyl cleavage enzyme plays an important role in controlling tissue levels of ether lipids. It should be pointed out, however, that 1-alkyl-sn-glycero-3-phosphorylethanolamine [117, 118] and even alkyl ethyleneglycols [118] are also split by the alkylglycerol cleavage system, but the enzyme will not attack alkylglycerolipids containing carbonyl, acyl, or phosphate groups substituted on the glycerol portion. In the whole animal, therefore, other catabolic enzymes seemingly work in concert with the alkyl cleavage system to regulate cellular levels of alkyl lipids. Enzymes that cleave the alk-1-enyl moiety of glycerolipids have not been investigated in cancerous tissues.

E. Factors Affecting the Formation of Ester- and Ether-Linked Lipids in Intact Cells

At the present time very little is known about the biochemical mechanisms that regulate the metabolism of ester- and ether-linked glycerolipids. SANSONE-BAZZANO et al. [86] have shown that when testosterone is administered to female mice, immature mice, or castrated adult male mice, it increases the levels of lipids derived from alcohol moieties in preputial glands (a rich source of ether lipids) and it also increases the incorporation of label from [6-^{14}C]-glucose into O-alkyl lipids. However, to our knowledge no other investigations on the hormonal control of ester-versus-ether pathways have been carried out.

LUMB and SNYDER [70] have illustrated that the quantitative contribution of the ether-lipid pathway can be determined, even in very complex systems, when [1-^{3}H]-fatty alcohols are used as precursors. The tritium is lost as water when the alcohol is oxidized to the acid, but it is retained when it is incorporated into O-alkyl lipids or disappears by half when incorporated into O-alk-1-enyl lipids.

Perhaps the most suitable way to investigate factors that regulate glycerolipids and their precursors in intact cells is to use tissue cultures (see chap. 5 in this book and the monograph edited by ROTHBLAT and KRITCHEVSKY [83], but one must still be careful in evaluating the results. For example, it is known that the turnover rates of glycerophospholipids greatly increase when 3T3 cultured mouse cells grow to confluency, even though there is no change in the rates of net synthesis [24]. Also, the nature of the lipids in the media (e.g., from the serum or detergents) can markedly influence lipid composition of the cells, although serum lipids did not markedly affect the ether-lipid content of L-M cells grown in suspension cultures [2]. However, STEELE and JENKIN [128] have reported that both isomeric forms (9-*cis* and 6-*cis*) of octadecenoic acids reduced the alkyl-diacylglycerols by about 10 % in Novikoff hepatoma cells grown in culture; the authors reasoned that this occurred because lower quantities of alcohols were formed from the monoenoic acids.

A recent publication from our laboratory [11] described a very simple cell culture system that should be extremely useful for investigating cellular regulatory mechanisms that affect the metabolism of ether-linked and ester-linked glycerolipids. In this system, L-M fibroblasts that grow in chemically defined media are easily manipulated by growth conditions, e.g., suspension-versus-monolayer cultures or pH, so that the relative proportions of ether- and ester-type glycerolipids are significantly altered. The fact that the altered cells show a great difference in their potential to develop tumors when they are reinjected into mice indicate that the biochemical changes in these cells might well be related to their malignancy.

V. Conclusion

It should be evident even from this rather concise and abbreviated overview of the literature that much more detailed and definitive work is required, especially in the area of enzyme regulation, to correlate and fully understand the multitude of alterations that have been reported in glycerolipids after neoplastic change. We do know that ether lipids accumulate in tumor tissues and that this build-up is caused by an elevation in enzymic activities for synthesis of ether lipids and their precursors, combined with a decrease in the activity of enzymes that degrade them. There is a trend toward unsaturation in the fatty acids at the sn-2-position of the ether lipids; this trend has also been described for many glycerolipids in reports of investigations that did not include subfractionation of the lipid classes. However, the reasons for these phenomena and the possible function of the high levels of ether lipids in cancer cells are still not known.

VI. References

1   AGRANOFF, B. W. and HAJRA, A. K.: The acyl dihydroxyacetone phosphate pathway for glycerolipid biosynthesis in mouse liver and Ehrlich ascites tumor cells. Proc. nat. Acad. Sci., Wash. 68: 411–415 (1971).
2   ANDERSON, R. E.; CUMMING, R. B.; WALTON, M., and SNYDER, F.: Lipid metabolism in cells grown in tissue culture: O-alkyl, O-alk-1-enyl, and acyl moieties of L-M cells. Biochim. biophys. Acta 176: 491–501 (1969).
3   ANGERVALL, L.; BJORNTORP, P., and STENER, B.: The lipid composition of hibernoma as compared with that of lipoma and of mouse brown fat. Cancer Res. 25: 408–409 (1969).
4   ANSELL, G. B. and SPANNER, S.: The magnesium-ion-dependent cleavage of the vinyl ether linkage of brain ethanolamine plasmalogen. Biochem. J. 94: 252–258 (1965).
5   BAER, E. and FISCHER, H. O. L.: Studies on acetone-glyceraldehyde, and optically active glycerides. IX. Configuration of the natural batyl, chimyl, and selachyl alcohols. J. biol. Chem. 140: 397–410 (1941).

6 BANDI, Z. L.; AAES-JORGENSEN, E., and MANGOLD, H. K.: Metabolism of unusual lipids in the rat. I. Formation of unsaturated alkyl and alk-1-enyl chains from orally administered alcohols. Biochim. biophys. Acta 239: 357–367 (1971).

7 BANDI, Z. L.; MANGOLD, H. K.; HOLMER, G., and AAES-JORGENSEN, E.: The alkyl and alk-1-enyl glycerols in the liver of rats fed long-chain alcohols or alkylglycerols. FEBS Letters 12: 217–220 (1971).

8 BERGELSON, L. D.; DYATLOVITSKAYA, E. V.; TORKHOVSKAYA, T. I.; SOROKINA, I. B., and GORKOVA, N. P.: Dedifferentiation of phospholipid composition in subcellular particles of cancer cells. FEBS Letters 2: 87–90 (1968).

9 BLANK, M. L. and SNYDER, F.: Long-chain fatty alcohols in normal and neoplastic tissues. Lipids 5: 337–341 (1970).

10 BLANK, M. L. and SNYDER, F.: Specificities of alkaline and acid phosphatases in the dephosphorylation of phospholipids. Biochemistry 9: 5034–5036 (1970).

11 BLANK, M. L. and SNYDER, F.: A tissue culture system for studies of the regulation of ether-linked and ester-linked aliphatic moieties in glycerolipids. Arch. Biochem. Biophys. 160: 100–105 (1974).

12 BLANK, M. L.; WYKLE, R. L.; ALPER, S., and SNYDER, F.: Microsomal synthesis of the ether analogs of triacylglycerols: Acyl CoA:alkylacylglycerol and acyl CoA:alk-1-enylacylglycerol acyltransferases in tumors and liver. Biochim. biophys. Acta 348: 397–403 (1974).

13 BLANK, M. L.; WYKLE, R. L.; PIANTADOSI, C., and SNYDER, F.: The biosynthesis of plasmalogens from labeled O-alkylglycerols in Ehrlich ascites cells. Biochim. biophys. Acta 210: 442–447 (1970).

14 BOLLINGER, J. N.: The isolation and tentative identification of diacylglyceryl ethers from the Walker 256 carcinoma of the rat and a human lymphosarcoma. Lipids 2: 143–148 (1967).

15 BOSCH, H. VAN DEN; GOLDE, L. M. G. VAN, and DEENEN, L. L. M. VAN: Dynamics of phosphoglycerides. Rev. Physiol. 66: 13–145 (1972).

16 BOXER, G. E. and SHONK, C. E.: Mitochondrial triose phosphate isomerase. Biochim. biophys. Acta 37: 194–196 (1960).

17 BREMER, J. and GREENBERG, D. M.: Methyl transferring enzyme system of microsomes in the biosynthesis of lecithin (phosphatidylcholine). Biochim. biophys. Acta 46: 205–216 (1961).

18 CARRUTHERS, C.: The fatty acid composition of the phosphatides of normal and malignant epidermis. Cancer Res. 27: 1–6 (1967).

19 CHAE, K.; PIANTADOSI, C., and SNYDER, F.: An alternate enzymic route for the synthesis of the alkyl analog of phosphatidic acid involving alkylglycerol. Biochem. biophys. Res. Commun. 51: 119–124 (1973).

20 CHAE, K.; PIANTADOSI, C., and SNYDER, F.: Reductase, phosphatase, and kinase activities in the metabolism of alkyldihydroxyacetone phosphate and alkyldihydroxyacetone. J. biol. Chem. 248: 6718–6723 (1973).

21 CHENG, S.; PIANTADOSI, C., and SNYDER, F.: Lipid droplets and glyceryl ether diesters in Ehrlich ascites cells grown in tissue culture. Lipids 2: 193–194 (1967).

22 CHRISTENSEN LOU, H. O. and CLAUSEN, J.: Polar lipids of oligodendrogliomas. J. Neurochem. 15: 263–264 (1968).

23 CHRISTIAN, J. C.; KANG, K. W., and POON, Y. C.: Phospholipids and cholesterol in plasma and erythrocyte membranes of adults with active malignancies. Clin. chim. Acta *21:* 289–290 (1968).

24 CUNNINGHAM, D. D.: Changes in phospholipid turnover following growth of 3T3 mouse cells to confluency. J. biol. Chem. *247:* 2464–2470 (1972).

25 DESNUELLE, P. and SAVARY, P.: Specificities of lipases. J. Lipid Res. *4:* 369–384 (1963).

26 DIPAOLO, J. A.; HEINING, A., and CARRUTHERS, C.: Isolation and lipid analysis of lipoid granules from Ehrlich ascites tumor cells. Proc. Soc. exp. Biol. Med. *113:* 68–70 (1963).

27 DOHR, H.; MANTECA, A., and SCHEITHAUER, E.: Phospholipids in brain tumors. Acta neurochir. *17:* 103–112 (1967).

28 DYATLOVITSKAYA, E. V.; TORKHOVSKAYA, T. I., and BERGEL'SON, L. D.: Phospholipids of cell membranes in the liver and hepatomas of rats (translation). Dokl. biol. Sci. *186:* 363–365 (1969).

29 DYATLOVITSKAYA, E. V.; TORKHOVSKAYA, T. I., and BERGEL'SON, L. D.: Tumor lipids. Investigation of phospholipids of Jensen sarcoma. Biochemistry (Russian) *34:* 144–148 (1969).

30 DYATLOVITSKAYA, E. V.; TORKHOVSKAYA, T. I.; SOROKINA, I. B.; GOR'KOVA, N. P., and BERGEL'SON, L. D.: Tumor lipids. Distribution of phospholipids in the mitochondria and microsomes of Jensen sarcoma. Biochemistry (Russian) *34:* 462–464 (1969).

31 EICHBERG, J.; GILBERTSON, J. R., and KARNOVSKY, M. L.: Neutral plasmalogens analogous to the neutral triglycerides. J. biol. Chem. *236:* PC 15–16 (1961).

32 ERBLAND, J. F. and MARINETTI, G. V.: The metabolism of lysolecithin in rat-liver particulate systems. Biochim. biophys. Acta *106:* 139–144 (1965).

33 FERRELL, W. J. and KESSLER, R. J.: Enzymic relationship of free fatty acids, aldehydes and alcohols in mouse liver. Physiol. Chem. Phys. *3:* 549–558 (1971).

34 FIGARD, P. H. and GREENBERG, D. M.: Phosphatides of some mouse ascites tumors and rat hepatomas. Cancer Res. *22:* 361–367 (1962).

35 GAITI, A.; GORACCI, G.; DE MEDIO, G. E., and PORCELLATI, G.: Enzymic synthesis of plasmalogen and O-alkyl glycerolipid by base-exchange reaction in the rat brain. FEBS Letters *27:* 116–120 (1972).

36 GERSTL, B.; HAYMAN, R. B.; TAVASTSTJERNA, M. G., and SMITH, J. K.: Fatty acids of white matter of human brain. Experimentia *18:* 131–133 (1962).

37 GIGG, J. and GIGG, R.: The preparation of unsymmetrical diglycerides. J. chem. Soc. *C:* 431–434 (1967).

38 GILBERTSON, J. R. and KARNOVSKY, M. L.: Nonphosphatide fatty acyl esters of alkenyl and alkyl ethers of glycerol. J. biol. Chem. *238:* 893–897 (1963).

39 GOLDE, L. M. G. VAN and DEENEN, L. L. M. VAN: The effect of dietary fat on the molecular species of lecithin from rat liver. Biochim. biophys. Acta *125:* 496–509 (1966).

40 GOLDE, L. M. G. VAN and DEENEN, L. L. M. VAN: Molecular species of extracellular phosphatidylethanolamine from *Escherichia coli*. Chem. Phys. Lipids *1:* 157–164 (1967).

41 GOPAL, K.; GROSSI, E.; PAOLETTI, P., and USARDI, M.: Lipid composition of human intracranial tumors. A biochemical study. Acta neurochir. *11:* 333–347 (1963).
42 GOTTFRIED, E. L.: Lipids of human leukocytes. Relation to cell type. J. Lipid Res. *8:* 321–327 (1967).
43 GRAY, G. M.: The lipid composition of tumour cells. Biochem. J. *86:* 350–357 (1963).
44 HAAS, G. H. DE; SARDA, L., and ROGER, J.: Positional specific hydrolysis of phospholipids by pancreatic lipase. Biochim. biophys. Acta *106:* 638–640 (1965).
45 HAJRA, A. K.: Biosynthesis of phosphatidic acid from dihydroxyacetone phosphate. Biochem. biophys. Res. Commun. *33:* 929–935 (1968).
46 HAJRA, A. K.: Biosynthesis of alkyl-ether containing lipid from dihydroxyacetone phosphate. Biochem. biophys. Res. Commun. *37:* 486–492 (1969).
47 HAJRA, A. K.: Acyl dihydroxyacetone phosphate: Precursor of alkyl ethers. Biochem. biophys. Res. Commun. *39:* 1037–1044 (1970).
48 HAJRA, A. K. and AGRANOFF, B. W.: Acyl dihydroxyacetone phosphate. Characterization of a ^{32}P-labeled lipid from guinea pig liver mitochondria. J. biol. Chem. *243:* 1617–1622 (1968).
49 HAVEN, F. L. and BLOOR, W. R.: Lipids in cancer. Adv. Cancer Res. *4:* 237–314 (1956).
50 HILF, R.; MICHEL, I.; GIBBS, C. C., and BELL, C.: NADP-Linked enzymes and lipids in normal and neoplastic mammary tissue. Effects of estrogen and dietary glucose. Biochim. biophys. Acta *116:* 589–592 (1966).
51 HOKIN, M. R. and HOKIN, L. E.: The synthesis of phosphatidic acid from diglyceride and adenosine triphosphate in extracts of brain microsomes. J. biol. Chem. *234:* 1381–1387 (1959).
52 HOWARD, B. V.; MORRIS, H. P., and BAILEY, J. M.: Ether-lipids, α-glycerol phosphate dehydrogenase, and growth rate in tumors and cultured cells. Cancer Res. *32:* 1533–1538 (1972).
53 JOHNSON, M. and MORA, P. T.: Lipids of the Rauscher mouse leukemia virus. Virology *31:* 230–237 (1967).
54 JOHNSON, R. C. and GILBERTSON, J. R.: Isolation, characterization, and partial purification of a fatty acyl coenzyme A reductase from bovine cardiac muscle. J. biol. Chem. *247:* 6991–6998 (1972).
55 JOHNSON, R. M. and DUTCH, P. H.: The composition of the lipides of resting and pregnancy-stimulated mammary glands and mammary carcinomas. Arch. Biochem. Biophys. *40:* 239–244 (1952).
56 KASAMA, K.; UEZUMI, N., and ITOH, K.: Characterization and identification of glyceryl ether diesters in harderian gland tumor of mice. Biochim. biophys. Acta *202:* 56–66 (1970).
57 KATES, M.: Techniques of lipidology. Isolation, analysis and identification of lipids; in WORK and WORK Laboratory techniques in biochemistry and molecular biology (American Elsevier, New York 1972).
58 KAWALEK, J. C. and GILBERTSON, J. R.: Enzymic reduction of free fatty aldehydes in bovine cardiac muscle. Biochem. biophys. Res. Commun. *51:* 1027–1033 (1973).

59 KAWANAMI, J.; TSUJI, T., and OTSUKA, H.: Lipid of cancer tissues. I. Lipid composition of Shionogi carcinoma 115 and Nakahara-Fukuoka sarcoma. J. Biochem. *59:* 151–159 (1966).

60 KENNEDY, E. P. and WEISS, S. B.: The function of cytidine coenzymes in the biosynthesis of phospholipides. J. biol. Chem. *222:* 193–214 (1956).

61 KIYASU, J. Y. and KENNEDY, E. P.: The enzymatic synthesis of plasmalogens. J. biol. Chem. *235:* 2590–2594 (1960).

62 KORNBERG, A. and PRICER, W. E., JR.: Enzymatic esterification of α-glycerophosphate by long chain fatty acids. J. biol. Chem. *204:* 345–357 (1953).

63 KUZ'MINA, S. N.; TROITSKAYA, L. P.; ZBARSKII, I. B.; DYATLOVITSKAYA, E. V.; TORKHOVSKAYA, T. I., and BERGEL'SON, L. D.: Tumor lipids. Phospholipid composition of cell nuclei from the liver, ascites hepatoma, and Jensen sarcoma. Biochemistry (Russian) *34:* 610–614 (1969).

64 LABELLE, E. F., JR. and HAJRA, A. K.: Biosynthesis of acyl dihydroxyacetone phosphate in subcellular fractions of rat liver. J. biol. Chem. *247:* 5835–5841 (1972).

65 LANDS, W. E. M.: Metabolism of glycerolipids. II. The enzymatic acylation of lysolecithin. J. biol. Chem. *235:* 2233–2237 (1960).

66 LEE, T-c.; STEPHENS, N.; MOEHL, A., and SNYDER, F.: Turnover of rat liver plasma membrane phospholipids. Comparison with microsomal membranes. Biochim. biophys. Acta *291:* 86–92 (1973).

67 LETNANSKY, K. and KLC, G. M.: Glycerolphosphate oxidoreductases and the glycerophosphate cycle in Ehrlich ascites tumor. Arch. Biochem. Biophys. *130:* 218–226 (1969).

68 LINDLAR, F. and WAGENER, H.: On the lipid and fatty acid composition of experimental tumors. Schweiz. med. Wschr. *94:* 243–250 (1964).

69 LUMB, R. H.: Private commun.

70 LUMB, R. H. and SNYDER, F.: A rapid isotopic method for assessing the biosynthesis of ether linkages in glycerolipids of complex systems. Biochim. biophys. Acta *244:* 217–221 (1971).

71 MIHAILESCU, E.; APOSTOLESCU, I.; ZAMFIRESCU-GHEORGHIU, M., and MICU, D.: Investigation of lymph node lipid fractions by means of thin-layer chromatography. Rev. roum. med. Interne *5:* 209–216 (1968).

72 NAYYAR, S. N.: A study of phosphate, deoxyribonucleic acid, and phospholipid fractions in neural tumors. Neurology, Minneap. *13:* 287–291 (1963).

73 PALTAUF, F. and HOLASEK, A.: Enzymatic synthesis of plasmalogens. Characterization of the 1-O-alkyl-2-acyl-sn-glycero-3-phosphorylethanolamine desaturase from mucosa of hamster small intestine. J. biol. Chem. *248:* 1609–1615 (1973).

74 PALTAUF, F. and POLHEIM, D.: Occurrence of C_{20} alk-1-enyl and alkyl glycerol ethers in phospholipids of the rat intestinal mucosa. Biochim. biophys. Acta *210:* 187–189 (1970).

75 PFLEGER, R. C.; PIANTADOSI, C., and SNYDER, F.: The biocleavage of isomeric glyceryl ethers by soluble liver enzymes in a variety of species. Biochim. biophys. Acta *144:* 633–648 (1967).

76 RAGHAVAN, S.; RHOADS, D., and KANFER, J.: *In vitro* incorporation of [14C]serine, [14C]ethanolamine, and [14C]choline into phospholipids of neuronal and glial-enriched fractions from rat brain by base exchange. J. biol. Chem. *247:* 7153–7156 (1972).

77 RAO, G. A. and ABRAHAM, S.: α-Glycerolphosphate dehydrogenase activity and levels of glyceride-glycerol precursors in mouse mammary tissues. Lipids *8:* 232–234 (1973).

78 RAPPORT, M. M. and ALONZO, N.: Chromatographic fractionation of tissue (lymphosarcoma) phospholipids. Proc. 2nd Int. Conf. Biochemical Problems of Lipids, Ghent 1955, pp. 69–72 (Interscience, New York 1956).

79 RAY, T. K. and ROY, S. C.: Studies in lipids of human normal and malignant cervix uteri. I. Lipid content and composition of whole tissue and subcellular fractions. Indian J. Biochem. *3:* 106–110 (1966).

80 REES, E. D.; SHUCK, A. E., and ACKERMANN, H.: Lipid composition of rat mammary carcinomas, mammary glands, and related tissues; endocrine influences. J. Lipid Res. *7:* 396–402 (1966).

81 REES, E. D.; SHUCK, A. E., and ACKERMANN, H.: Hormonal influence on fatty acid composition of sterol ester and phospholipid fractions of experimental mammary carcinomas. Proc. Soc. exp. Biol. Med. *128:* 666–670 (1968).

82 RENKONEN, O.: Chromatographic separation of plasmalogenic, alkyl-acyl, and diacyl forms of ethanolamine glycerophosphatides. J. Lipid Res. *9:* 34–39 (1968).

83 ROTHBLAT, G. H. and KRITCHEVSKY, D. (eds): Lipid metabolism in tissue culture cells. Wistar Inst. Symp. Monogr. No. 6 (Wistar Institute Press, Philadelphia 1967).

84 RUGGIERI, S. and FALLANI, A.: Lipid composition of cytoplasmic fractions from rat liver and from Yoshida ascites hepatoma. Sperimentale *118:* 313–335 (1968).

85 SAND, D. M.; HEHL, J. L., and SCHLENK, H.: Biosynthesis of wax esters in fish. Reduction of fatty acids and oxidation of alcohols. Biochemistry *8:* 4851–4854 (1969).

86 SANSONE-BAZZANO, G.; BAZZANO, G.; REISNER, R. M., and HAMILTON, J. G.: The hormonal induction of alkyl glycerol, wax and alkyl acetate synthesis in the preputial gland of the mouse. Biochim. biophys. Acta *260:* 35–40 (1972).

87 SCHMID, H. H. O.; BAUMANN, W. J., and MANGOLD, H. K.: Alkoxylipids. III. Naturally occurring D(+)-1-*O*-*cis*-alk-1′-enyl-diglycerides. Biochim. biophys. Acta *144:* 344–354 (1967).

88 SCHMID, H. H. O. and MANGOLD, H. K.: Neutral plasmalogens and alkoxy diglycerides in human perinephric fat. Biochem. Z. *346:* 13–25 (1966).

89 SCHMID, H. H. O.; TUNA, N., and MANGOLD, H. K.: The composition of *O*-alkyl diglycerides of human subcutaneous adipose tissue. J. physiol. Chem. *348:* 730–732 (1967).

90 SCHOGT, J. C. M.; BEGEMANN, P. H., and KOSTER, J.: Nonphosphatide aldehydogenic lipids in milk fat, beef tallow, and ox heart. J. Lipid Res. *1:* 446–449 (1960).

91 SLAGEL, D. E.; DITTMER, J. C., and WILSON, C. B.: Lipid Composition of human glial tumour and adjacent brain. J. Neurochem. *14:* 789–798 (1967).

92 SLOT, E.: On the content and significance of some lipids in ovarian tumors. Acta Un. int. Cancr. *19:* 1154–1155 (1963).
93 SLOTBOOM, A. J.; HAAS, G. H. DE, and DEENEN, L. L. M. VAN: On the synthesis of plasmalogens. Chem. Phys. Lipids *1:* 192–208 (1967).
94 SMITH, S. W.; WEISS, S. B., and KENNEDY, E. P.: The enzymatic dephosphorylation of phosphatidic acids. J. biol. Chem. *228:* 915–922 (1957).
95 SMITH, R. R. and WHITE, H. B., JR.: Neutral lipid patterns of normal and pathologic nervous tissue. Studies by thin-layer chromatography. Arch. Neurol. *19:* 54–59 (1968).
96 SNYDER, C.; MALONE, B.; NETTESHEIM, P., and SNYDER, F.: Urethan-induced pulmonary adenoma as a tool for the study of surfactant biosynthesis. Cancer Res. *33:* 2437–2443 (1973).
97 SNYDER, F.: Ether-linked lipids in neoplasms of man and animals: Methods of measurement and the occurrence and nature of alkyl and alk-1-enyl moieties. Adv. exp. Med. Biol. *4:* 609–621 (1969).
98 SNYDER, F.: Lipids and cancer. A bibliography: 1947–1970. United States Atomic Energy Commission Report, ORAU 111 (1970).
99 SNYDER, F.: Glycerolipids in the neoplastic cell. Methodology, metabolism, and composition; in BUSCH Methods in cancer research, vol. 6, pp. 399–436 (Academic Press, New York 1971).
100 SNYDER, F.: The chemistry, physical properties, and chromatography of lipids containing ether bonds; in NIEDERWIESER and PATAKI Progress in thin-layer chromatography and related methods, vol. 2, chap. 4, pp. 105–141 (Ann Arbor Science Publishers, Ann Arbor 1971).
101 SNYDER, F. (ed.): Ether lipids. Chemistry and biology (Academic Press, New York 1972).
102 SNYDER, F.: Nomenclature; in SNYDER Ether lipids. Chemistry and biology, pp. XVII–XIX (Academic Press, New York 1972).
103 SNYDER, F.: Ether-linked lipids and fatty alcohol precursors in neoplasms; in SNYDER Ether lipids. Chemistry and biology, chap. 10, pp. 273–295 (Academic Press, New York 1972).
104 SNYDER, F.: Enzymatic systems that synthesize and degrade glycerolipids possessing ether bonds. Adv. Lipid Res. *10:* 233–259 (1972).
105 SNYDER, F.: Thin-layer chromatographic behavior of glycerolipid analogs containing ether, ester, hydroxyl, and ketone groupings. J. Chromat. *82:* 7–14 (1973).
106 SNYDER, F.: Analysis of alkyl and alk-1-enyl ether lipids and their derivatives by chromatographic techniques; in MARINETTI Lipid chromatographic analysis (Marcel Dekker, New York, in press).
107 SNYDER, F.; BLANK, M. L., and MALONE, B.: Requirement of cytidine derivatives in the biosynthesis of *O*-alkyl phospholipids. J. biol. Chem. *245:* 4016–4018 (1970).
108 SNYDER, F.; BLANK, M. L.; MALONE, B., and WYKLE, R. L.: Identification of *O*-alkyldihydroxyacetone phosphate, *O*-alkyldihydroxyacetone, and diacyl glyceryl ethers after enzymic synthesis. J. biol. Chem. *245:* 1800–1805 (1970).

109 SNYDER, F.; BLANK, M. L., and MORRIS, H. P.: Occurrence and nature of O-alkyl and O-alk-1-enyl moieties of glycerol in lipids of Morris transplanted hepatomas and normal rat liver. Biochim. biophys. Acta 176: 502–510 (1969).

110 SNYDER, F.; BLANK, M. L., and WYKLE, R. L.: The enzymic synthesis of ethanolamine plasmalogens. J. biol. Chem. 246: 3639–3645 (1971).

111 SNYDER, F.; CRESS, E. A., and STEPHENS, N.: An unidentified lipid prevalent in tumors. Lipids 1: 381–386 (1966).

112 SNYDER, F.; CLARK, M., and PIANTADOSI, C.: Biosynthesis of alkyl lipids. Displacement of the acyl moiety of acyldihydroxyacetone phosphate with fatty alcohol analogs. Biochem. biophys. Res. Commun. 53: 350–356 (1973).

113 SNYDER, F. and MALONE, B.: Enzymic interconversion of fatty alcohols and fatty acids. Biochem. biophys. Res. Commun. 41: 1382–1387 (1970).

114 SNYDER, F.; MALONE, B., and BLANK, M. L.: Enzymic synthesis of O-alkyl bonds in glycerolipids. J. biol. Chem. 245: 1790–1799 (1970).

115 SNYDER, F.; MALONE, B.; BLANK, M. L.; CLEVENGER, M., and GOSWITZ, F. A.: in 1971 Research report, Medical Division, Oak Ridge Associated Universities, pp. 127–128. United States Atomic Energy Commission Report ORAU-116 (1972).

116 SNYDER, F.; MALONE, B., and CUMMING, R. B.: Synthesis of glyceryl ethers by microsomal enzymes derived from fibroblasts (L-M) cells grown in suspension cultures. Canad. J. Biochem. 48: 212–215 (1970).

117 SNYDER, F.; MALONE, B., and PIANTADOSI, C.: Tetrahydropteridine-dependent cleavage enzyme for O-alkyl lipids. Substrate specificity. Biochim. biophys. Acta 316: 259–265 (1973).

118 SNYDER, F.; MALONE, B., and PIANTADOSI, C.: Enzymic studies of glycol and glycerol lipids containing O-alkyl bonds in liver and tumor tissues. Arch. Biochem. Biophys 161: 402–407 (1974).

119 SNYDER, F.; MALONE, B., and WYKLE, R. L.: The biosynthesis of alkyl ether bonds in lipids by a cell-free system. Biochem. biophys. Res. Commun. 34: 40–47 (1969).

120 SNYDER, F. and PIANTADOSI, C.: Deacylation of isomeric diacyl-1-^{14}C-alkoxyglycerols by pancreatic lipase. Biochim. biophys. Acta 152: 794–797 (1968).

121 SNYDER, F.; PIANTADOSI, C., and MALONE, B.: The participation of 1- and 2-isomers of O-alkylglycerols as acyl acceptors in cell-free systems. Biochim. biophys. Acta 202: 244–249 (1970).

122 SNYDER, F.; RAINEY, W. T., JR.; BLANK, M. L., and CHRISTIE, W. H.: The source of oxygen in the ether bond of glycerolipids: ^{18}O studies. J. biol. Chem. 245: 5853–5856 (1970).

123 SNYDER, F. and WOOD, R.: The occurrence and metabolism of alkyl and alk-1-enyl ethers of glycerol in transplantable rat and mouse tumors. Cancer Res. 28: 972–978 (1968).

124 SNYDER, F. and WOOD, R.: Alkyl and alk-1-enyl ethers of glycerol in lipids from normal and neoplastic human tissues. Cancer Res. 29: 251–257 (1969).

125 SNYDER, F.; WYKLE, R. L., and MALONE, B.: A new metabolic pathway. Biosynthesis of alkyl ether bonds from glyceraldehyde-3-phosphate and fatty

alcohols by microsomal enzymes. Biochem. biophys. Res. Commun. *34:* 315–321 (1969).

126 SOODSMA, J. F.; PIANTADOSI, C., and SNYDER, F.: The biocleavage of alkyl glyceryl ethers in Morris hepatomas and other transplantable neoplasms. Cancer Res. *30:* 309–311 (1970).

127 SOODSMA, J. F.; PIANTADOSI, C., and SNYDER, F.: Partial characterization of the alkylglycerol cleavage enzyme system of rat liver. J. biol. Chem. *247:* 3923–3929 (1972).

128 STEELE, W. and JENKIN, H. M.: The effect of two isomeric octadecenoic acids on alkyl diacyl glycerides and neutral glycosphingolipids of Novikoff hepatoma cells. Lipids *7:* 556–559 (1972).

129 STEIN, A. A.; OPALKA, E., and ROSENBLUM, I.: Fatty acid analysis of two experimental transmissible glial tumors by gas-liquid chromatography. Cancer Res. *25:* 201–205 (1965).

130 STEIN, A. A.; OPALKA, E., and ROSENBLUM, I.: Hepatic lipids in tumor-bearing (glioma) mice. Cancer Res. *25:* 957–961 (1965).

131 STEIN, A. A.; OPALKA, E., and SERRONE, D.: Sequential hepatic triglycerides in tumor-bearing mice. Cancer Res. *26:* 1707–1710 (1966).

132 STEIN, O. and STEIN, Y.: Utilization of lysolecithin by Landschütz ascites tumor *in vivo* and *in vitro*. Biochim. biophys. Acta *137:* 232–239 (1967).

133 STOFFEL, W.; LEKIM, D., and HEYN, G.: Sphinganine (dihydrosphingosine), an effective donor of the alk-1'-enyl chain of plasmalogens. Z. Physiol. Chem. *351:* 875–933 (1970).

134 TABAKOFF, B. and ERWIN, G. V.: Purification and characterization of a reduced nicotinamide adenine dinucleotide phosphate-linked aldehyde reductase from brain. J. biol. Chem. *245:* 3263–3268 (1970).

135 TAKAHASHI, T. and SCHMID, H. H. O.: Long-chain alcohols in mammalian tissues. Chem. Phys. Lipids *4:* 243–246 (1970).

136 THOMPSON, G. A., JR.: The biosynthesis of ether-containing phospholipids in the slug, *Arion ater*. III. Origin of the vinylic ether bond of plasmalogens. Biochim. biophys. Acta *152:* 409–411 (1968).

137 THOMPSON, G. A., JR. and LEE, P.: Studies of the α-glyceryl ether lipids occurring in molluscan tissues. Biochim. biophys. Acta *98:* 151–159 (1965).

138 TIETZ, A.; LINDBERG, M., and KENNEDY, E. P.: A new pteridine-requiring enzyme system for the oxidation of glyceryl ethers. J. biol. Chem. *239:* 4081–4090 (1964).

139 UEZUMI, N.; HASEGAWA, S.; KASAMA, K., and YAMADA, T.: Studies on the lipids of neoplastic tissues. I. Fatty acid compositions of the lipids from several induced and transplanted tumors of mice. Mie med. J. *12:* 159–170 (1962).

140 VEERKAMP, J. H.; MULDER, I., and DEENEN, L. L. M. VAN: Comparative studies on the phosphatides of normal rat liver and primary hepatoma. Z. Krebsforsch. *64:* 137–148 (1961).

141 WALLACH, D. F. H.; SODERBERG, J., and BRICKER, L.: The phospholipids of Ehrlich ascites carcinoma cells. Composition and intracellular distribution. Cancer Res. *20:* 397–402 (1960).

142 Warner, H.R. and Lands, W.E.M.: The metabolism of plasmalogen. Enzymatic hydrolysis of the vinyl ether. J. biol. Chem. *236:* 2404–2409 (1961).
143 Weber, G. and Cantero, A.: Phospholipid content in Novikoff hepatoma, regenerating liver and in liver of fed and fasted normal rats. Exp. Cell Res. *13:* 125–131 (1957).
144 Weiss, S.B.; Kennedy, E.P., and Kiyasu, J.Y.: The enzymatic synthesis of triglyerides. J. biol. Chem. *235:* 40–44 (1960).
145 Wood, R. and Healy, K.: Tumor lipids. Biosynthesis of plasmalogens. J. biol. Chem. *245:* 2640–2648 (1970).
146 Wood, R. and Snyder, F.: Characterization and identification of glyceryl ether diesters present in tumor cells. J. Lipid Res. *8:* 494–500 (1967).
147 Wood, R. and Snyder, F.: Tumor lipids. Metabolic relationships derived from structural analyses of acyl, alkyl, and alk-1-enyl moieties of neutral glycerides and phosphoglycerides. Arch. Biochem. Biophys. *131:* 478–494 (1969).
148 Wood, R.; Walton, M.; Healy, K., and Cumming, R.B.: Plasmalogen biosynthesis in Ehrlich ascites cells grown in tissue culture. J. biol. Chem. *245:* 4276–4285 (1970).
149 Wykle, R.L.: Private commun.
150 Wykle, R.L.; Blank, M.L.; Malone, B., and Snyder, F.: Evidence for a mixed-function oxidase in the biosynthesis of ethanolamine plasmalogens from 1-alkyl-2-acyl-*sn*-glycero-3-phosphorylethanolamine. J. biol Chem. *247:* 5442–5447 (1972).
151 Wykle, R.L.; Blank, M.L., and Snyder, F.: The biosynthesis of plasmalogens in a cell-free system. FEBS Letters *12:* 57–60 (1970).
152 Wykle, R.L.; Blank, M.L., and Snyder, F.: The enzymic incorporation of arachidonic acid into ether-containing choline and ethanolamine phosphoglycerides by deacylation-acylation reactions. Biochim. biophys. Acta *326:* 26–33 (1973).
153 Wykle, R.L.; Piantadosi, C., and Snyder, F.: The role of acyldihydroxyacetone phosphate, NADH, and NADPH in the biosynthesis of *O*-alkyl glycerolipids by microsomal enzymes of Ehrlich ascites tumor. J. biol. Chem. *247:* 2944–2948 (1972).
154 Wykle, R.L. and Schremmer, J.M.: A lysophospholipase D pathway in the metabolism of ether-linked lipids in brain microsomes. J. biol. Chem. *249:* 1742–1746 (1974).
155 Wykle, R.L. and Snyder, F.: The glycerol source for the biosynthesis of alkyl glyceryl ethers. Biochem. biophys. Res. Commun. *37:* 658–662 (1969).
156 Wykle, R.L. and Snyder, F.: Biosynthesis of an *O*-alkyl analogue of phosphatidic acid and *O*-alkylglycerols via *O*-alkyl ketone intermediates by microsomal enzymes. J. biol. Chem. *245:* 3047–3058 (1970).
157 Yamakawa, T. and Ueta, N.: Biochemistry of lipids of neoplastic tissue. II. Variability of the fatty acid composition of mouse ascites tumor cells. Jap. J. exp. Med. *32:* 591–598 (1962).
158 Yamakawa, T.; Ueta, N., and Irie, R.: Biochemistry of lipids of neoplastic tissue. I. Lipid composition of ascites tumor cells of mice. Jap. J. exp. Med. *32:* 289–296 (1962).

159 YAMASHITA, S.; HOSAKA, K., and NUMA, S.: Resolution and reconstitution of the phosphatidate-synthesizing system of rat-liver microsomes. Proc. nat. Acad. Sci., Wash. *69:* 3490–3492 (1972).
160 YAVIN, E. and GATT, S.: Oxygen-dependent cleavage of the vinyl-ether linkage of plasmalogens. I. Cleavage by rat-brain supernatant. Europ. J. Biochem. *25:* 431–436 (1972).
161 YAVIN, E. and GATT, S.: Oxygen-dependent cleavage of the vinyl-ether linkage of plasmalogens. II. Identification of the low molecular weight active component and the reaction mechanism. Europ. J. Biochem. *25:* 437–446 (1972).

Authors' address: FRED SNYDER and CATHY SNYDER, Oak Ridge Associated Universities, P.O. Box 117, *Oak Ridge, TN 37830* (USA)

Fatty Acid Metabolism in Tumors[1]

ARTHUR A. SPECTOR[2]

Departments of Biochemistry and Internal Medicine, University of Iowa, Iowa City, Iowa

Contents

I. Introduction	43
II. Sources of Tumor Fatty Acids	44
A. Fatty Acid Biosynthesis in Tumors	44
1. *De novo* Biosynthesis	45
2. Desaturation and Chain Elongation	46
a) Essential Fatty Acids	46
3. Regulation of Fatty Acid Biosynthesis	48
4. Glucose Incorporation into Lipids	49
5. Relative Contributions of Biosynthesis and Host Lipids	50
B. Lipids Supplied by the Host	51
1. Blood Plasma Lipids	51
a) Role of Dietary Lipids	52
2. Ascites Plasma Lipids	52
a) Comparison of Cell and Extracellular Fluid Fatty Acid Composition	54
III. Free Fatty Acid Utilization by Tumors	54
A. Free Fatty Acid Utilization *in vivo*	55
B. Mechanism of Free Fatty Acid Utilization in Tumors	56
1. Uptake	57
2. Oxidation	58
3. Esterification	60
IV. Esterified Fatty Acid Utilization by Tumors	62
A. Lipoprotein Triglycerides	62
V. Turnover of Esterified Fatty Acids in Tumors	64
A. Free Fatty Acid Release from Tumor Cells	65

1 This review was aided by research grants from the National Institutes of Health (HL 14,388 and HL 14,781).
2 Recipient of a Research Career Development Award from the National Institutes of Health (K 04-HL 20,338).

VI. Summary and Comment	66
A. Future Directions	67
VII. Acknowledgments	69
VIII. References	69

I. Introduction

Tumors require fatty acids as oxidative substrates, for replacement of membrane lipid components and to manufacture the new membranes needed for growth and cell division. The fatty acids necessary for these purposes can be obtained either by synthesis from a nonlipid precursor, such as glucose, or they can be supplied preformed by the host. In spite of the central role of fatty acids for tumor nutrition and growth, relatively little work has been done in this area during the last 15 years. Two factors appear to be responsible for this paucity of new information concerning fatty acid metabolism in tumors. First, other approaches appear more attractive in terms of providing key insights into the etiology of cancer. Second, the currently available information suggests that the metabolic pathways for fatty acids are, in general, the same in tumors and nonmalignant tissues. This has led to a commonly held viewpoint that chemotherapy aimed at fatty acid metabolism would not be a fruitful approach to cancer control. On the other hand, much new information about fatty acid metabolism has been acquired in the last 15 years, for the apparent relationship between hyperlipidemia and atherosclerosis has stimulated renewed interest in this area. Drugs that inhibit fatty acid biosynthesis, depress free fatty acid release from adipocytes or lower plasma lipoprotein concentrations are becoming available for clinical usage [41, 43, 47]. Therefore, it seems appropriate to review the presently available data on fatty acid metabolism in tumors, in the hope of gaining new perspectives concerning the possible applicability of these recent advances to cancer control. In addition to summarizing the current state of knowledge in this field, I shall try to point out the gaps that exist in our present information and to highlight those areas which seem to hold the greatest potential for future study.

Some of the most interesting aspects related to fatty acid metabolism in tumors either have been omitted entirely from this review or are treated only superficially. One such area is glycerophosphatide metabolism, particularly the synthesis of the alkyl ethers and plasmalogens. Others include fatty acid metabolism in nonmalignant cells grown in culture and the control of fatty acid biosynthesis in tumors. These subjects are covered in detail in other sections of this book.

II. Sources of Tumor Fatty Acids

MEDES et al. [50] incubated slices from a number of transplanted rodent tumors with [2-^{14}C]acetate or [U-^{14}C]glucose. These included hepatoma 98/10, mammary adenocarcinoma TA3, rhabdomyosarcoma MC-1A and sarcoma 37. In each case, radioactivity was recovered in the fatty acids of the tumor. The rates of fatty acid synthesis, however, were judged to be too slow to satisfy the lipid requirements of these rapidly growing tumors. It appeared that these tumors obtained a large portion of their fatty acid requirements preformed from the host animal. MEDES et al. [49] then fed [1-^{14}C]palmitate dissolved in olive oil through an intragastric tube to mice bearing these tumors. Most of the administered palmitate was oxidized to CO_2 and H_2O during the 24 h following feeding. During the first 6 h, the majority of the labeled fatty acid remaining in the animal was recovered in the liver. At 6 h, the specific radioactivity of the liver fatty acids was 5–20 times greater than those of the tumors. The specific radioactivity of the tumor fatty acids increased and reached a peak in 48–72 h. Slices of these tumors oxidized some of the incorporated radioactive fatty acids during in vitro incubation, and the specific radioactivity of the evolved CO_2 was approximately the same as that of the tumor fatty acids. These findings were interpreted as supporting the viewpoint that tumors derive a large part of their fatty acid requirement from the host.

Studies with the Ehrlich ascites carcinoma have led to similar conclusions; namely, that although fatty acid synthesis from glucose does occur, most of the fatty acids present in the cells are derived from the ascites plasma in the form of free fatty acids [8, 53, 75, 84] and triglycerides contained in very low density lipoproteins [14, 84].

A. Fatty Acid Biosynthesis in Tumors

Many investigators have confirmed the observations of MEDES et al. [50] that tumors are able to incorporate radioactive water-soluble substrates into fatty acids. A series of transplanted mouse ovarian tumors and a hepatoma incorporated [1-^{14}C]acetate into fatty acids [26]. The rate of fatty acid synthesis was 5–10 times higher in the hepatoma than in the ovarian tumors. Glucose stimulated the incorporation of labeled acetate into fatty acids in these tumors [27]. Fructose produced less of a stimulatory effect. Since the carbohydrate-induced increase was not accompanied by any decrease in acetate oxidation, this effect is not due simply to a diversion of acetate from the oxidative to the synthetic pathway. It was concluded that fatty acid synthesis was stimulated because the carbohydrates were providing energy through glycolysis [27]. Subsequent work with the Ehrlich ascites tumor suggests that glucose exerts the stimulatory effect either by providing NADPH through the pentose phosphate pathway [40] or by being converted to the triose acceptors, L-glycerol 3-phosphate or dihydroxyacetone phosphate, for incorporation of the newly formed fatty acids into lipid esters [95]. A tumor induced by the Rous sarcoma virus in chorioallantoic membranes of chicken eggs also was found to incorporate labeled acetate, pyruvate and glucose into fatty acids [29].

Hepatomas synthesize fatty acids, but the rate at which this occurs varies widely. For example, the Novikoff hepatoma incorporated labeled glucose into fatty acids at a rate which was 11 % of that observed in normal liver, whereas in the slow-growing Morris hepatoma 5123, the rate was less than 1 % of that seen in normal liver [103]. Slightly more pyruvate than glucose was converted to fatty acids by the Morris hepatoma 5123. The slow rate of fatty acid synthesis is not characteristic of all Morris hepatomas, for the minimal deviation hepatoma 9121 incorporated [1-^{14}C]acetate into fatty acids at the same rate as normal liver and somewhat faster than host liver [24].

Fatty acids also are synthesized from glucose by the Barrett mammary adenocarcinoma [1]. As compared with mammary tissue from C3H mice, however, the tumor incorporated much smaller amounts of labeled glucose into fatty acids. Less lactate also was converted to fatty acids by the tumor than by normal mammary tissue. The slow rate of fatty acid synthesis in this mammary adenocarcinoma may have been due to lack of NADPH, for the pentose phosphate pathway, which was active in the normal mammary tissue, was inoperative in the tumor.

1. De Novo Biosynthesis

The incorporation of glucose and acetate into fatty acids by Ehrlich ascites carcinoma cells is stimulated by addition of bicarbonate to the incubation medium or by increasing the CO_2 tension of the gas phase [40, 62]. Fatty acid synthesis takes place in the extramitochondrial compartment of the cells and is dependent on an NADPH-generating system. Separation of the individual fatty acids by gas-liquid chromatography demonstrated that carbon atoms from glucose were incorporated into palmitate [40]. Taken together, these findings indicate that the malonylcoenzyme A pathway for *de novo* fatty acid biosynthesis is operative in the Ehrlich cell. The enzymes that comprise this pathway, acetylcoenzyme A carboxylase and the fatty acid synthetase complex, are present in homogenates of the Morris hepatomas 7777 and 9618A [45]. Physical measurements and chemical assays indicate that the hepatoma acetylcoenzyme A carboxylase is the same enzyme as that present in normal rat liver. The hepatomas, however, were found to contain increased amounts of this enzyme.

2. Desaturation and Chain Elongation

In addition to palmitate, glucose carbon atoms were recovered in palmitoleate, stearate, oleate, and surprisingly, linoleate [40]. How these carbon atoms entered linoleate is unexplained, for all other available evidence indicates that mammalian cells cannot synthesize essential fatty acids. The presence of carbon atoms from glucose in palmitoleate and oleate indicates that the fatty acid desaturase enzyme system is operative in Ehrlich cells. Subsequent work has shown that Ehrlich cells can elongate fatty acids [22]. Chain elongation also occurs in Morris hepatomas [68]. Two elongation pathways have been described in mammalian cells; a mitochondrial system in which acetylcoenzyme A is the donor of the two carbon atom fragment that adds to the fatty acid carboxyl group, and a microsomal system in which malonylcoenzyme A is the two carbon atom donor. The extent to which each of these pathways contributes to chain elongation in tumor cells remains to be elucidated.

a) Essential Fatty Acids

Some interesting observations concerning the chain elongation system and the role of essential fatty acids in tumor cells have been made by BAILEY and DUNBAR [4]. Ehrlich ascites cells and sarcoma 180 were grown in essential fatty acid deficient mice. Essential fatty acids comprised only 2 % of the plasma fatty acids in these mice, whereas they account for 30 % of the plasma fatty acid in normal mice. No change in the mean generation time of the cells was noted, and they grew normally when transplanted into normal mice. There was no change in the lipid content of the tumors grown in the fatty acid deficient mice, but their fatty acid composition was altered. Oleate increased from 24 to 42 % of the total fatty acids, and eicosatrienoate accounted for 14 % of the fatty acids in these cells. Concurrently, the essential fatty acids decreased from 36 to 3 % of the cell total fatty acids.

The main trienoic acids that accumulate in the tumors under these conditions are 20:3ω9 and 22:3ω9 [4]. In order to learn whether these acids were synthesized by the tumor cells or taken up from the essential fatty acid deficient host, the Ehrlich cells were grown in a lipid-free medium [22]. After 2 months, 75 % of the cell fatty acids were composed of palmitoleate, oleate and a 20 carbon atom acid containing a single unsaturated bond. Negligible amounts of polyunsaturated fatty acid was recovered from the cells. When [1-^{14}C]oleate was added to these cultures, radioactive 20:1ω9 was recovered. Hence, these essential fatty acid deficient cells elongated but did not further desaturate oleate. In addition [1-^{14}C]linoleate was converted to 20:2ω6 and 22:2ω6, indicating that it was elongated but not further desaturated. These results indicate that the trienoic acids that accumulated in the Ehrlich cells grown in essential fatty acid deficient mice were derived from the host, not synthesized by the tumor.

These findings raise a number of interesting questions. Is the inability to desaturate fatty acids of the oleate series a general property of Ehrlich cells, or does it result from some change produced by prolonged growth in a fat-free culture medium? Does the fatty acid desaturase system of other tumor cells exhibit similar substrate specificity? Furthermore, how general is the observation that cells grow normally if their essential fatty acids are replaced by long-chain trienoic acids? Other cell lines, including tumor cells, can replicate in the absence of essential fatty acids [3, 28, 71]. By contrast, CHD3 Chinese hamster cells and macrophages in culture require the addition of linoleic acid for growth and replication [21, 35]. HeLa S_3 cells cultured in a lipid-deficient medium develop the fatty acid pattern seen in essential fatty acid deficiency [32]. They fail to grow, and mitochondrial function is impaired as evidenced by changes in oxidative phosphorylation and respiratory control. These defects can be partially or totally prevented by inclusion of albumin-bound linoleate or arachidonate in the culture medium. Conversely, mitochondrial respiratory control and the tightness of coupling were normal in essential fatty acid deficient Ehrlich cells [5]. However, succinate dehydrogenase appears to be increased in mitochondria from these cells, and they were more resistant to osmotic shock than normal Ehrlich cells [5].

3. Regulation of Fatty Acid Biosynthesis

Fatty acid biosynthesis in the liver is depressed when animals are placed on a high fat diet. This regulatory mechanism is not exhibited by intact hepatomas. SABINE et al. [66] observed no change in acetate incorporation into fatty acids by slices of hepatoma BW 7756 in male mice who were fed a high fat diet. The same diet reduced acetate incorporation into fatty acids in liver slices from normal mice. Fasting reduced acetate incorporation in liver slices but not in the hepatoma [67]. Refeeding a high carbohydrate diet to previously fasted mice produced a large increase in acetate incorporation in the normal liver slices but not in those from the hepatoma. It was suggested that the hepatoma did not respond to these dietary manipulations because an activator or inhibitor could not enter the hepatoma cells. This hypothesis was supported by studies with homogenates of hepatoma BW 7756 [69]. The malonylcoenzyme A pathway for *de novo* fatty acid synthesis was active in the supernatant fraction of the homogenate. As in normal liver homogenates, this pathway in the hepatoma homogenate was inhibited by palmitate and palmitylcoenzyme A. Long-chain fatty acids and the coenzyme A thioesters are thought to mediate the diet induced changes in fatty acid synthesis in normal liver. The experimental data indicated that the hepatoma enzymes responded to these inhibitors in the same manner as the enzymes of normal liver. This supports the idea that the failure of hepatomas to respond to dietary variations is due to an inability of the inhibitors to regulate the synthesis of the enzyme rather than to any difference in the enzyme system itself.

Lack of dietary regulation of fatty acid biosynthesis also has been observed in Morris hepatomas 5123C, 7793, 7795 and 7800 [68], 7777 and 9618A [45] and 9121 [24]. As in hepatoma BW 7756 [69], the acetylcoenzyme A carboxylase and fatty acid synthetase in Morris hepatomas 7777 and 9618A were inhibited by addition of palmitylcoenzyme A [45]. The acetylcoenzyme A carboxylases of these hepatomas appear to be identical to that of normal liver. Taken together, these findings support the view of SABINE et al. that the loss of dietary control in hepatomas is not due to any fundamental difference in the responsiveness of the fatty acid synthetic enzymes.

Dietary regulation of cholesterol synthesis also does not occur in certain hepatomas, indicating that aberrant metabolic control is not confined to the fatty acid pathways. For example, when rats were placed on a fat-deficient diet, cholesterol synthesis was not increased in Morris hepatoma 9121 [24]. On the other hand, cholesterol synthesis in HTC cells, derived from Morris hepatoma 7288C, does respond to the lipid content of the culture medium [102]. Cholesterol synthesis was stimulated in cells that were incubated in lipoprotein-poor serum, and it was depressed by about 80 % when lipoproteins were added to these lipid-deficient media. The lipid content of the medium also has an effect on fatty acid synthesis in HTC cells, but this is considerably smaller than the effect produced on cholesterol synthesis [102]. Fatty acid synthesis was reduced by about 20 % when lipoproteins were added to cells growing in a culture medium containing lipoprotein-poor serum. The failure to note a profound decrease in this pathway may be due to the fact that free fatty acids, the modulators of the fatty acid synthesis, are not removed from lipoprotein-poor serum.

4. Glucose Incorporation into Lipids

When Ehrlich ascites tumor cells are incubated with [U-^{14}C]glucose *in vitro*, a considerable quantity of radioactivity is present in cell lipids [95]. After saponification of these lipids, however, 65–80 % of the radioactivity is recovered in water-soluble materials. Moreover, the addition of long-chain fatty acids, known inhibitors of *de novo* biosynthesis, to the incubation medium markedly increased the incorporation of glucose into Ehrlich cell lipids [84]. Much of this increase was accounted for by incorporation into the lipid material that was water-soluble following saponification. These results indicate that most of the glucose incorporated into lipids was in the glycerol backbones of lipid esters rather than in fatty acids. Furthermore, they indirectly suggest that most of the fatty acid incorporated into cell lipid esters was supplied by extracellular fatty acids. Similar data were obtained when [U-^{14}C]glucose was injected intraperitoneally in mice bearing the Ehrlich ascites tumor [75]. After saponification, 94 % of the glucose radioactivity incorporated into cell lipids was water soluble. Similar results also were observed with human leukemic lymphocytes. Most of the radioactive glucose was incorporated into the glycerol backbones of the glycerolipids and the hexose moieties of glycosphingolipids, not fatty acids [56]. Concurrent studies demonstrated that the Ehrlich ascites tumor cells readily utilized long chain fatty acids that were added to the extracellular fluid [30, 51, 77, 96]. Many other tumors also utilize fatty acids when they are available in the incubation medium [49, 52, 72, 105]. Based upon these findings, it appears that many tumors derive a large fraction of their fatty acids from the host [37]. Therefore, it is important to learn how the host transfers lipids to a tumor.

5. Relative Contributions of Biosynthesis and Host Lipids

Before dismissing *de novo* fatty acid biosynthesis in tumors as being unimportant, however, some additional information should be considered. In at least a few experimental tumor systems, the rates of fatty acid biosynthesis are equal to or greater than those seen in normal liver [37]. Furthermore, certain tumor cells that derive fatty acids from the extracellular fluid under ordinary conditions can adapt to a lipid-free environment. For example, HeLa cells and the JTC-16 rat ascites hepatoma, as well as other cell lines, grow and replicate in long-term culture in a lipid-free medium [5, 28, 100]. When lipid is removed from the culture medium, an induction of one or more of the fatty acid synthetic enzymes occurs [64]. Another factor that should be considered in assessing the role of fatty acid synthesis in tumors is the substrate that is employed. Either radioactive glucose or acetate have been used for most studies of fatty acid synthesis in tumors. In the light of this, it is interesting to note that in HTC cells, the rates of synthesis are higher when measured with ^3H_2O than with glucose. In addition, acetate is utilized poorly by tumors, apparently because they contain very low activities of acetylcoenzyme A synthetase [38]. Moreover, certain of the biosynthetic rates that have been reported may be low because insufficient amounts of bicarbonate were added to the incubation medium. Fatty acid synthesis in intact tumor cells is strongly dependent on bicarbonate concentration [40, 62], a fact that probably is not generally appreciated. Finally, the activities of the biosynthetic enzymes in homogenates of Ehrlich cells are higher than might be expected from the rates of fatty acid synthesis observed in the intact cells [McGee and Spector, unpublished observations]. These findings suggest that under certain conditions, fatty acid synthesis may play a larger role in supplying tumor cells with lipid than might be predicted from the currently available reports. They also suggest that tumors probably can adapt to a reduced fatty acid supply from the host by accelerating their biosynthetic rate. The crucial question, which cannot be answered at this time, is whether the extent of such compensation would be sufficient to maintain the same rate of growth as when lipids are abundantly available from the host.

B. Lipids Supplied by the Host

The host responds to the presence of a tumor by releasing increased amounts of lipids into the circulation. In rodents, carcass lipids are depleted, and frank hyperlipidemia occurs. Although hyperlipidemia apparently is not associated with cancer in man, there is some evidence to indicate that increased free fatty acid turnover does occur in certain forms of the disease.

1. Blood Plasma Lipids

The lipid content of the carcass of rats bearing the Walker carcinosarcoma 256 decreases as the tumor grows, and the total fatty acid content of the blood plasma increases [36, 55]. Extreme hyperlipidemia is noted when rats bearing this tumor are fed a high fat diet [99], and this occurs as soon as the tumor begins to grow [31]. In Swiss mice bearing the Krebs-2-carcinoma, there is a 50-percent loss of body weight seven days after transplantation, that is, at a time when the tumor is still small [18]. This early fat loss also is produced by injection of nonviable tumor. During the phase of acute fat loss, there is increased acetate incorporation into host lipids.

In mice bearing the Ehrlich ascites carcinoma, there is a large decrease in the high density lipoproteins and an elevation in low density lipoproteins in the blood plasma [19]. Although hyperlipidemia is noted during the early stages of tumor growth, the plasma lipid levels decrease to less than normal values by the tenth day following transplantation. Hyperlipidemia also was noted in Buffalo-strain rats bearing the Morris hepatoma 7777 [59], but the type of hyperlipidemia differed considerably from that observed in the mice bearing the Ehrlich ascites tumor. Hyperlipidemia was not observed at 2 weeks after the tumor was transplanted, but it was evident by the fourth week. The serum total lipids, phospholipids and cholesterol were elevated. A large increase was noted in plasma high density lipoproteins, particularly the HDL_2 subfraction, and a slow-moving high-density lipoprotein band appeared on disc gel electrophoresis. A small increase in plasma low-density lipoproteins was noted, but the triglyceride-rich very low-density lipoproteins were decreased considerably. By contrast, the high-density lipoprotein subfraction HDL_2

was decreased in rats bearing carcinogen-induced mammary tumors or the Walker carcinosarcoma 256 [10].

Cancer in humans is not associated with hyperlipidemia [48], but plasma free fatty acid elevation has been reported [58]. The loss of lipid from the carcass of cancer victims is due to excessive free fatty acid mobilization from the adipose tissue depots [48]. In patients with cancer, higher plasma glucose concentrations are required to suppress free fatty acid mobilization [23].

Taken together, these findings suggest that the presence of a tumor causes the host to release more lipid into the blood plasma. The mechanism of this response is unknown at present. It is likely that this may vary with the type of tumor and the animal species in which the tumor grows.

a) Role of Dietary Lipids

The work of Littman et al. [44] indicates that the growth rate of rodent tumors is dependent on an adequate supply of lipid in the diet. Feeding of a fat-free diet retarded the growth of sarcoma 180, the Ehrlich ascites carcinoma, adenocarcinoma 755 and the Novikoff hepatoma. Previous studies demonstrated that dietary fatty acids can be incorporated into tumor lipids [49, 108]. It is tempting to speculate that the protective effect of a fat-free diet resulted from lower plasma lipid concentrations, but no plasma lipid measurements are reported in these studies. Furthermore, since the diet also was cholesterol-free, it is possible that the slower growth rate was due to a reduced sterol supply to the tumors and actually was unrelated to fatty acid deprivation. This possibility is supported by the observation that administration of hypocholesterolemic drugs to these animals also reduced the growth rate of the tumors.

2. Ascites Plasma Lipids

Ascites tumors grow in the peritoneal cavity of the tumor bearing host. The tumor cells are suspended in an ascitic plasma which appears to be a transudate of the blood plasma. Therefore, the ascites plasma actually is the extracellular fluid of these tumors. In an attempt to understand better the interaction between a tumor cell and the lipids in its immediate environment, studies of the content and turnover of various lipid fractions in the ascites plasma have been performed in my laboratory.

The Ehrlich ascites tumor plasma contains both free and esterified fatty acids [75]. In our initial studies, 12 CDF_1 mice were sacrificed at varying time intervals after inoculation of the tumor. The total volume of the tumor ranged from 1.5 to 18 ml. The ascites plasma free fatty acid concentration was 0.069–0.36 µEq/ml, and the albumin concentration was about 0.3 µmol/ml. Therefore, the molar ratio of free fatty acid to albumin, an important parameter in determining the rate of free fatty acid utilization [80], was between 0.2 and 1.2. About 85 % of the fatty acids making up the ascites plasma free fatty acid fraction was composed of oleate, palmitate, linoleate and stearate. There was 28 times more esterified than free fatty acid in these samples of Ehrlich ascites plasma.

These studies were repeated in Ehrlich ascites tumor plasma obtained from male CBA mice, the tumors being obtained either 10 or 12 days after transplantation [84]. The free fatty acid concentration in these ascites plasmas was 0.7–0.8 µEq/ml, and the esterified fatty acid concentration was 3.3–5.9 µEq/ml. For the most part, the esterified fatty acids were contained in triglycerides, although phospholipids and cholesteryl esters also were present. Three lipoprotein fractions were isolated, very low density (d < 1.006), low density (d = 1.006–1.063) and high density (d = 1.063–1.21). These lipoproteins all had α-mobility on electrophoresis at pH 8.8 in a barbital buffer. Some chylomicron-like particles were isolated by ultracentrifugation, but this material was not visualized clearly on electrophoresis. Most of the ascites plasma lipids were contained in the very low density lipoproteins. The very low and low density lipoproteins were precipitated by addition of heparin and manganese, whereas the high density lipoproteins were not. Immunoelectrophoresis revealed that the ascites plasma very low density lipoproteins cross reacted with mouse blood plasma antiserum, mouse blood plasma lipoprotein antiserum and mouse blood plasma high density lipoprotein antiserum. Likewise, the ascites plasma high density lipoproteins cross reacted with these three antisera. No cross-reactivity was noted, however, with the ascites plasma low density lipoproteins. None of these ascites plasma lipoproteins cross reacted with rat blood plasma lipoprotein antisera.

These results were obtained using mice who had access to food and water up to the time of sacrifice. However, we found no reduction in the amount of very low density lipoproteins present in the ascites plasma when mice were fasted for 18 h before sampling. Therefore, the ascites plasma triglycerides appear to be endogenous in origin rather than being derived from dietary fat.

In additional experiments, the tumor cells were incubated *in vitro*, and the incubation medium was examined for lipoproteins. No release of lipoproteins from the cells to the medium was observed under these conditions. Based upon this, it was concluded that the ascites plasma lipoproteins were derived from the host, not from the tumor cells.

MERMIER and BAKER [54] have made similar measurements of ascites plasma lipid concentrations in the Lettre-Ehrlich tumor grown in Swiss Webster mice. Food was removed from these animals 2 h prior to sacrifice. The free fatty acid concentration of the ascites plasma was 0.28 ± 0.06 µEq/ml [54], in good agreement with our earlier finding of 0.24 ± 0.09 µEq/ml [75]. On the other hand, the total fatty acid content of the ascites plasma was only 0.43 µEq/ml in the Swiss Webster mice. Therefore, in marked contrast to our findings, more than 50 % of the total fatty acid present in the ascites plasma of this strain of Ehrlich ascites tumor was in the form of free fatty acids.

a) Comparison of Cell and
Extracellular Fluid Fatty Acid Composition

When strain L fibroblasts in culture are grown in a medium that contains serum, the fatty acid composition of the cells is quite similar to that of the serum [33]. Profound changes in the fatty acid composition of the L cells occurred when they were transferred to a serum-free medium. The fatty acid composition of the L cells also could be altered markedly by addition of a single free fatty acid to the culture medium. Similar findings have been made with tumor cells in culture. When HeLa and bovine lymphosarcoma cells are grown in the presence of bovine serum, their fatty acid compositions resemble that of the culture medium [13]. The fatty acid composition of the HeLa cell changes considerably when it is grown in a medium containing albumin rather than fetal calf serum [32]. When Ehrlich cells grow in the mouse peritoneal cavity, their fatty acid composition resembles that of the ascites plasma free fatty acids [75]. These findings are compatible with the view that tumor cells obtain a large fraction of their fatty acids preformed from the host. Therefore, it is of interest to learn how tumor cells utilize the lipids that are available in their immediate environment.

III. Free Fatty Acid Utilization by Tumors

Free fatty acids are excellent substrates for tumors during *in vitro* incubation [30, 51, 96]. The finding that free fatty acids are present in

Ehrlich ascites tumor plasma and are elevated or turn over more rapidly in the blood plasma of cancer patients led to the question of the importance of this substrate for tumors *in vivo*. Recent studies using the Ehrlich ascites tumor as a model indicate that free fatty acids indeed may be a very important source of lipids for tumor cells in the intact animal.

A. Free Fatty Acid Utilization *in vivo*

Tracer amounts of labeled fatty acids complexed with bovine serum albumin were injected intraperitoneally into mice bearing the Ehrlich ascites tumor [75]. The tumor was sampled from 1.5 to 15 min after injection, and the flux of the labeled fatty acids was followed. With [1-^{14}C]palmitate, the decrease in labeled free fatty acid in the ascites plasma obeyed first-order kinetics during the first 6 min. The first order rate constant was 0.23 ± 0.04 min^{-1} and the half-life was 3.9 ± 0.7 min. After 10 min, 51 % of the injected radioactivity was associated with the cells. With [1-^{14}C]oleate, 37 % of the injected radioactivity was recovered from the cells after 10 min. With either [1-^{14}C]palmitate or [1-^{14}C]linoleate, 12 % of the radioactivity taken up by the cells was in the water-soluble materials and 88 % was in lipids. Of the radioactivity present in cells lipids at 15 min after injection, 72–82 % was in phospholipids and 18–27 % in neutral lipids. Based upon these data, it was concluded that in the Ehrlich ascites tumor, the ascites plasma free fatty acids turned over rapidly and that most of the turnover was accounted for by incorporation into the tumor cells. Assuming that the cell doubling time was 24 h and that the rate of fatty acid oxidation was 0.1 μEq/h $\times$ 10^8 cells, we estimated that about 80 % of the lipid requirement of the tumor cells was supplied by the ascites plasma free fatty acids. Because there is considerable uncertainty attached to these values and substantial variation was observed among individual mice, the value of 80 % must be taken only as a gross estimate. Although not tested experimentally, it was assumed that the ascites plasma free fatty acids were derived from the blood plasma of the host.

Recently, these studies were extended by BAKER and his co-workers [8, 53, 54]. Their findings also demonstrated that ascites plasma free fatty acids turn over very rapidly and that most of the turnover is accounted for by uptake into the tumor cells. Several very important additional observations were made. These workers noted that the fraction of injected free fatty acid remaining in the tumor was dependent on the volume of the tumor [53]. In small tumors, as much as 50 % of the injected [9,10-³H]palmitate was taken up by the host, whereas in tumors of 8 ml or greater, all of the injected radioactivity remained in the tumor. The total fatty acid increase in the tumor between days 5 and 12 was 100 µEq [8]. From this measurement, they concluded that our estimate of the fatty acid requirements of Ehrlich ascites cells was much too large. Tracer studies with [9,10-³H]palmitate indicated that 38 % of the ascites plasma free fatty acid was removed each minute, a large portion of which entered the tumor cells [54]. Free fatty acids were incorporated into cell lipid esters at a rate that was 26 times greater than the net accumulation of esterified fatty acid in the cells. These calculations indicate that the ascites plasma free fatty acid flux easily accounts for all of the fatty acids required for lipid ester formation in the Ehrlich cells. The most surprising finding was that although free fatty acids were rapidly entering the ascites plasma, less than 2 % of the influx was derived from the free fatty acids of the host's blood plasma. It was suggested that free fatty acids are transferred directly from the host into the ascites plasma without entering the blood plasma. MERMIER and BAKER suggest three potential sources for the ascites plasma free fatty acids; the lipid esters of the tumor cells themselves (which cannot, of course, account for the net accumulation of lipids in the tumor cells), the lipid esters of host tissues in the peritoneal cavity, or the lipid esters of host tissues which have lymphatics that drain into the peritoneal cavity [54]. Another potential source is the lipid esters of the ascites plasma lipoproteins.

In summary, these *in vivo* studies indicate that free fatty acids present in the extracellular fluid are a very important lipid substrate for tumor cells, at least in the case of the only tumor in which such studies have been performed. Therefore, it is important to understand the mechanism through which free fatty acids are utilized by tumor cells.

B. Mechanism of Free Fatty Acid Utilization in Tumors

Under the usual physiological conditions, more than 99 % of the circulating long-chain free fatty acids are bound to serum albumin [34, 86, 88]. The binding of fatty acids to albumin occurs through physical interactions, and no covalent bonds are involved. Most of the work on the mechanism of free fatty acid utilization by tumors has been done with washed Ehrlich ascites tumor cells. Radioactive free fatty acids bound to serum albumin are rapidly taken up by Ehrlich cells during *in vitro* incubation [30, 51, 98]. The fatty acid is transferred from albumin to the cell; that is, the fatty acid-albumin complex is not taken up intact. Fatty acid uptake actually is not dependent on the presence of albumin, for rapid uptake occurs from a protein-free medium [30, 98] as well as when the fatty acid is bound to β-lactoglobulin [85] or plasma lipoproteins [90]. In fact, it appears that the fatty acid dissociates from the carrier protein prior to uptake by the cell, but this as yet has not been proven rigorously [77].

1. Uptake

The first step in the free fatty acid uptake is binding to the cell in unesterified form [30, 98]. Two pools of cell free fatty acid exist, one that is rapidly exchangeable with the extracellular fluid and another that is more slowly exchangeable [93]. The rapidly exchangeable pool appears to be located at or near the cell surface, probably associated with plasma membrane receptors [98]. Studies with normal human erythrocytes indicate that the slowly exchangeable cell free fatty acid pool also may be located in the cell membrane; that is, the membrane contains at least two classes of fatty acid binding sites [82]. Whether this is also the case in tumor cells remains to be determined. The anionic form of the fatty acid probably is the species that is taken up by the cell [77]. The presence of an anionic group, however, does not appear to be an absolute requirement, for hexadecanol can be taken up just as readily as palmitate [92]. Uptake in the Ehrlich cell is not dependent on metabolic energy, and it probably occurs by diffusion through the lipid phase of the cell membrane [77, 98]. By contrast, entry of fatty acids into nonmalignant cells appears to involve a lecithin-lysolecithin cycle in the cell membrane [107]. Additional investigation is required to determine whether this actually represents a true difference between a normal and malignant cell.

One factor that regulates the amount of free fatty acid that is taken up by Ehrlich cells is the molar ratio of fatty acid to albumin [98]. Uptake increases in an exponential fashion as the molar ratio is raised. Another factor is pH, for uptake increases considerably as the pH of an albumin-containing medium is reduced below 7.4 [78]. This may provide Ehrlich cells with an advantage in terms of ability to utilize available free fatty acid, for the pH of the tumor extracellular fluid is 6.9 throughout the growth period [16]. Fatty acid structure also is important in determining the amount of free fatty acid that is taken up by the Ehrlich cell. Uptake increases as the fatty acid chain length increases, and at a given chain length, it decreases as the degree of unsaturation increases [96]. In other words, the relative uptakes are stearate > palmitate > myristate > laurate > decanoate, and stearate > oleate > linoleate. Short-chain and medium-chain acids such as acetate and octanoate are taken up very poorly. These differences cannot be explained entirely by differences in fatty acid binding to serum albumin. For example, palmitate is bound more tightly than laurate by albumin; yet, at a given molar ratio of fatty acid to albumin, more palmitate than laurate is taken up by the Ehrlich cell. Therefore, a major determinant in free fatty acid uptake appears to be the affinity of the Ehrlich cell membrane receptors for the particular fatty acid. The nature of these receptors is unknown, but it has been suggested that they are membrane lipids [92]. The relative affinities for the various fatty acids are not unique to the Ehrlich cell, for similar relative uptakes have been obtained with rat liver mitochondria and human platelets [83, 87].

More free fatty acid is utilized by Ehrlich cells when glucose is present in the incubation medium [95]. This is not due, however, to any increase in the amount of free fatty acid that is bound to the cell in unesterified form. It results entirely from an increase in the amount of free fatty acid that is incorporated into the lipid esters of the cell. In other words, glucose increases the turnover of the cell-free fatty acid pools and thereby increases the rate of free fatty acid uptake indirectly.

Free fatty acid uptake by the Ehrlich cell was increased by addition of chlorophenoxyisobutyrate to incubation media containing albumin [89]. This compound is the circulating form of a commonly used hypolipidemic drug, clofibrate. The increase in uptake appears to be due to weakening of the bonds between free fatty acid and albumin. Esterification also was enhanced when chlorophenoxyisobutyrate was present. It is thought that the stimulation of esterification is indirect, being due to an increase in the size of the cell-free fatty acid pools.

2. Oxidation

The studies from WEINHOUSE's laboratory demonstrated that slices prepared from a variety of tumors were able to oxidize fatty acids to CO_2 and H_2O [105]. As compared with slices of normal liver, the rates of oxidation by the tumors were low, particularly with short-chain fatty acid substrates. In ascites hepatoma 98/15, the addition of glucose was found to reduce the oxidation of decanoate, laurate and palmitate, but not butyrate [72]. High concentrations of long-chain fatty acids overcame the respiratory depression produced by glucose, and they uncoupled oxidative phosphorylation. These studies, however, were done in the absence of albumin or any other fatty acid carrier, so that the cells were exposed to very high concentrations of unbound fatty acid. Under ordinary conditions the total free fatty acid concentration to which cells are exposed may be as high as 2 mM, but most of this is bound to albumin, and the unbound concentration rarely exceeds 0.01 mM. Long-chain fatty acids produce very little increase in O_2 consumption by Ehrlich cells when albumin is present [97].

Addition of glucose reduced both O_2 consumption and palmitate oxidation in Ehrlich cells [51, 95]. Galactose also decreased palmitate oxidation, but lactate and acetate had little effect [51]. When palmitate was available, it accounted for one third of the O_2 consumption of the Ehrlich cell. Glucose was the preferred respiratory substrate, however, and two thirds of the cellular O_2 consumption was accounted for by glucose oxidation. This differs from what occurs in tissues such as the heart, where fatty acids reduce glucose oxidation, but glucose does not influence the amount of free fatty acid that is oxidized [74].

The ability of homogenates of hepatomas to oxidize fatty acids to CO_2 and H_2O or acetoacetate varies considerably [11]. Large quantities of palmitate and butyrate were oxidized by the minimal deviation Morris hepatomas 5123, 7787 and 7793. Less oxidation was observed with 7288C and the Reuber hepatoma H35. Very little oxidation was noted with 3683, 3924 and the Novikoff hepatoma. It should be noted that carnitine was not added in these homogenate incubations, and palmitate was added without a protein carrier. Those hepatomas that exhibited low rates of fatty acid oxidation have a very high glycolytic capacity and have high levels of ATP-glucose phosphotransferase [11, 104]. As in other tissues, fatty acid oxidation occurs in the mitochondria of hepatomas and is mediated by the usual tricarboxylic acid cycle [15, 25].

The presence of long-chain fatty acylcoenzyme A synthetase activity has been demonstrated in homogenates of Ehrlich ascites cells [76]. Carnitine palmityltransferase activity also was found in these homogenates, and palmitate oxidation was stimulated 2.5 times by addition of carnitine. By contrast, addition of carnitine did not stimulate fatty acid oxidation in intact cell incubations.

The rate at which a fatty acid was oxidized was dependent upon its structure [96]. Very little acetate or octanoate was oxidized by the Ehrlich ascites cells. The rates of oxidation of the long-chain acids were: linoleate > palmitate and oleate > stearate > laurate. Each of these acids was tested with a different preparation of Ehrlich cells. Therefore, these comparative rates of oxidation are open to some question.

3. Esterification

Free fatty acids are incorporated into phospholipids, glycerides and cholesterol esters of Ehrlich cells [96]. Under most conditions, more fatty acid is incorporated into phospholipids than other lipid esters [84]. Leukemic leukocytes also incorporate more fatty acid into phospholipids than glycerides, whereas the reverse is true for normal leukocytes [46].

When fatty acids are injected intraperitoneally into mice bearing the Ehrlich ascites tumor, from 71 to 83 % of the fatty acid incorporated into phospholipids is recovered in lecithin [97]. With either palmitate or oleate, approximately equal amounts of fatty acid were incorporated into the 1- and 2-positions of the lecithins. This is surprising because in Ehrlich cells, as in most other mammalian tissues, the 1-position of the choline-containing phosphatides contains predominantly saturated fatty acids and the 2-position predominantly polyunsaturated fatty acids [106]. It is likely that some of the radioactive fatty acid inserted into the 2-position of lecithin was elongated or desaturated prior to incorporation. The extent of such transformations was minimal, however, for 90 % of the radioactive palmitate that was incorporated in cell lipids remained as palmitate, and 94 % of the labeled oleate remained as oleate.

Fatty acid incorporation into the esterified lipids of the Ehrlich cell is increased greatly by the presence of glucose in the incubation medium [95]. Most of the glucose is incorporated into the glycerol backbones of phospholipids, particularly lecithin, and triglycerides [84]. Initially, this was puzzling, for it was throught that glucose had to be converted to L-glycerol 3-phosphate in order to enter the lipid glycerol moiety, and tumors were known to be deficient in glycerolphosphate dehydrogenase [12, 17, 70]. This dilemma appears to be resolved by the finding that glucose is incorporated into the glycerolipids of Ehrlich cells predominantly through dihydroxyacetone phosphate, not glycerol 3-phosphate [2]. This obviates the need for an active glycerolphosphate dehydrogenase.

Since glycerophosphatides and acylglycerols are composed of mixtures of fatty acids, one would suspect that the rates of free fatty acid incorporation into esterified lipids might be slowed if only a single fatty acid were available to the cell. Yet, the incorporation of [1-^{14}C]palmitate into the lipid esters of Ehrlich cells was not increased when mixtures of free fatty acids were available as compared with only palmitate alone [79]. Likewise, the incorporation of labeled glucose into cell lipids was only slightly enhanced when mixtures of fatty acids were available as compared with a single fatty acid. Since only a small percentage of palmitate or oleate is converted to other fatty acids by Ehrlich cells [97], the cell apparently does not make a mixture of fatty acids from the one that is provided *in vitro*. Two explanations are suggested for the ability of a single fatty acid to support a high rate of esterification. One is that the Ehrlich cell may be able to synthesize lipid esters, if necessary, using only a single species of fatty acid. The other is that a mixture of acids is needed to support maximal esterification but that the mixture is provided by continual turnover of intracellular lipid esters [97]. A practical consequence of these findings is that, in the Ehrlich cell, reasonably accurate rates of fatty acid esterification can be measured using labeled palmitate alone. It would be much more complicated to obtain rates of esterification if it were necessary to work with mixtures of labeled fatty acids.

The amount of labeled fatty acid incorporated into the lipid esters of the Ehrlich cell increases markedly as the molar ratio of free fatty acid to albumin increases [84]. Raising the molar ratio increases incorporation into both phospholipids and glycerides. The percentage increase for incorporation into triglycerides, however, is greater than that in phospholipids. For example, at a molar ratio of 1.0, about 5 times more palmitate is incorporated into phospholipids than triglycerides. Although much more palmitate is incorporated into both ester fractions at molar ratio of 4.5, only 1.3 times more palmitate is incorporated into phospholipids than glycerides. Similar results were obtained with oleate. Glucose incorporation into lipid esters also increases as the molar ratio of free fatty acid to albumin is raised, and a higher percentage of the total glucose incorporation also is channeled into glycerides [95]. These findings suggest that the Ehrlich cells store fatty acids in the form of triglycerides when large quantities are available, whereas they direct fatty acids primarily into phospholipids for membrane formation when the fatty acid supply is limited.

The rate of esterification also is dependent on the structure of the fatty acid [96]. Octanoate is not incorporated to any appreciable extent into the lipid esters of the Ehrlich cell. Laurate is esterified at a rate that is only 10–20 % that of palmitate. Oleate is esterified at a slightly slower rate than palmitate by the Ehrlich cell, and the ratio of incorporation into triglycerides as compared to phospholipids is somewhat higher with oleate [84]. The differences in amounts of palmitate and oleate esterified are based on a given fatty acid to albumin molar ratio. It should be remembered, however, that more palmitate than oleate is taken up in unesterified form at each molar ratio. If the rate of esterification is calculated relative to the cell unesterified fatty acid content, there is almost no difference between palmitate and oleate [79].

IV. Esterified Fatty Acid Utilization by Tumors

The finding of a triglyceride-rich lipoprotein in the ascites plasma of the Ehrlich ascites tumor raised the question of whether the cells are able to utilize esterified fatty acid contained in the extracellular fluid. Studies with nonmalignant cells in culture indicated that triglycerides present in serum lipoproteins could be utilized [6]. These triglycerides are hydrolyzed in the extracellular fluid, and the released fatty acids are the species actually taken up [39]. In addition, phospholipids contained in lipoproteins are transferred to erythrocytes [65] and chick embryo fibroblasts [63].

In an attempt to learn whether Ehrlich cells also were capable of utilizing exogenous esterified fatty acid, we incubated the cells with a model compound that was relatively simple to use, a fatty acid methyl ester bound to albumin [42]. Some methyl ester was taken up intact by the cells. This was due to transfer of the methyl ester from albumin to the cell, not uptake of the intact methyl ester-albumin complex. The fatty acids that were hydrolyzed within the cells were readily oxidized and esterified into phospholipids and triglycerides. Homogenates of the Ehrlich cell were found to contain a methyl ester hydrolase which was firmly bound to membrane [91].

A. Lipoprotein Triglycerides

Additional studies demonstrated that Ehrlich cells also are able to utilize triglycerides present in the ascites plasma very low density lipoproteins [14, 84]. A depletion of chemically measured triglycerides from the incubation medium was observed during *in vitro* incubation of the ascites plasma very low density lipoproteins with the Ehrlich cells. Labeled tripalmitin and triolein were incorporated into these lipoproteins, and they were subsequently incubated with the cells. Uptake of labeled triglycerides was noted. Some of the labeled fatty acid was hydrolyzed and then either oxidized or esterified by the cells. In addition, a considerable amount of the triglyceride was taken up intact and remained as triglyceride in the cells. The possibility that some or all of this material was bound to the cell membrane cannot be excluded. Although some of the triglyceride uptake could be accounted for by binding of intact lipoproteins to the Ehrlich cell, concomitant measurements of lipoprotein cholesterol and protein indicated that binding of intact lipoproteins could not account for all of the intact triglyceride uptake. Therefore, as with methyl esters, it appears that some triglycerides are transferred intact to the lipid phase of the cell membrane under these conditions of incubation. We believe that hydrolysis takes place after the triglyceride molecule is associated with the cell.

Such a mechanism for lipoprotein triglyceride utilization, although different from the currently accepted viewpoint, actually does have some additional experimental support. First, the studies with methyl esters in Ehrlich cells are compatible with such an interpretation [42]. Second, triglycerides are transferred from very low density to high density lipoproteins during *in vitro* incubation [60], so it is not unreasonable to postulate that similar transfer might occur between these lipoproteins and cell membranes. Third, electron microscopic evidence has been presented which suggests that triglycerides are hydrolyzed inside the endothelial cells of the capillary, not in the capillary lumen [73]. Finally, evidence for intact triglyceride uptake has been obtained in the perfused rabbit aorta [101] and with L strain mouse fibroblasts in tissue culture [7].

On the other hand, the possibility that the mechanism of triglyceride uptake *in vivo* differs from what we have observed during *in vitro* incubation must be considered. The currently available studies on the interaction of the ascites lipoproteins with Ehrlich cells were done in a simplified system in which no other proteins were present [14, 84]. It is entirely possible that in the intact ascites tumor *in vivo,* the lipoprotein triglycerides are hydrolyzed to fatty acids prior to uptake. In this context, it is interesting to note that the Walker carcinosarcoma 256 contains a lipase which hydrolyzes lipoprotein triglycerides [9]. Like other lipoprotein lipases, this enzyme is released from the cells by heparin. Studies were not done, however, to determine whether this enzyme is associated with the tumor cells or the vasculature of the tumor mass. If triglyceride hydrolysis actually takes place in the ascites plasma, it might explain the origin of some of the ascites plasma free fatty acids. The results of MERMIER and BAKER indicate that the host releases free fatty acids into the ascites plasma but that they are not derived from free fatty acids in the host's blood [54]. A possible explanation is that very low density lipoproteins are transferred from the host's plasma to the ascites and that free fatty acids are formed from the lipoprotein triglycerides within the ascites plasma.

V. Turnover of Esterified Fatty Acids in Tumors

The esterified fatty acids contained in the phospholipids and glycerides of tumors are continuously turning over. MEDES et al. [49] incorporated labeled fatty acids into the lipid esters of a variety of transplantable tumors by feeding [1-^{14}C]palmitate to tumor-bearing mice. After 24 h, the tumors were removed and sliced, and the slices were allowed to respire in the absence of added substrate. Radioactivity was recovered in the collected CO_2. With hepatoma 98/15 and TA3 carcinoma, the specific activity of the respiratory CO_2 was about the same as that of the tumor fatty acids. By contrast, with the Ehrlich ascites carcinoma, rhabdomyosarcoma MC-1A, sarcoma 37 and TA3 ascites carcinoma, the specific activity of the CO_2 was greater than that of the tumor fatty acids. These studies demonstrated that the endogenous fatty acids of tumors were an important oxidative substrate, and they helped to explain the low respiratory quotients that had been reported earlier for tumor slices [20]. Addition of glucose to media containing the fatty acid-labeled Ehrlich cells reduced radioactive CO_2 production by as much as 70 % [52]. At high concentrations of glucose, the specific activity of the radioactive CO_2 was reduced by almost 50 %, and the O_2 uptake was lowered by 60 %. Similar effects were produced by fructose and lactate, but very little inhibition of labeled CO_2 production occurred when acetate was added. When the concentration of the added substrate was low, there was little or no change in O_2 consumption, and the amount of substrate that was oxidized roughly corresponded to the amount of suppression of endogenous fatty acids oxidation. These results indicate that the extent to which endogenous fatty acids are oxidized is dependent on the availability of other substrates in the extracellular fluid.

These observations were extended recently using Ehrlich cells [97]. The cells were labeled by injection of albumin-bound [1-^{14}C]palmitate, oleate or linoleate. From 90 to 94 % of the radioactivity that was present in Ehrlich cell lipids 1 h after the injection remained in the same fatty acid as was administered. From 74 to 86 % of the radioactivity was recovered in phospholipids and, of this, 70–83 % was in lecithin. When the labeled cells were incubated *in vitro* without any exogenous substrate, the cell lipid radioactivity was reduced by 10–17 % in 2 h. By contrast the lipid ester content of the cells decreased only about 3 % during the same period. When unlabeled palmitate and glucose were added to the incubation medium, the cell lipid ester content increased by 13 % in 2 h, but the lipid ester radioactivity still decreased from 3 to 9 %. These findings indicate the presence of one or more small lipid ester pools in the Ehrlich cell that turn over much faster than the remainder of the cell esterified fatty acids. Moreover, these esterified fatty acid pools continue to turn over in spite of a net esterified fatty acid accumulation in the cells. Much of the rapidly turning over esterified fatty acid pool was present in lecithin.

Addition of unlabeled fatty acids reduced the oxidation of cell-esterified fatty acids, whereas the O_2 consumption of the cells increased slightly [97]. The suppression was not specific for a particular fatty acid; that is, palmitate, oleate, stearate and linoleate all suppressed the oxidation of endogenous fatty acid in palmitate-labeled cells to about the same degree. As noted with the carbohydrate substrates [52], the amount of suppression of endogenous fatty acid oxidation was about equal to the amount of added fatty acid that was oxidized, indicating that the extracellular fatty acids were replacing endogenous esterified fatty acids as the oxidative substrate [81].

The ability of the Ehrlich cell to utilize its esterified fatty acids also has been observed using cytochemical techniques [16]. Lipid droplets accumulate in the cytoplasm of Ehrlich ascites cells during growth in the peritoneal cavity. If the cells are made to grow more rapidly by aspirating most of the tumor contents from the peritoneal cavity, the cytoplasmic lipid droplets disappear. This is associated with an increase in cell lipase activity as measured cytochemically.

A. Free Fatty Acid Release from Tumor Cells

In addition to oxidation, the endogenous esterified fatty acids of the Ehrlich cell also can be released into the incubation medium as free fatty acid [94]. Plasma albumin or another fatty acid acceptor is required in the extracellular fluid for free fatty acid release to occur. When free fatty acid is present in the medium, the release manifests itself as fatty acid exchange. In the absence of extracellular fatty acid, however, a net efflux of free fatty acid can occur. Less free fatty acid release occurred when glucose was present in the incubation medium, probably because of a glucose-stimulated increase in reesterification. Unlike adipose tissue, most of the released fatty acids are derived from phospholipids of the Ehrlich cell.

The *in vivo* free fatty acid turnover studies of MERMIER and BAKER poignantly demonstrate the considerable degree to which esterified fatty acid turnover occurs in Ehrlich ascites cells [54]. Using [9,10-^{3}H]palmitate as the tracer, they found that 96% of the ascites plasma free fatty acid that was incorporated into cell lipid esters was subsequently hydrolyzed and released as free fatty acid. Data from *in vitro* studies also support continual turnover of the esterified fatty acids in Ehrlich cells. When the cells were incubated with albumin-bound [9,10-^{3}H]palmitate, a large net fatty acid uptake occurred in the first 2 min of incubation [94]. Subsequently, the medium-free fatty acid content remained constant in spite of continued utilization of the labeled palmitate by the cells, and the specific activity of the medium-free fatty acids decreased by 28%. These findings indicate a continuous release of unlabeled fatty acids from the cells concomitant with the [9,10-^{3}H]palmitate utilization, and they support the concept that the fatty acid esters of the Ehrlich cell are turning over rapidly.

Preliminary results indicate that free fatty acid release from Ehrlich cells is not stimulated by certain fat-mobilizing hormones; epinephrine, nonepinephrine and adrenocorticotropin [SPECTOR, unpublished results]. This may be explained by the fact that most of the released fatty acid is derived from phospholipids, not triglycerides [97]. Much more work is required, however, before definitive statements can be made about hormonal regulation of lipid ester hydrolysis is Ehrlich cells.

VI. Summary and Comment

Tumor cells exhibit a very active metabolism of fatty acids. They can utilize fatty acids that are available in the extracellular fluid in either free or esterified form. Fatty acids that are taken up by tumor cells are either oxidized or incorporated into cell lipid esters. Certain of the esterified fatty acids in tumor cells, particularly those in lecithin, turn over rapidly. Tumor cells also are able to release free fatty acid into the surrounding medium. The fatty acid composition of most tumors reflects that of the extracellular fluid. Tumor cells do contain the enzymes required for fatty acid *de novo* biosynthesis, chain elongation and desaturation. Because of the ready availability of lipids in the surrounding fluid, however, fatty acid synthesis usually is turned off. In general, it appears that the fatty acid metabolic pathways are the same in tumors as in nonmalignant tissues. Therefore, a chemotherapeutic approach aimed at a specific reaction in the fatty acid metabolic pathway is unlikely to be fruitful.

Such a general statement is open to serious question, however, because of the very limited number of tumors in which fatty acid metabolism has been studied in detail. Almost all of the recent work on fatty acid turnover and utilization has been done with the Ehrlich ascites carcinoma. Most of the mechanistic studies on fatty acid biosynthesis have been done with Morris hepatomas. There is some reason to think that many of the findings made with these two experimental systems probably are representative of tumors as a class. On the other hand, definite differences in lipid metabolism have been observed, and it is unclear at this time whether these are due to the fact that the tumors studied were different or the animal species in which they grew differed. For example, Buffalo rats bearing a Morris hepatoma exhibit elevations in plasma high density lipoproteins [59], whereas CBA mice bearing the Ehrlich ascites carcinoma have increases in very low density lipoproteins [84]. These differences clearly demonstrate the potential danger of extrapolating such observations from one tumor to another, at least until a wider spectrum of tumors has been studied.

A. Future Directions

In spite of these limitations, a few potentially useful observations have been made which, in turn, lead to some intriguing questions. Much of the fatty acid requirement of the Ehrlich ascites carcinoma is supplied by the host as free fatty acids [54, 75]. Is this also true for other tumors? Are the structural specificities that regulate fatty acid uptake and utilization in Ehrlich cells exhibited by other tumors? An intracellular fatty acid binding protein has been found recently in intestine and liver [57, 61].

Do tumor cells that depend on large amounts of free fatty acids for energy and growth contain such a protein and, if so, does it have any special properties to facilitate fatty acid transport or utilization? Finally, is the mechanism of fatty acid uptake really different in tumor cells; that is, do normal cells regulate fatty acid uptake through an integrated series of enzymatic reactions [106] whereas entry into tumor cells is gained simply through diffusion [92]?

The area of lipoprotein metabolism also holds potential promise. Ehrlich cells can utilize the triglycerides contained in lipoproteins as a source of fatty acids [84]. Therefore, under *in vivo* conditions, what are the relative contributions of lipoprotein triglycerides and free fatty acids in terms of supplying lipids to the Ehrlich cell? Can other tumors also utilize the fatty acids contained in plasma lipoprotein triglycerides? Some lipoprotein triglycerides appear to be taken up intact by Ehrlich cells and fibroblasts in culture [7, 14]. Is the mechanism of triglyceride uptake in these cells really different from that in nonmalignant cells as is suggested by the currently held views concerning the role of lipoprotein lipase in normal tissues? If so, this may afford a means of getting drugs with a triglyceride-like structure specifically into tumor cells.

Another crucial point in this regard is whether a reduction in the supply of lipids from the host might slow tumor growth. The dietary studies of LITTMAN *et al.* [44] indirectly support such a proposition. Since tumors can synthesize fatty acids and can grow in culture in a fat-free medium [5], it is unlikely that reduced lipid supply from the host, *per se*, would inhibit growth. The key point, however, is whether biosynthesis could completely compensate and provide fatty acids fast enough to support the rate of growth that ordinarily occurs in a lipid-rich environment. Another aspect of this problem that warrants further study is the effect of changes in the fatty acid composition of tumors on their function. The phospholipid-fatty acid composition of tumors can be altered by changing the fat in the diet [37]. Nothing is known about whether such changes affect the properties of cell membranes. It is conceivable that the membrane lipid composition could be altered in such a way as to enhance the effectiveness of radiation or chemotherapy. Likewise, the fatty acid composition of tumor cells can be altered by placing the host on an essential fatty acid deficient diet [5]. Neither growth rate nor transplantability of the tumors were inhibited by these changes. It is possible, however, that such changes in fatty acid composition may produce more subtle effects on membrane properties that could prove to be useful in terms of therapy. It is hoped that the answers to these questions will be forthcoming, for they could provide insights that may lead eventually to more effective means for the treatment of cancer.

VII. Acknowledgments

I wish to express my deep appreciation to Dr. NOME BAKER, Department of Medicine, University of California at Los Angeles and the Veterans Administration Hospital, Los Angeles, Calif. and Dr. J. MARTYN BAILEY, Department of Biochemistry, George Washington University School of Medicine, Washington, DC for sending me copies of their manuscripts prior to publication.

VIII. References

1 ABRAHAM, S. and CHAIKOFF, I. L.: Metabolism of Barrett mammary adenocarcinoma. Glucose utilization, fatty acid synthesis, and terminal oxidative patterns and glutamine synthesis. Cancer Res. 25: 647–655 (1965).
2 AGRANOFF, B. W. and HAJRA, A. K.: The acyl dihydroxyacetone phosphate pathway for glycerolipid biosynthesis in mouse liver and Ehrlich ascites tumor cells. Proc. nat. Acad. Sci., Wash. 68: 411–415 (1971).
3 BAILEY, J. M.: Cellular lipid nutrition and lipid transport; in ROTHBLAT and KRITCHEVSKY Lipid metabolism in tissue culture cells, pp. 85–109 (Wistar Institute Press, Philadelphia 1967).
4 BAILEY, J. M. and DUNBAR, L. M.: Lipid metabolism in cultured cells. Growth of tumor cells deficient in essential fatty acids. Cancer Res. 31: 91–97 (1971).
5 BAILEY, J. M. and DUNBAR, L. M.: Essential fatty acid requirements of cells in tissue culture. Exp. molec. Path. 18: 142–161 (1973).
6 BAILEY, J. M.; HOWARD, B. V.; DUNBAR, L. M., and TILLMAN, S. F.: Control of lipid metabolism in cultured cells. Lipids 7: 125–134 (1972).
7 BAILEY, J. M.; HOWARD, B. V., and TILLMAN, S. F.: Lipid metabolism in cultured cells. XI. Utilization of serum triglycerides. J. biol. Chem. 248: 1240–1247 (1973).
8 BAKER, N.; MERMIER, P., and WILSON, L.: Net increase of lipid fatty acid in Ehrlich ascites carcinoma during growth. Lipids 8: 433–436 (1973).
9 BARCLAY, M.; GARFINKEL, E.; TEREBUS-KEKISH, O.; SHAH, E. B.; GUIA, M. DE; BARCLAY, R. K., and SKIPSKI, V. P.: Properties of lipoprotein lipase extracted from livers of normal rats and livers and tumors of rats bearing Walker carcinosarcoma 256. Arch. Biochem. Biophys. 98: 397–405 (1962).
10 BARCLAY, M.; SKIPSKI, V. P.; TEREBUS-KEKISH, O.; MERKER, P. L., and CAPPUCCINO, J. G.: Serum lipoproteins in rats with tumors induced by 9,10-dimethyl-1,2-benzanthracene and with transplanted carcinosarcoma 256. Cancer Res. 27: 1158–1167 (1967).

11 BLOCH-FRANKENTHAL, L.; LANGAN, J.; MORRIS, H. P., and WEINHOUSE, S.: Fatty acid oxidation and ketogenesis in transplantable liver tumors. Cancer Res. 25: 732–736 (1965).
12 BOXER, G. E. and SHONK, C. E.: Low levels of soluble DPN-linked α-glycerophosphate dehydrogenase in tumors. Cancer Res. 20: 85–91 (1960).
13 BOYLE, J. J. and LUDWIG, E. H.: Analysis of fatty acids of continuously cultured mammalian cells by gas-liquid chromatography. Nature, Lond. 196: 893–894 (1962).
14 BRENNEMAN, D. E. and SPECTOR, A. A.: Uptake of ascites plasma very low density lipoprotein triglycerides by Ehrlich cells. Fed. Proc. 32: 2587 (1973).
15 BROWN, G. W., JR.; KATZ, J., and CHAIKOFF, I. L.: The oxidative metabolic pattern of mouse hepatoma C 954 as studied with C^{14}-labelled acetate, propionate, octanoate and glucose. Cancer Res. 16: 509–519 (1956).
16 BURNS, E. R. and SOLOFF, B. L.: Cytoplasmic lipid in Ehrlich ascites tumor cells before and during recurrent growth. Oncology 25: 283–288 (1970).
17 CIACCIO, E. I.; KELLER, D. L., and BOXER, G. E.: The production of L-α-glycerolphosphate during anerobic glycolysis in normal and malignant tissues. Biochim. biophys. Acta 37: 191–193 (1960).
18 COSTA, G. and HOLLAND, J. F.: Effects of Krebs-2-carcinoma on the lipids metabolism of male Swiss mice. Cancer Res. 22: 1081–1083 (1962).
19 CREININ, H. L. and NARAYAN, K. A.: Effect of Ehrlich ascites tumor cells on mouse plasma lipoproteins. Z. Krebsforsch. 75: 93–98 (1971).
20 DICKENS, F. and SIMER, F.: The metabolism of normal and tumor tissue. III. The respiratory quotient and the relationship of respiration to glycolysis. Biochem. J. 24: 1301–1326 (1930).
21 DUBIN, I. N.; CZERNOBILSKY, B., and HERBST, B.: Effect of albumin and linoleic acid on growth of macrophages in tissue culture. J. nat. Cancer Inst. 34: 43–51 (1965).
22 DUNBAR, L. M. and BAILEY, J. M.: Essential fatty acid requirement of tumor cells. Fed. Proc. 31: 674 (1972).
23 EDMONSON, J. H.: Fatty acid mobilization and glucose metabolism in patients with cancer. Cancer 19: 277–280 (1966).
24 ELWOOD, J. C. and MORRIS, H. P.: Lack of adaptation in lipogenesis by hepatoma 9121. J. Lipid Res. 9: 337–341 (1968).
25 EMMELOT, P. and BOS, C. J.: Factors influencing the fatty acid oxidation of tumor mitochondria with special reference to changes in spontaneous mouse hepatomas. Experientia 11: 353–354 (1955).
26 EMMELOT, P. and BOSCH, L.: The metabolism of neoplastic tissue. Synthesis of fatty acids from acetate by transplanted mouse tumors *in vitro* and *in vivo*. Brit. J. Cancer 9: 327–338 (1955).
27 EMMELOT, P. and BOSCH, L.: The metabolism of neoplastic tissues. The relation of carbohydrate utilization to cholesterol and fatty acid synthesis in tumor tissue slices. Brit. J. Cancer 9: 339–343 (1955).
28 EVANS, V. J.; BRYANT, J. C.; KERR, H. A., and SCHILLING, E. L.: Chemically defined media for cultivation of long-term cell strains from four mammalian species. Exp. Cell Res. 36: 439–474 (1965).

29 FIGARD, P. H. and LEVINE, A. S.: Incorporation of labelled precursors into lipids of tumors induced by Rous sarcoma virus. Biochim. biophys. Acta *125:* 428–434 (1966).
30 FILLERUP, D. L.; MIGLIORI, J. C., and MEAD, J. F.: The uptake of lipoproteins by ascites tumor cells. J. biol. Chem. *233:* 98–101 (1958).
31 FREDERICK, G. L. and BEGG, R. W.: A study of hyperlipidemia in the tumor-bearing rat. Cancer Res. *16:* 548–552 (1956).
32 GERSCHENSON, L. E.; MEAD, J. F.; HARARY, I., and HAGGERTY, D. F., JR.: Studies on the effects of essential fatty acids on growth rate, fatty acid composition, oxidative phosphorylation and respiratory control of HeLa cells in culture. Biochim. biophys. Acta *131:* 42–49 (1967).
33 GEYER, R. P.: Uptake and retention of fatty acids by tissue culture cells; in ROTHBLAT and KRITCHEVSKY Lipid metabolism in tissue culture cells, pp. 33–44 (Wistar Institute Press, Philadelphia 1967).
34 GOODMAN, D. S.: The interaction of human serum albumin with long-chain fatty acid anions. J. amer. chem. Soc. *80:* 3892–3898 (1958).
35 HAM, R. G.: Albumin replacement by fatty acids in clonal growth of mammalian cells. Science *140:* 802–803 (1963).
36 HAVEN, F. L.; BLOOR, W. R., and RANDALL, C.: Lipids of the carcass, blood plasma and adrenals of the rat in cancer. Cancer Res. *9:* 511–514 (1949).
37 HENDERSON, J. F. and LEPAGE, G. A.: The nutrition of tumors. A review. Cancer Res. *19:* 887–902 (1959).
38 HEPP, D.; PRUSSE, E.; WEISS, H. und WIELAND, O.: Essigsäure als Endprodukt des aeroben Krebsstoffwechsels. Biochem. Z. *344:* 87–102 (1966).
39 HOWARD, B. V. and KRITCHEVSKY, D.: The source of cellular lipid in the human diploid cell strain W1-38. Biochim. biophys. Acta *187:* 393–401 (1969).
40 KIMURA, Y.; NIWA, T.; WADA, E., and KOMEIJI, T.: Incorporation of labelled glucose carbon into different fractions of Ehrlich ascites tumor cells, with special reference to lipogenesis from glucose. Jap. J. exp. Med. *34:* 267–291 (1964).
41 KRITCHEVSKY, D.: Newer hypolipidemic agents. Fed. Proc. *30:* 835–840 (1971).
42 KUHL, W. E. and SPECTOR, A. A.: Uptake of long-chain fatty acid methyl esters by mammalian cells. J. Lipid Res. *11:* 458–465 (1970).
43 LEES, R. S. and WILSON, D. E.: The treatment of hyperlipidemia. New Engl. J. Med. *284:* 186–195 (1971).
44 LITTMAN, M. L.; TAGUCHI, T., and MOSBACH, E. H.: Effect of cholesterol-free, fat-free diet and hypocholesterolemic agents on growth of transplantable animal tumors. Cancer Chemother. Rep. *50:* 25–45 (1966).
45 MAJERUS, P. W.; JACOBS, R.; SMITH, M. B., and MORRIS, H. P.: The regulation of fatty acid biosynthesis in rat hepatomas. J. biol. Chem. *243:* 3588–3595 (1968).
46 MALAMOS, B.; MIRAS, C.; LEVIS, G., and MANTOZOS, J.: *In vitro* incorporation of acetate-1-^{14}C into leukemic and normal leukocyte lipids. J. Lipid Res. *3:* 222–228 (1962).
47 MARAGOUDAKIS, M. E.: On the mode of action of lipid-lowering agents. VI.

Inhibition of lipogenesis in rat mammary gland cell culture. J. biol. Chem. *246:* 4046–4052 (1971).

48 MAYS, E. T.: Serum lipids in human cancer. J. surg. Res. *9:* 273–277 (1969).

49 MEDES, G.; PADEN, G., and WEINHOUSE, S.: Metabolism of neoplastic tissues XI. Absorption and oxidation of dietary fatty acids by implanted tumors. Cancer Res. *17:* 127–133 (1957).

50 MEDES, G.; THOMAS, A. J., and WEINHOUSE, S.: Metabolism of neoplastic tissues. IV. A study of lipid synthesis in neoplastic tissue slices *in vitro*. Cancer Res. *13:* 27–29 (1953).

51 MEDES, G.; THOMAS, A. J., and WEINHOUSE, S.: Metabolism of neoplastic tissue. XV. Oxidation of exogenous fatty acids in Lettré Ehrlich ascites tumor cells. J. nat. Cancer Inst. *24:* 1–12 (1960).

52 MEDES, G. and WEINHOUSE, S.: Metabolism of neoplastic tissues. XIII. Substrate competition in fatty acid oxidation in ascites tumor cells. Cancer Res. *18:* 352–359 (1958).

53 MERMIER, P. and BAKER, N.: Volume-dependent transfer of free fatty acids from Ehrlich ascites carcinoma to host tissues. Lipids *8:* 534–535 (1973).

54 MERMIER, P. and BAKER, N.: Flux of free fatty acids between host tissues, ascites fluid and Ehrlich ascites carcinoma cells. J. Lipid Res. (in press).

55 MIDER, G. B.; SHERMAN, C. D., jr., and MORTON, J. J.: The effect of Walker carcinoma 256 on the total lipid content of rats. Cancer Res. *9:* 222–227 (1949).

56 MIRAS, C. J.; LEGAKIS, N. J., and LEWIS, G. M.: Conversion of glucose to lipids by normal and leukemic lymphocytes. Cancer Res. *27:* 2153–2158 (1967).

57 MISHKIN, S.; STEIN, L.; GATMAITAN, Z., and ARIAS, I. M.: The binding of fatty acids to cytoplasmic proteins. Binding to Z protein in liver and other tissues of the rat. Biochim. biophys. Res. Commun. *47:* 997–1003 (1972).

58 MUELLER, P. S. and WATKIN, D. M.: Plasma unesterified fatty acid concentrations in neoplastic disease. J. Lab. clin. Med. *57:* 95–108 (1961).

59 NARAYAN, K. A. and MORRIS, H. P.: Serum lipoproteins of rats bearing transplanted Morris hepatoma 7777. Int. J. Cancer *5:* 410–414 (1970).

60 NICHOLS, A. V.: Functions and interrelationships of different classes of plasma lipoproteins. Proc. nat. Acad. Sci., Wash. *64:* 1128–1137 (1969).

61 OCKNER, R. K.; MANNING, J. A.; POPPENHAUSEN, R. B., and HO, W. K. L.: A binding protein for fatty acids in cytosol of intestinal mucosa, liver, myocardium and other tissues. Science *177:* 56–58 (1972).

62 PEDERSEN, B. N.; GROMEK, A., and DAEHNFELDT, J. L.: Extramitochondrial fatty acid synthesis in Ehrlich ascites tumor cells propagated *in vitro* and *in vivo*. Proc. Soc. exp. Biol. Med. *141:* 506–509 (1972).

63 PETERSON, J. A. and RUBIN, H.: The exchange of phospholipids between cultured chick embryo fibroblasts and their growth medium. Exp. Cell Res. *58:* 365–378 (1969).

64 RAFF, R. A.: Induction of fatty acid synthesis in cultured mammalian cells. Effects of cycloheximide and X-rays. J. cell. comp. Physiol. *75:* 341–352 (1970).

65 REED, C. F.: Phospholipid exchange between plasma and erythrocytes in man and the dog. J. clin. Invest. *47:* 749–760 (1968).

66 SABINE, J. R.; ABRAHAM, S., and CHAIKOFF, I. L.: Lack of feedback control of fatty acid synthesis in a transplantable hepatoma. Biochim. biophys. Acta *116:* 407–409 (1966).
67 SABINE, J. R.; ABRAHAM, S., and CHAIKOFF, I. L.: Control of lipid metabolism in hepatomas. Insensitivity of the rate of fatty acid and cholesterol synthesis by mouse hepatoma BW 7756 to fasting and to feedback control. Cancer Res. *27:* 793–799 (1967).
68 SABINE, J. R.; ABRAHAM, S., and MORRIS, H. P.: Defective dietary control of fatty acid metabolism in four transplantable rat hepatomas; numbers 5123C, 7793, 7795 and 7800. Cancer Res. *28:* 46–51 (1968).
69 SABINE, J. R. and CHAIKOFF, I. L.: Control of fatty acid synthesis in homogenate preparations of mouse hepatoma BW 7756. Austr. J. exp. Biol. med. Sci. *45:* 541–548 (1967).
70 SACKTOR, B. and DICK, A. R.: Alpha glycerophosphate and lactate dehydrogenases of hematopoietic cells from leukemic mice. Cancer Res. *20:* 1408–1412 (1960).
71 SAVCHUCK, W. B.; LOCKHARD, W. L., and LONG, H. W.: Proliferation of cultured liver cells in the presence of lysine and arginine salts of fatty acids. Exp. Cell Res. *37:* 169–174 (1965).
72 SCHOLEFIELD, P. G.; SATO, S., and WEINHOUSE, S.: The metabolism of fatty acids by ascites hepatoma 98/15. Cancer Res. *20:* 661–668 (1960).
73 SCOW, R. O.; HAMOSH, M.; BLANCHETTE-MACKIE, E. J., and EVANS, A. J.: Uptake of blood triglycerides by various tissues. Lipids *7:* 497–505 (1972).
74 SHIPP, J. C.: Interrelation between carbohydrate and fatty acid metabolism of isolated perfused rat heart. Metabolism *13:* 852–867 (1964).
75 SPECTOR, A. A.: The importance of free fatty acid in tumor nutrition. Cancer Res. *27:* 1580–1586 (1967).
76 SPECTOR, A. A.: Effect of carnitine on free fatty acid utilization in Ehrlich ascites tumor cells. Arch. Biochem. Biophys. *122:* 55–61 (1967).
77 SPECTOR, A. A.: The transport and utilization of free fatty acid. Ann. N.Y. Acad. Sci. *149:* 768–783 (1968).
78 SPECTOR, A. A.: Influence of pH of the medium on free fatty acid utilization by isolated mammalian cells. J. Lipid Res. *10:* 207–215 (1969).
79 SPECTOR, A. A.: Free fatty acid utilization by mammalian cell suspensions. Comparison between individual fatty acids and fatty acid mixtures. Biochim. biophys. Acta *218:* 36–43 (1970).
80 SPECTOR, A. A.: Metabolism of free fatty acids. Progr. biochem. Pharmacol., vol. 6, pp. 130–176 (Karger, Basel 1971).
81 SPECTOR, A. A.: Fatty acid, glyceride and phospholipid metabolism; in ROTHBLAT and CRISTOFALO Growth, nutrition and metabolism of cells in culture, vol. 1, pp. 257–296 (Academic Press, New York 1972).
82 SPECTOR, A. A.; ASHBROOK, J. D.; SANTOS, E. C., and FLETCHER, J. E.: Quantitative analysis of the uptake of free fatty acid by mammalian cells: lauric acid and human erythrocytes. J. Lipid Res. *13:* 445–451 (1972).
83 SPECTOR, A. A. and BRENNEMAN, D. E.: Effect of free fatty acid structure on binding to rat liver mitochondria. Biochim. biophys. Acta *260:* 433–438 (1972).

84 SPECTOR, A. A. and BRENNEMAN, D. E.: Role of free fatty acid and lipoproteins in the lipid nutrition of tumor cells; in WOOD Tumor lipids: biochemistry and metabolism, pp. 1–13 (Amer. Oil Chem. Soc. Press, Champaign 1973).
85 SPECTOR, A. A. and FLETCHER, J. E.: Binding of long-chain fatty acids to β-lactoglobulin. Lipids 5: 403–411 (1970).
86 SPECTOR, A. A.; FLETCHER, J. E., and ASHBROOK, J. D.: Analysis of long-chain free fatty acid binding to bovine serum albumin by determination of stepwise equilibrium constants. Biochemistry 10: 3229–3232 (1971).
87 SPECTOR, A. A.; HOAK, J. C.; WARNER, E. D., and FRY, G. L.: Utilization of long-chain free fatty acids by human platelets. J. clin. Invest. 49: 1489–1496 (1970).
88 SPECTOR, A. A.; JOHN, K., and FLETCHER, J. E.: Binding of long-chain fatty acids to bovine serum albumin. J. Lipid Res. 10: 56–67 (1969).
89 SPECTOR, A. A. and SOBOROFF, J. M.: Effect of chlorophenoxyisobutyrate on free fatty acid utilization by mammalian cells. Proc. Soc. exp. Biol. Med. 137: 945–947 (1971).
90 SPECTOR, A. A. and SOBOROFF, J. M.: Utilization of free fatty acids complexed to human plasma lipoproteins by mammalian cell suspensions. J. Lipid Res. 12: 545–552 (1971).
91 SPECTOR, A. A. and SOBOROFF, J. M.: Long-chain fatty acid methyl ester hydrolase activity in mammalian cells. Lipids 7: 186–190 (1972).
92 SPECTOR, A. A. and SOBOROFF, J. M.: Studies on the cellular mechanism of free fatty acid uptake using an analogue, hexadecanol. J. Lipid Res. 13: 790–796 (1972).
93 SPECTOR, A. A. and STEINBERG, D.: The utilization of unesterified palmitate by Ehrlich ascites tumor cells. J. biol. Chem. 240: 3347–3753 (1965).
94 SPECTOR, A. A. and STEINBERG, D.: Release of free fatty acids from Ehrlich ascites tumor cells. J. Lipid Res. 7: 649–656 (1966).
95 SPECTOR, A. A. and STEINBERG, D.: Relationship between fatty acid and glucose utilization in Ehrlich ascites tumor cells. J. Lipid Res. 7: 657–663 (1966).
96 SPECTOR, A. A. and STEINBERG, D.: The effect of fatty acid structure on utilization by Ehrlich ascites tumor cells. Cancer Res. 27: 1587–1594 (1967).
97 SPECTOR, A. A. and STEINBERG, D.: Turnover and utilization of esterified fatty acids in Ehrlich ascites tumor cells. J. biol. Chem. 242: 3057–3062 (1967).
98 SPECTOR, A. A.; STEINBERG, D., and TANAKA, A.: Uptake of free fatty acids by Ehrlich ascites tumor cells. J. biol. Chem. 240: 1032–1041 (1965).
99 STEWART, A. G. and BEGG, R. W.: Systemic effects of tumors in force-fed rats. III. Effect on the composition of the carcass and liver and on the plasma lipids. Cancer Res. 13: 560–565 (1953).
100 TAKAOKA, T. and KATSUTA, H.: Long-term cultivation of mammalian cell strains in protein- and lipid- free chemically defined synthetic media. Exp. Cell Res. 67: 295–304 (1971).
101 VOST, A.: Uptake and metabolism of circulating chylomicron triglyceride by rabbit aorta. J. Lipid Res. 13: 695–704 (1972).
102 WATSON, J. A.: Regulation of lipid metabolism in *in vitro* cultured minimal deviation hepatoma 7288C. Lipids 7: 146–155 (1972).

103 WEBER, G.; MORRIS, H. P.; LOVE, W. C., and ASHMORE, J.: Comparative biochemistry of hepatomas. II. Isotope studies of carbohydrate metabolism in Morris hepatoma 5123. Cancer Res. *21:* 1405–1411 (1961).

104 WEINHOUSE, S.: Glycolysis, respiration and anomalous gene expression in experimental hepatomas: G. H. A. Clowes memorial lecture. Cancer Res. *32:* 2007–2016 (1972).

105 WEINHOUSE, S.; ALLEN, A., and MILLINGTON, R. H.: Metabolism of neoplastic tissue. V. Fatty acid oxidation in slices of transplanted tumors. Cancer Res. *13:* 367–371 (1953).

106 WOOD, R. and SNYDER, F.: Tumor lipids. Metabolic relationships derived from structural analyses of acyl, alkyl, and alk-1-enyl moieties of neutral glycerides and phosphoglycerides. Arch. biochem. Biophys. *131:* 478–494 (1969).

107 WRIGHT, J. D. and GREEN, C.: The role of the plasma membrane in fatty acid uptake by rat liver parenchymal cells. Biochem. J. *123:* 837–844 (1971).

108 YASUDA, M.: Lipid metabolism of tumors. III. Influence of food lipids upon the nature of tumor phospholipids. Proc. Soc. exp. Biol. Med. *28:* 1074–1075 (1931).

Author's address: ARTHUR A. SPECTOR, Departments of Biochemistry and Internal Medicine, University of Iowa, *Iowa City, IA 52242* (USA)

Lipoproteins in Relation to Cancer

Marion Barclay and Vladimir P. Skipski

Memorial Sloan-Kettering Cancer Center, New York, N.Y.

Contents

I. Introduction	76
II. Lipoproteins in Serum or Plasma, Determinations and Quantitation	77
III. Selection and Description of Normal Subjects and Patients with Cancer	79
IV. Lipoproteins in Serum from Normal Subjects and Patients with Cancer	81
A. Classes of Lipoproteins	81
1. Very Low-Density Lipoproteins (VLDL) – Chylomicra or Particles	81
2. Low-Density Lipoproteins (LDL or β-Lipoproteins)	86
3. High-Density Lipoproteins (HDL_2 and HDL_3) or α-Lipoproteins	89
4. Albumin-Bound Free Fatty Acid Complexes ('Ultracentrifugal Residue'; Very High-Density Lipoproteins)	96
5. Commentary	97
V. Lipoproteins in Serum or Plasma from Animals with Experimental Tumors	100
A. Classes of Lipoproteins	100
1. Chylomicra or Particles, and the VLDL	100
2. Low-Density Lipoproteins (LDL or β-Lipoproteins)	101
3. High-Density Lipoproteins (HDL_2 and HDL_3) or α-Lipoproteins	102
VI. Summary	106
VII. Acknowledgments	107
VIII. References	107

I. Introduction

It can no longer be doubted that lipid (and probably associated protein) metabolism is deranged when an animal has cancer. Lipemic serum from patients and animals with cancer was seen early in clinical observations and cancer research. The full significance of 'lipemic serum' has been revealed relatively recently by the recognition that lipids are transported in serum or plasma as lipoproteins, in which form they are soluble in aqueous media containing salts. The lipids being transported are complexed in one or another form with globulins but free fatty acids are associated also with albumins.

At present the numerous methods available for separating lipoproteins have promoted abundant research efforts into the lipoprotein complexes. Their lipid components have been studied thoroughly and related to several diseases, and work on their protein components is going forward. It is possible that the final answers to why certain lipoproteins are affected by cancer may reside in the proteins (or apoproteins), in the lipid moieties, or the complex as a whole.

II. Lipoproteins in Serum or Plasma, Determinations and Quantitation

Figure 1 illustrates the presently known spectrum of lipoproteins in serum from normal fasting women aged 25–50 years, and the values for the chemical constituents were derived from these samples [60, 61]. Serum from other clinically normal fasting subjects, i.e. men and children, has essentially the same spectrum of lipoproteins but the quantities will vary [9–11]. At present there is no evidence that the lipoprotein fractions from various groups of normal fasting subjects have remarkably different amounts or kinds of lipids (or proteins).

The lipoprotein fractions illustrated were obtained by taking advantage of the rather specific hydrated density properties of lipoprotein macromolecules, as listed in figure 1 and classified by the terminology used (i.e., VLDL, LDL, HDL_2, HDL_3, $VHDL_1$ and $VHDL_2$ and legend to fig. 1). By narrowing the preparative density certain fractions can be separated even further, if desired. For most purposes, sufficiently accurate values can be calculated for these subfractions or classes by means of the analytical ultracentrifuge (Beckman-Spinco Model E). The values in most tables in this review were obtained in this manner [6].

In many early experiments with lipoproteins, electrophoresis (originally moving boundary in buffer) was used. This separation technique, based upon the different electric charges on the lipoproteins, was adapted to various solid media. Many, probably most, studies on serum lipoproteins use these relatively simple procedures, including at present polyacryl-

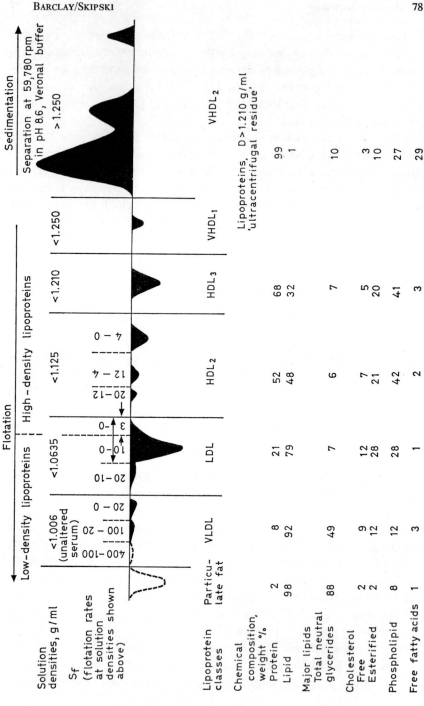

amide gel (disc) electrophoresis [46]. Although presumably simpler, these techniques do not provide means for accurate calculations of the amounts of lipoproteins, and sometimes the separations are not clear-cut. Electrophoretic procedures result in separation of two major groups, β-lipoproteins and α-lipoproteins. These correspond to some extent to LDL and total HDL obtained by ultracentrifugation. A pre-β-lipoprotein fraction or band can correspond to VLDL. Details of these different procedures may be obtained by consulting LINDGREN et al. [38].

III. Selection and Description of Normal Subjects and Patients with Cancer

It has been demonstrated that the levels of lipoproteins in serum are influenced by a variety of factors in health as well as in disease [3]. Therefore, in order to learn what effect cancer has upon serum lipoproteins, it is essential to be exceedingly strict in the selection of subjects. Most of the presumably normal subjects, especially the women who comprise the majority of the subjects in the authors' experiments, were given physical examinations. No subjects, including the patients with cancer, had been receiving recent medication (especially hormones), and all subjects had fasted at least 18 h before providing the blood specimens. All the

Fig. 1. Spectra of lipoproteins in human serum. Solid vertical lines enclose fractions, obtained at the solution densities listed, by preparative ultracentrifugation [6]. Broken vertical lines separate classes of lipoproteins based upon the calculated S_f (or flotation rate) limitations at each solution density. It is recognized that the S_f limitations, especially in VLDL and LDL, for example S_f10–20 in LDL, comprise a spectrum of macromolecules. S_f0–3 in the LDL is commonly designated HDL$_1$, because its hydrated density, 1.05 g/ml, is on the borderline density when separations are made at solution density 1.0635 g/ml, which is used conventionally to separate LDL from the HDLs [38]. The values under 'Particulate fat' are from SKIPSKI [60]; all other values for the constituents listed are from SKIPSKI et al. [61]. It should be noted that all the lipid constituents are present, in varying amounts, in all the fractions of lipoproteins, and the total quantity in serum reflects the amounts of the lipoproteins in serum. For example, if the LDL is greatly elevated, the total amount of cholesterol will be elevated.

women were still menstruating regularly and were in the 25- to 50-year age range. Some of the men were older, but none was receiving medication and all provided fasting samples. Two normal children had a light breakfast.

It was fortunately possible for us to study a well-integrated group of women living under the same conditions of diet and activities. At the beginning of the study these subjects provided weekly samples of fasting serum over a 12-week period. They filled out extensive questionnaires at this time and also 5 and 10 years later. They were examined by a physician at the 0- and 5-year periods. Any disease occurring at the 5- and 10-year periods was noted.

From these observations and from those of the men, it was discerned that the subjects who reported a high incidence (grand-parent(s), parent(s), sibling(s) or two or more uncles or aunts) of relatives with cancer had unusually and consistently low values for one of the high-density lipoprotein components, HDL_2. The consistency of the low values in these subjects over a period of several weeks (even at ovulation when HDL_2 is usually elevated [5]) strengthens these observations. Henceforth all normal subjects were classified into those with positive or negative (at most one uncle or aunt) family histories of cancer. Subsequent experiments have shown that most persons with a positive family history have unusually low values for HDL_2. Thus, an additional factor influencing levels of serum lipoproteins in clinically normal persons was revealed.

Both tables I and II were compiled from data published previously [9–11] and augmented with recent aquisitions. In both tables I and II the quantities of only the lipoprotein classes in which one or more components were unusually different from those of the normal subjects with a negative family history are shown. The values for the means which are significantly different are printed in italics for emphasis.

Table I is concerned with women subjects: normal; with carcinoma of breast, both primary operable and advanced inoperable; and with other types of cancer. The patients with primary operable disease were studied before mastectomy when this surgery was performed [10]. These patients did not have extensive disease with metastasis. The patients with advanced inoperable disease were studied before palliative surgery (e.g. bilateral oophorectomy) was done. They had received no medication [10, 11]. Some had bone and/or soft tissue metastases [31]. Serum samples from the patients with other types of cancer were obtained before treatment and after an 18-hour fast, the routine procedure with all patients.

IV. Lipoproteins in Serum from Normal Subjects and Patients with Cancer

A. Classes of Lipoproteins

1. Very Low-Density Lipoproteins (VLDL) – Chylomicra or Particles

The two groups of normal women (table I) have practically the same values for total VLDL and these are quite low. The low standard deviations indicate that most values are grouped about the means, 12 and 7 mg per 100 ml serum.

On the other hand, normal men with a positive family history of cancer have very elevated values for total VLDL, and there is obviously more variability, ± 91 (table II). The levels of total VLDL in men with a negative family history of cancer were remarkably consistent and significantly lower than the other men. However, all the men with positive family histories of cancer had greater quantities of total VLDL than any of the women. The two normal children sampled after a light breakfast had quite high values for total VLDL (209 and 216 mg/100 ml serum). Of the fasted children, one had no total VLDL, the other had no $S_f100–400$ lipoproteins, but a total of 150 mg/100 ml serum in the total VLDL (table II). The mean or average (75 mg/100 ml serum) for the children without breakfast should be compared with that for the children with cancer, a few of whom did not have greatly elevated total VLDL (consult table VI for details).

All the patients with cancer had significantly higher levels of total VLDL. In figure 1 it can be seen that the total fraction, when removed as an entity from unaltered serum (D <1.006 g/ml [6]), shows two or three boundaries in the analytical ultracentrifuge. The elevations in total VLDL (tables I, II, V and VI) result primarily from the components with $S_f20–100$ and $S_f0–20$. VLDL with $S_f100–400$ were generally absent except for small amounts in the group of normal men with a positive family history of cancer, and in men and women with cancer. The levels of $S_f100–400$ were elevated in children with chronic reticuloendotheliosis, acute leukemia, lymphoma and myelomonocytic leukemia (table VI).

HIGAZI et al. [28] reported significant increases in total lipids and the 'nonmobile' fraction, separated with paper electrophoresis, in serum of patients with acute myeloblastic leukemia. The 'nonmobile' fraction probably includes chylomicra (particles and/or some VLDL). In contrast, in a case of acute lymphoblastic leukemia there was a marked decrease in

Table I. Lipoproteins in serum from clinically normal women, in patients with two stages of carcinoma of the breast, and in patients with other types of cancer

Lipoproteins[1], mg/100 ml serum	Normal women		Women with carcinoma of breast		Women with other types of cancer
	fam. hist. neg. (25–50 years)	fam. hist. pos. (25–59 years)	primary opera. (32–40 years)	advanced inop. (33–49 years)	(30–60 years)
Number of analyses	86	72	5	25	10
VLDL					
Mean	12	7	*103*[2]	103	84
Range	0–55	0–28	63–132	0–260	0–161
SD	± 11	± 9	± 36	± 60	± 35
p[3]			<0.001	<0.001	<0.001
LDL S_f10–20					
Mean	12	8	214	46	15
Range	0–84	0–46	0–433	0–208	2–40
SD	± 25	± 18	± 90	± 90	± 15
S_f0–3 (HDL$_1$)					
Mean	45	49	*18*	74	95
Range	5–143	5–134	13–22	15–162	38–123
SD	± 30	± 29	± 5	± 48	± 33
p			<0.05	<0.02	<0.001

Table I. Continuation

Lipoproteins[1], mg/100 ml serum	Normal women		Women with carcinoma of breast		Women with other types of cancer
	fam. hist. neg. (25–50 years)	fam. hist. pos. (25–59 years)	primary opera. (32–40 years)	advanced inop. (33–49 years)	(30–60 years)
HDL$_2$ (S$_f$0-4)					
Mean	120	52	84	51	61
Range	45–275	10–134	59–126	8–98	21–93
SD	± 56	± 28	± 37	± 30	± 29
p		<0.001	<0.05	<0.001	<0.001
HDL$_3$					
Mean	126	126	87	111	134
Range	62–255	53–270	62–115	49–165	85–188
SD	± 30	± 36	± 27	± 34	± 36
p			<0.02		

1 Lipoproteins in serum: VLDL = very low-density lipoproteins with D <1.006 g/ml; LDL = low-density lipoproteins with D <1.0635 g/ml – S$_f$10–20 and S$_f$0-3 (HDL$_1$) are subfractions or classes of LDL; HDL$_2$ = high-density lipoproteins with D <1.125 g/ml – S$_f$0-4 is the major component; HDL$_3$ = high-density lipoprotein with D <1.210 g/ml.

2 Italic numbers are significantly different from those of normal subjects with negative family histories of cancer.

3 p is the probability that the difference observed between the means of 2 groups could be attributable to chance alone; if p is equal to or less than 0.05, the difference is considered significant.

Table II. Lipoproteins in serum from clinically normal men and children, and in men and children with cancer

Lipoproteins[1], mg/100 ml serum	Normal men fam. hist. neg. (40–70 years)	Normal men fam. hist. pos. (30–50 years)	Men with cancer (30–60 years)	Normal children (0.5–10 years)	Children with cancer (1–11 years)
Number of analyses	7	10	8	4	11
VLDL					
Mean	43	*197*[2]	230	75	114
Range	40–46	73–296	156–418	0–150	10–264
SD	± 2	± 91	± 95	± 99	± 99
p[3]		<0.001	<0.001		
LDL S$_f$10–20					
Mean	27	31	43	4	23
Range	0–58	0–77	5–90	0–13	1–68
SD	± 28	± 33	± 36	± 5	± 28
0–3 S$_f$(HDL$_1$)					
Mean	42	50	124	64	53
Range	10–110	0–105	15–276	59–68	2–100
SD	± 30	± 39	± 83	± 5	± 32
p			<0.05		

Table II. Continuation

Lipoproteins[1], mg/100 ml serum	Normal men		Men with cancer	Normal children	Children with cancer
	fam. hist. neg. (40–70 years)	fam. hist. pos. (30–50 years)	(30–60 years)	(0.5–10 years)	(1–11 years)
HDL$_2$ (S$_f$0–4)					
Mean	97	26	23	65	*19*
Range	70–122	16–46	15–53	60–72	0–48
SD	± 26	± 14	± 13	± 6	± 17
p		<0.001	<0.001		<0.001
HDL$_3$					
Mean	148	142	120	217	*106*
Range	112–187	91–212	51–173	118–314	61–155
SD	± 33	± 42	± 39	± 98	± 37
p					<0.05

1 Lipoproteins in serum: VLDL = very low-density lipoproteins with D <1.006 g/ml; LDL = low-density lipoproteins with D <1.0635 g/ml – S$_f$10–20 and S$_f$0–3 (HDL$_1$) are subfractions or classes of LDL; HDL$_2$ = high-density lipoproteins with D <1.125 g/ml – S$_f$0–4 is the major component; HDL$_3$ = high density lipoprotein with D <1.210 g/ml.

2 Italic numbers are significantly different from those of normal subjects with negative family histories of cancer.

3 p is the probability that the difference observed between the means of 2 groups could be attributable to chance alone; if p is equal to or less than 0.05, the difference is considered significant.

total lipids and the 'nonmobile' fraction. The authors [28] rightly state that additional evidence is required to prove the interesting differences observed in the effects on lipids and lipoproteins of various leukemias. The changes in levels of lipid and lipoproteins also were in acute, rather than chronic, forms of leukemia.

2. Low-Density Lipoproteins (LDL or β-Lipoproteins)

Figure 1 shows that the lipoprotein fraction with the next highest density (D >1.006 <1.0635 g/ml) can be composed of at least three classes with S_f values ranging from 0 to 20 when it is separated from HDL, after the complete previous removal of VLDL [6]. The majority of the components have an S_f range from 3 to 10. Most of the serum cholesterol is associated with LDL. When serum or plasma lipoproteins are separated by the heavy-metal, cold ethanol procedure [17], once rather extensively used for cancer research in our laboratory [7, 8, 31], LDL can be estimated from the cholesterol content of fraction III and HDL (α-lipoproteins) from the phospholipid content of fraction IV. The LDL are comparable to β-lipoproteins in electrophoresis [33, 38].

Tables I and II show that with the ultracentrifugal techniques used [6], the amounts of only two of the classes of LDL in serum from patients are different from those in normal subjects. The values in the two groups of normal subjects are remarkably similar and rather low. Women with primary operable carcinoma of the breast have the highest values for S_f10–20, but data from additional subjects might bring this average down. Women with advanced carcinoma of the breast may also have high values for S_f10–20. The high standard deviations in the cases with cancer of the breast indicate the wide spread of the values and result in no statistical significance with these numbers of cases. This is true also for the men (table II).

The S_f0–3 component, however, was significantly increased in women with advanced carcinoma of the breast and in women with different types of cancer (table I). However, women with primary operable disease had unusually low values. The highest values were observed in patients with lung cancer, neuroblastoma, and all cases of ovarian carcinoma. In women with cancer, other than breast, it was lowest in choriocarcinoma.

In men with various types of cancer (table II), S_f0–3 was also high, especially in late myeloma, leiomyosarcoma, colon carcinoma, chronic myelocytic leukemia, and highest in epidermoid bronchogenic carcinoma. It was not especially high in early myeloma, parotid and gastrointestinal cancer (table V). The high levels of S_f0-3 and low levels of HDL_2, somewhat similar in density properties (fig. 1), are suggestive of a metabolic shift toward production of molecules with lighter weight.

Children with various kinds of cancer did not usually have values for S_f0-3 that were abnormally high (table II). The most elevated values occurred in malignant and chronic reticuloendotheliosis, acute leukemia and Hodgkin's disease.

The only significant elevation (659 mg/100 ml serum) in the class with S_f3-10, occurred in a child with Hodgkin's disease [9]. The S_f3-10 components (except in Hodgkin's disease) are least affected by cancer, even though they comprise the bulk of the total serum lipoproteins. This implies that the effects of cancer upon serum lipoproteins are quite selective as suggested by certain well-defined effects discussed below.

A number of relatively early reports presented information on the lipoproteins in myelomatosis. Decreased values were observed for certain of the β-lipoproteins (−S 25–40) [37]. When the serum proteins were separated and estimated with paper electrophoresis and the lipoproteins with ultracentrifugal analyses, a male patient with a combination of myelomatosis and xanthomatosis had excessively high values for a β-globulin which had the mobility of a β_2-globulin [51]. NEUFELD et al. [51] consider this β_2-globulin (or lipoprotein) to be an abnormal molecule synthesized by the plasma cells in this disease. It stained intensely with lipid stain and coincided with greatly elevated values for esterified cholesterol and phospholipids.

Ultracentrifugal analysis of the serum at density 1.0635 g/ml (thus serving to float both VLDL and LDL) produced the data in table III compiled from NEUFELD et al. [51] and from our data. These results and those published previously by LENNARD-JONES [35] are significant in that they are pioneer observations on the suspected production, by abnormal cells, of 'abnormal' or unusual lipoprotein macromolecules in disease in man. In addition to being present in greatly elevated quantities, these low density β-lipoproteins also had altered mobilities in an electrical field. After a prolonged period on a diet very low in animal fat and containing only 30 g vegetable fat per day there was an increase in the S_f0-12 class of the LDL. This is not so sensitive to diet as the less dense $S_f20-400$ class (VLDL), which decreased. LENNARD-JONES [35] observed no change on this type of diet. Evidently, in spite of dietary and chemotherapeutic treatment (1 g urethane daily), the S_f0-12 (primarily β-lipoprotein) continued to be synthesized in large amounts (or not catabolized), and

Table III. Lipoproteins in sera from patients with a combination of myelomatosis and xanthomatosis

Lipoproteins with densities <1.0635 g/ml (VLDL and LDL)	Patients[1], mg/100 ml serum	Control males[2], mg/100 ml serum
S_f 0–12	250–350	450
S_f 12–20	100–150	29
S_f 20–100	1,100–1,500	38
S_f 100–400	1,200–2,000	9

1 Values when patients were first studied, compiled from NEUFELD *et al.* [51].
2 Average values are from the present authors' data.

NEUFELD *et al.* [51] suggested that this was the primary abnormal component.

A serum factor believed to cause alterations in the character of the β-lipoprotein as cited above was described by SPIKES *et al.* [62] in other patients with the rare combination of myelomatosis and hyperlipoproteinemia. Serum from these patients contained β-lipoproteins with an altered migration rate in starch gel. When this 'abnormal' serum from the patients was mixed with normal serum, it caused the β-lipoproteins of normal serum also to have altered or unusual migration rates. The serum component producing these effects was termed 'serum lipoprotein-altering factor' (SLAF). It was separated most efficiently by ultracentrifugation and was concentrated in the fraction with density > 1.21 g/ml (the ultracentrifugal residue). It was demonstrable only in serum from patients who had both myeloma and hyperlipoproteinemia. At present, the nature of the relationship between SLAF, β-lipoproteins and myeloma remains unclear.

Elevated levels of β-lipoproteins were noted when paper electrophoresis was used to detect lipoproteins in serum from patients with other types of cancer [45]. Patients with breast carcinoma had especially high levels which decreased after surgery but tended to increase with time. Significantly higher turbidometric indices were obtained for the β-lipoproteins in 54 patients with breast carcinoma and also in 16 patients with sarcomatosis [32] when compared with control indices.

In patients with most types of leukemia, however, the β-lipoproteins are not significantly increased [28].

Certain other procedures have, for the most part, shown that LDL is elevated in serum of patients with cancer. Serum from patients with Ewing's sarcoma was shown by immunoelectrophoresis to contain prominent precipitation arcs for β-lipoprotein when this serum was tested against rabbit-anti-human serum. This pronounced pattern was not observed in other malignant diseases nor in apparently healthy controls [24].

3. High-Density Lipoproteins (HDL_2 and HDL_3) or α-Lipoproteins

From Figure 1 it can be seen that serum may contain at least two lipoprotein fractions currently termed high-density because these lipoproteins have hydrated densities greater than 1.0635 g/ml, the density originally used to separate the two main large fractions, VLDL + LDL from the HDL components [34]. It is now generally accepted that HDL_2 (D $<$ 1.125 g/ml) and HDL_3 (D $<$ 1.210 g/ml) are different [59]. The HDL_2 fraction was shown to have two components in addition to the principal and usually most abundant one, HDL_2 proper [6, 13]. To distinguish these, S_f boundaries were calculated for them according to their flotation rates in a density 1.125 g/ml.

In numerous reports dealing with serum lipoproteins the data were obtained with techniques other than ultracentrifugation. In, for example, electrophoretic [33, 38] and precipitation techniques [17], the two major HDLs are separated from VLDL and LDL as α-lipoproteins or fraction IV, each encompassing both HDLs.

With respect to cancer, it appears that the most pronounced effect upon serum lipoproteins resides in or is associated with HDL_2 (fig. 1, $<$ 1.125, S_f4–0).

The relationship between levels of HDL_2 in normal women and their pronounced family history of cancer is shown in table IV, which contains only the HDL_2 values from previously published and more detailed data [11]. These subjects were described earlier and are presented here because the consecutive samples from them show considerable consistency. The most important point is that there is no overlap between the means of the two groups, lending credence to the postulate that a pronounced history of cancer in first degree relatives can influence the levels of HDL_2, not only in patients who have cancer, but in their relatives. Those with posi-

Table IV. Lipoproteins in serum from normal women related to incidence of cancer in their immediate families

Subjects	HDR$_2$ (S$_f$0–4)[1], mg/100 ml serum			Relatives with cancer (periods, years)		
	mean	range	SD	0	5	10
Women with positive family histories of cancer						
M.P.D. (10)[2]	86	54–107	± 22	none	father	no additional
R.M.H. (9)	70	49–89	± 16	maternal grandmother, 5 maternal aunts	no additional	2 first cousins
M.L. (4)	43	23–49	± 13	2 maternal aunts, 1 uncle		propositus
M.G.G. (10)	42	23–63	± 19	mother, father	no additional	no additional
T.M.K. (11)	27	15–45	± 11	father, mother paternal uncle	brother, sister, maternal uncle	propositus
W.M.M. (11)	26	15–35	± 6	none	none	sister
Average	49					
Women with little or no family history of cancer						
P.M. (10)	168	137–215	± 23	paternal aunt	no additional	no additional
R.M.O. (9)	135	108–176	± 36	none	none	none
M.J. (10)	124	103–147	± 20	maternal uncle	no additional	no additional
B.M.B. (11)	105	79–128	± 19	none	none	none
M.P.F. (5)	105	89–113	± 21	maternal aunt	no additional	no additional
M.J.D. (11)	93	69–118	± 21	none	none	none
Average	122					

1 HDL$_2$ (S$_f$0–4): Major class of the high-density lipoprotein fraction with D <1.125 g/ml. Subjects are listed in decreasing order of this component.
2 Numbers in parentheses represent weeks of samples; 1 blood specimen per week.

tive family histories have very low values for HDL_2, averaging 26–86 mg/100 ml serum compared with averages of 93–168 mg/100 ml serum for the subjects with negative family histories. There is also little overlap in the individual values, and the standard deviations are relatively low.

It can be seen also that two of the women with positive family histories reported cancer in themselves at the 10-year survey period, and note that they had consistently low values for HDL_2.

The other lipoproteins, i.e., VLDL, LDL and HDL_3 were not different between these two groups of relatively young, still menstruating women (table I).

Men with a positive history of cancer present the same picture with regard to HDL_2: those with a positive family history have average values for HDL_2 in serum, 16–46 mg/100 ml (mean 26 mg); those with a negative family history have average values ranging from 70 to 122 mg/100 ml (mean 97 mg). However, men with cancer in their first degree relatives have a pronounced elevation in VLDL. The S_f100–400 class ranged from 0 to 72 mg (mean 19 mg), the other two components in VLDL ranged from 73 to 296 mg (mean 197 mg). The men with negative family histories had no S_f100–400 (most normal fasting persons have none [3]) and only 40–46 mg (mean 43 mg) for the remainder of the VLDL.

Two men with the highest VLDL and lowest HDL_2 developed cancer about 5 years after this survey was made.

The HDL_3 values in the serum from these men were remarkably similar regardless of family history of cancer (table II).

Thus, normal fasting subjects, clinically free from disease, taking no drugs, eating a usual normal diet, having an active busy life, etc., tend to have low levels of HDL_2 if they are members of families with a high cancer incidence. In men, but not in women, there are also elevations in the VLDL.

In women with cancer (table I), the values for HDL_2 in serum are quite low, especially in patients with more advanced disease. Although it was unusual for a patient with minimal disease (e.g. primary operable carcinoma of the breast) to have normal values, a few did, and it might be assumed that their prognosis was better. In fact, in a few cases followed for several years, those who had more normal values for HDL_2 when first examined made more satisfactory responses to the mastectomy procedures, and maintained higher serum levels of HDL_2 during the postoperative period in which serum was analyzed several times for lipoproteins [7, 8, 10, 42].

Women with cancer other than carcinoma of the breast also have low values for HDL_2. Very low values were observed in patients with neuroblastoma (21 mg/100 ml), lung cancer (34 mg/100 ml), and in some patients with ovarian carcinoma (52 mg/100 ml). The only type in which the values approached normal was choriocarcinoma (93 mg/100 ml).

Men with cancer (table II) have very low values for HDL_2 (but remarkably similar to the normal men with positive family histories of cancer). They are proportionately much lower (25 %) than in normal men with a negative family history than are HDL_2 values in the women with cancer (43 %), compared with women who have a negative family history.

The data in table V show the unusually low values for HDL_2 in serum from men with the different types of cancer we have investigated. Compared with the normal averages listed, $S_f100-400$ is elevated and the total remaining VLDL is exceedingly elevated. There is also an unusually high average for S_f0-3, 1.0635 (HDL_1) in the several types of cancer. It may be significant that HDL_2 is consistently low whereas the other components are present in elevated but quite variable amounts.

Table V. Lipoproteins in serum from men with different types of cancer

Subjects and cancer type	Lipoprotein fractions, mg/100 ml serum		
	D <1.006 g/ml	D <1.0635 g/ml	D <1.125 g/ml
	$S_f100-400$[1] VLDL[2]	S_f0-3 (HDL_1)	(S_f0-4, HDL_2)
B. Parotid and gastrointestinal cancer	4 296	38	17
G. Melanoma	72 200	45	20
M. Melanoma	46 418	93	22
R. Leiomyosarcoma	59 205	151	53
A. Colon carcinoma	42 174	92	16
L.A. Chronic myelocytic leukemia	38 156	115	15
E.B. Epidermoid bronchogenic carcinoma	0 165	276	19
Average	37 231	116	23
Average, normal men[3]	0 43	42	97

1 Spectrum of components with the lowest densities in the VLDL fraction.
2 $S_f100-400$ components.
3 Normal men with a negative family history of cancer.

Table VI shows the relationship between the levels of serum HDL_2 (and the VLDL) in children with a variety of malignancies. An unusual observation in some children was the quite low or absent HDL_2 concurrently with an increase in one of the two other components seen in this lipoprotein fraction (with $D < 1.125$ g/ml). This is suggestive only (in so few cases) of a shift toward synthesis of a less-dense component. However, the presence of $S_f 4$–12 (1.125) along with no HDL_2 ($S_f 0$–4, 1.125) is quite unusual.

As with the men who have cancer, the $S_f 100$–400 and remaining VLDL (fig. 1) are greatly elevated, but in children with certain types of cancer only.

The results cited above, obtained by ultracentrifugal separations and analyses of serum lipoproteins, suggest that the most important effect of cancer is upon one of the high-density lipoproteins: HDL_2. Admittedly, the VLDL are frequently elevated, but these are more affected by other factors, especially diet [3], less easily controlled in clinical research.

There have been numerous reports showing or suggesting that these

Table VI. Lipoproteins in serum from children with different types of cancer

Subjects and cancer type	Lipoprotein fractions, mg/100 ml serum			
	$S_f 100$–400[1]	VLDL[2]	$D < 1.125$ g/ml	
			$S_f 4$–12	$S_f 0$–4 (HDL_2)
M.G. Malignant reticuloendotheliosis	0	36	24	15
M.B. Chronic reticuloendotheliosis	45	174	0	48
A.F. Acute leukemia	56	264	17	0
J.S. Lymphoma	33	209	0	27
D.M. Lymphangiomatosis	0	10	23	0
K.M. Hodgkin's disease[3]	0	34	3	14
S.H. Myelomonocytic leukemia[4]	34	69	0	30
Average	24	114	10	19
Average, normal children	0	75	0	65

1 Spectrum of components with the lowest densities in the VLDL fraction.
2 $S_f 100$–400 components.
3 Means of two experiments on the same subject, K.M.
4 Means of four experiments on the same subjects, S.H.

HDLs, either as Cohn's fraction IV [7, 8, 31] or as α-lipoproteins [15, 29, 30, 39, 41, 42, 54, 64, 66] are abnormally low in patients with cancer. Only a few of these can be dealt with in any detail.

Values for the α-lipoproteins in serum of women with various types of gynecological cancer have been reported by ALVAREZ and GOODELL [2]. The patients were compared not only with normal women but with women in the three different trimesters of normal pregnancy and up to six weeks postpartum. These are interesting associations because of the increased growth, presumably controlled and uncontrolled, occurring in these different subjects.

The mean value for α-lipoproteins (expressed as percent of $\alpha + \beta$ lipoproteins) in control subjects was 31%. In pregnant women the percentages were: first trimester, 32%; second trimester, 29%; third trimester, 24%; postpartum 1–5 days and 6 weeks, 27%. Values from women with cancer ranged from 13 to 24%, the highest in women with invasive carcinoma of the cervix. With the exception of the latter, all values for patients with cancer are well below controls and pregnant women. The β-lipoproteins in patients with gynecological cancer were also higher than in pregnant women, but again the percentage in the third trimester was the same as that in invasive carcinoma of the cervix.

ALVAREZ and GOODELL [2] report that serum total cholesterol is significantly increased in patients with carcinoma of the cervix *in situ*, but not in those with invasive carcinoma of the cervix or in any other invasive tumors. Thus, the low value, 18% for α-lipoproteins in carcinoma of the cervix *in situ*, can reflect a proportional decrease resulting from an increase in β-lipoprotein which transports most of the serum cholesterol. However, the decreases in α-lipoproteins in the other tumors may be real since the values for serum phospholipids in these patients with cancer are somewhat below normal. The only value for phospholipids which is near normal (still some 20 mg/100 ml serum below normal) is that for carcinoma of the cervix *in situ*.

In ovarian carcinoma the values for α-lipoproteins were 13%, significantly ($p < 0.001$) below all others. This is approximately 42% of the value obtained in controls. On a percentage basis, $HDL_2 + HDL_3$ (α-lipoproteins) in other studies [11] on patients with ovarian carcinoma was also somewhat below control levels, 57% of the total HDL in controls (table I). Considering the difference in techniques, these are lower values than normal. Invasive carcinoma of the cervix and metastatic carcinoma had values for α-lipoproteins closest to normal.

These data corroborate the assumption that low levels of α-lipoproteins (or HDL) are related to cancerous growth rather than to the growth of pregnancy.

Results from an increasing number of laboratories throughout the world are lending credence to the earlier reports showing that α-lipoproteins are decreased or quite low in patients with different types of cancer. Studies in Japan have shown changes in correlation (ratio) between serum total cholesterol and α-lipoproteins in patients with different types of cancer [44, 58]. The authors observed low concentrations of α-lipoproteins, especially in patients with prostate gland and gastric cancer. During the course of the disease in 'control' patients in whom there were initially abnormally low values and who received no treatment (radical or simple surgery, irradiation, or chemotherapy) there was a gradual decrease in serum α-lipoproteins. In cases where radical procedures were employed to remove or inhibit the cancer there was an increase in α-lipoproteins which remained for 3–6 months, especially in patients with stomach, breast, or rectal cancer. However, in patients treated by less radical procedures, the increases in α-lipoproteins were not pronounced. It had been noticed previously [7, 8, 42] that the α-lipoproteins increased in patients following radical mastectomy and in whom there was no recurrence of disease.

Recently, the levels of β- and α_1-lipoproteins were investigated with a rather different technique in patients with a variety of types of cancer and after different kinds of treatment [53]. NYDEGGER and BUTLER [53] used the single radial immunodiffusion method [40] adapted for plasma lipoproteins [25]. The results corroborate previous observations by others on a wide spectrum of cancers using various techniques [11, 44, 58].

NYDEGGER and BUTLER [53] studied a wide range of age groups, from below 30 to 90 years. The most outstanding observation was the consistently lower value for α_1-lipoproteins in patients with cancer in all age groups. Age *per se* seemed not to be a causative factor. In fact, values for α_1-lipoproteins tended to increase with age in their healthy controls. Since the sexes are not delineated in the report by NYDEGGER and BUTLER, a preponderance of values from males would offset those from women and possibly result in an increase in values for α_1-lipoproteins [1, 3, 4].

The β-lipoproteins were not so dramatically affected, increasing in only two age groups and decreasing in four age groups; there seemed to be no correlation with age.

When type of cancer was considered, the greatest decreases in serum α_1-lipoproteins were in patients with tumors in the gastrointestinal tract and in patients with gynecological cancer. This had been observed previously [11, 44, 58] and some of the lowest values for HDL_2 were in subjects with intestinal and colon cancer and in ovarian carcinoma [11].

Interesting and provocative are the observations that values for a_1-lipoproteins were actually lower in 45 patients with primary tumors only, compared with values from 41 patients who had metastatic cancer. These data contrast somewhat with previous reports [8, 31] in which it was shown that patients with advanced metastatic carcinoma of the breast had significantly less ($p < 0.001$) a-lipoproteins (measured as total phospholipid) than controls.

The values presented by NYDEGGER and BUTLER [53] presumably are averages from a variety of types of cancer; this may explain why they differ from cases of breast carcinoma in which levels of a-lipoproteins were noted to decrease as the severity and metastatic nature of the disease increased [7]. It is possible that in certain kinds of cancer the invasive or noninvasive nature may be immaterial so far as levels of serum lipoproteins are concerned. In line with NYDEGGER and BUTLER's observations [53] is the following case from our experience. In a child with myelomonocytic leukemia whose serum was analyzed four times during her illness, the values for both HDL_2 and HDL_3 (both included in a-lipoproteins) rose and fell over a period of 1.5 years. In this progressing disease the values did not become normal [9]. Data from a case such as this and others like it would, however, greatly influence average values from numerous types; thus it seems advisable to study and evaluate cancer types individually rather than as cancer *per se*.

NYDEGGER and BUTLER [53] reported no increases in levels of a_1-lipoproteins in patients treated with drugs, X-ray and surgery. Several other reports [3, 7, 44, 58] have suggested that a-lipoproteins or HDL_2 are increased after treatment, especially in patients who respond favorably. More extensive subsequent examination may reveal that the HDL again decrease to pretreatment levels. Increases following estrogen therapy, are of course, canonical [3].

4. Albumin-Bound Free Fatty Acid Complexes
('Ultracentrifugal Residue'; Very High-Density Lipoproteins)

The free fatty acids are complexed with proteins of which albumins transport over 50% [60]. The albumin complexes have high densities (fig. 1) and thus remain in the final infranatant when serum lipoproteins are separated by density gradient procedures. Because of their comparatively smaller size and lower molecular weight they migrate more rapidly than other lipid-protein complexes in most electrophoretic systems using solid support media. When lipid stains are utilized, the lipid nature of the bands formed is obvious. However, in most work in which the effects of cancer upon levels of fatty acids were investigated, these lipids were measured as fatty acids rather than as complexes with albumins.

Data from a few typical experiments dealing with human patients will be presented here; considerably more work has been carried out with experimental animals and animal tumors and has been considered in a previous chapter in this volume. ALVAREZ and GOODELL [2] showed that in patients with gynecologic malignancies the average concentrations of free fatty acids in their serum were somewhat higher than in control serum [2]. The values given were 13.6 mg/100 ml (530 μEq/l) serum in controls compared with 20.4 mg/100 ml (796 μEq/l) serum in patients. The above observations confirm a previous report by MUELLER and WATKIN [43] in which they demonstrated that most cancer patients, those with leukemias, solid tumors, lymphomas and myelomas, had significantly elevated values for free fatty acids.

However, even more important may be the observation by ALVAREZ and GOODELL [2] that in serum from patients with malignancies, the different fatty acids were present in varying amounts depending upon the type of disease and the lipid category. In general, there was a notable decrease in linoleic acid (18:2), along with marked increases in stearic (18:0) and palmitic (16:0) acid levels. It is suggested that most gynecologic cancers utilize preferentially one of the essential fatty acids (linoleic).

5. Commentary

There is a definite relationship between the levels of the lipoproteins in serum and cancer in humans. There are certain salient points which should be emphasized. It should be realized that abnormal quantities of lipoproteins in serum from normal subjects and/or patients with cancer, who have been selected so as to eliminate other factors that influence these quantities, do not represent superficial observations. As our knowledge of these important macromolecules evolves, it becomes more obvious that increased or decreased levels of lipoproteins in serum merely represent the tip of the iceberg. This has certainly been demonstrated in the presently extensive developments on the hyperlipoproteinemias [23] and with the 'lipoidoses' [16] now acknowledged to be heritable enzyme deficiencies. Although the precise metabolic defect(s) resulting in the abnormal malignant cell is not presently elucidated, the increasing volume of evidence implicating lipid-containing complexes in relation to cancer should be considered important. In addition to lipids and proteins, these complexes transport a multitude of vitally important co-factors, factors, vitamins, hormones, and carbohydrates in the form of glycosphingolipids. Alterations in any one or several of these, in relation to the whole complex, could possibly result in cancer. The following chapter on the proteolipids (neoproteolipids in cancer) will be concerned with one aspect of this.

The most striking observation that will probably offer a clue to the cancer riddle is the abnormally low quantity of the HDL, especially HDL_2 in patients. (Apropos of previous remarks, it should be recalled (fig. 1) that most of the phospholipids are transported by these complexes.) It seems a firm observation because it occurs regardless of technique and type of disease (except, of course, in liver cancer).

It is also intriguing that many 'normal' subjects who are members of families (first degree relatives) in which there is or has been a high cancer incidence, have values for HDL (HDL_2) as low as those in patients with overt disease. In view of the obvious genetic associations in other diseases involving lipid-protein complexes [16, 23], one cannot ignore, or possibly even deny, these relationships in cancer.

Details of the metabolic interrelationships between the different lipoproteins are outside the scope of this review. However, certain consistent observations deserve some comment. In normal men but not in normal women, both fasting, with positive family histories of cancer, high levels of VLDL coexist with the low levels of HDL_2. In men these reciprocal relationships between all the classes in the VLDL and HDL_2 have a rather high coefficient of correlation ($r = -0.8$).[1] This reciprocal relationship is observed in both men and women with cancer.

Reciprocal relationships between pre-β-lipoproteins (VLDL) and α-lipoproteins (HDL) were also observed in patients with hyperlipemia (hyperlipoproteinemia) and severe α-lipoprotein deficiency (Tangier disease), both inherited disorders [23, 36].

Possible conversions between VLDL and HDL *in vivo* and *in vitro* have been investigated both in man and in experimental animals, but some doubts still remain about these conversions [21]. There are several parameters involved: delipidation of the lipoproteins may be required before

---

[1] The authors wish to thank Dr. ISABEL MORGAN MOUNTAIN for doing this statistical work.

the protein moieties (mainly apo LP-Glu and apo LP-Ala) are either transferred to or exchanged between VLDL, LDL, and principally, HDL. Since there are several apoproteins in the lipoproteins, these may exchange at different rates or not at all. The distribution of these apoproteins (from VLDL) between the other lipoproteins may be dependent, in part, upon the ratio of lipoproteins in serum initially [19].

It has been shown that certain serum factors which cause inhibition of growth of HeLa cells in tissue culture are bound to the a_1-lipoproteins [56, 57]. It can be suggested that a decrease in a_1-lipoprotein (as a carrier) would reduce the availability of these inhibitory factors if indeed they require the a_1-lipoproteins as carriers.

The tumor proteolipids, or neoproteolipids described in the following chapter, were first discovered associated with the HDL from serum of patients with carcinoma of the breast. They were not seen in the other serum lipoproteins with the analytical procedures used. Their presence has since been observed in serum of patients with various types of cancer. Why is serum neoproteolipid associated with or bound either only or mainly to the HDL? Does this complex, which can be isolated in significant amounts from tumor tissue, inhibit the synthesis of the HDL by the liver, or hasten the degradation of HDL so that serum contains significantly decreased amounts? Or does its binding change the physicochemical properties of the HDL so that some quantities elude detection based upon these properties? Unfortunately, the presence or absence of neoproteolipid in serum of normal subjects with positive family histories of cancer has not been systematically investigated.

There are at least two other intriguing implications to the relationship of decreased HDL_2 in patients with cancer and in men and women with positive family histories of cancer. The observations that levels of a-lipoproteins were increased by the administration of thymus extracts [55] and that 70% of the immunoglobulins of the lymphyocyte were reported to be lipoproteins with density < 1.125 g/ml [65], increase the possible connection between HDL and the immune competence of the individual.

It would seem that to have 'adequate' quantities of a-lipoproteins or HDL_2 is a desirable state for man and animals [3]. These have been increased therapeutically by the administration of estrogens but this indirect approach suffers from the side effects which result. If the HDL can be shown to have important roles either as carriers of growth inhibitors, or as potentiators of immunity, these avenues should be explored in attempts to understand the cancer process, prevent it, or cure it. Since it is now obvious that the HDL are deficient in the patient with cancer, it would seem logical to explore their use, e.g. Cohn's fraction IV [17], as replacement therapy.

V. Lipoproteins in Serum or Plasma from Animals with Experimental Tumors

Although hyperlipemia was noted early and quite generally in humans with cancer, the reasons for this phenomenon and its characteristics were sought in experimental animals in whom tumors had either been induced chemically or implanted. This problem was discussed quite extensively from the points of view that hyperlipemia can be 'attributed to mobilization of fat from the stores, both to meet the increased energy demands of the tumor-bearing host and to supply unsaturated fatty acids for maintenance of the high concentration of phospholipids within the tumor' [27]. On the other hand, BEGG and co-workers [14, 22, 63] suggested alternatives such as a 'combination of impaired entrance and unimpaired exit of lipid to and from the fat stores, together with continued absorption of dietary lipid' [14]. Additional experiments from this group confirmed that labeled fatty acids fed by gavage were indeed taken up by body lipids [22, 63].

Many factors may be operative: hormones, enzymes, protein deficiencies (or an abnormal apoprotein with unusual affinity for lipids). It may be significant that, regardless of experimental design the neutral lipid fraction seems to be elevated rather than the phospholipid in many but not all types of tumors. This, in terms of lipoproteins, suggests elevated VLDL and possibly LDL along with decreased HDL, as seen in experiments on man.

Of the neutral lipids, the fatty acids (as albumin-bound free fatty acid complexes) especially, and triglycerides (as chylomicra, particles or VLDL) are elevated in serum of animals bearing certain tumors [26], notably the transplanted Walker carcinosarcoma 256 in the rat.

A. Classes of Lipoproteins

1. Chylomicra or Particles, and the VLDL

In the nonfasting normal animal there will be a degree of lipemia associated with chylomicra or primary particles which are composed principally of exogenous triglyceride. The values for VLDL, which are composed mainly of endogeneous glycerides and some other lipids, are also elevated in nonfasting normal animals, just as they can be in man. Experiments on fasting female rats of the Sprague-Dawley strain, ranging in age from 110 to 210 days, indicate that the total VLDL tends to increase with age, but the values are still well below those of nonfasting 50-day-old female rats. Even under controlled dietary conditions there is considerable variation. This variation extends to the effects of the tumor induced in fasting female rats by 9,10-dimethyl-1,2-benz(α)-anthracene (DMBA) as may be seen in table VII, adapted from previously published work [12]. The values for VLDL obtained over a period of 160 days in two separate sequential experiments are shown. Some of the variation in both control and treated rats may be seasonal, since these experiments were carried out at the different dates listed. One can say only that there is a general tendency for total VLDL to be higher at 100–160 days after induction of the tumor, but there is no well-defined pronounced effect of this tumor upon VLDL levels.

The effects of a transplanted tumor, Walker carcinosarcoma 256 (W256), upon the levels of serum VLDL were also not significant in any one of the four different experiments carried out, and the variability was similar to that observed with the DMBA experiments.

Experiments performed by VON EULER et al. [20], in which serum from rats bearing unspecified tumors was studied by thin-layer electrophoresis, showed wide bands remaining at the origin. In a series of experiments NARAYAN and co-workers [46–50] used polyacrylamide gel (disc) electrophoresis to separate proteins and lipoproteins from serum of rats fed chemical carcinogens. The densitometric tracings of the gels showed quite dense areas of stained lipid either in the stacking gel or at the top of the separating gel. These 'peaks' (representing molecules which obviously did not enter, or barely enterred the gel) were higher in the rats which had received the larger quantity of the carcinogen used, N-2-fluorenylacetamide.

The same phenomenon was observed when the rats were fed this carcinogen while on a diet deficient in essential fatty acids [47]. All these 'origin bands' could result from a combination of chylomicra plus VLDL.

2. Low-Density Lipoproteins (LDL or β-Lipoproteins)

The authors have observed that the lowest density class of the total LDL, $S_f 10$–20 (fig. 1) is routinely absent in serum of female rats and is only occasionally present in trace amounts in serum of male rats. This is in some contrast to the observations in normal men and women who may have small but significant quantities.

As in humans, the $S_f 0$–10 class is the most abundant in the LDL of serum from rats, but the quantities are relatively much lower than in man. Usually, this class appears to be less heterogeneous than the other classes or lipoproteins in serum. However, in serum from rats who have developed tumors from DMBA feeding, the total quantity of $S_f 0$–10 is not elevated, but the usual 'single' component is now definitely two. In table VII it may be seen that at day 60 and later, in animals in which palpable tumors occurred, the total $S_f 0$–10 was divided into two components. These appeared in the ultracentrifuge as two separated boundaries or 'peaks' [12]. These components, especially that with $S_f 6$–15 (1.0635), obviously were composed of lipoproteins with lower densities, and thus faster flotation rates. Their total quantities, however, were almost precisely the same as the amounts in control rat serum. The lower densities of these two components suggest differences in lipid content. Rats that had received DMBA but failed to develop palpable tumors did not have these two components in this lipoprotein fraction. These two lipoproteins may be comparable to the 'abnormal' β-lipoprotein(s) seen in patients with a combination of myeloma and hyperlipoproteinemia [35, 51, 62].

Alterations in LDL may occur in other types of animal tumors. On the disc electrophoretic patterns from serum of rats with the Morris hepatoma 7777, NARAYAN and MORRIS [50] observed an additional band(s) in the areas where the LDL or β-lipoproteins should be.

Table VIII, which gives results for measurements of LDL, HDL_2 and HDL_3 only, shows increased levels in serum LDL ($S_f 0$–10) in three of the four experiments on serum from rats bearing the W256 tumor. Although the control rats in these experiments with W256 were males of a different strain, their serum contained quantities of LDL similar to those in female controls used for the DMBA-induced tumor experiments. However, unlike the tumors induced by DMBA, the W256 tumors did not 'produce a split $S_f 0$–10'. These quite different effects upon LDL in serum of rats with these two tumors may be explained by a combined effect of DMBA plus tumor, since LDL did not appear divided in rats which had received DMBA, but in rats which tumors were not detectable.

3. High-Density Lipoproteins (HDL_2 and HDL_3) or α-Lipoproteins

The fraction with density between 1.0635 and 1.125 g/ml (HDL_2) usually contains only one component in rats in contrast to humans in

Table VII. Lipoprotein values (mg/100 ml serum) in serum from control rats and those fed DMBA[1] in two separate sequential experiments

Lipoproteins[2] and treatment	Day 0		Day 60		Day 100		Day 160	
	I, May 16	II, Oct. 24	I, July 18	II, Dec. 26	I, Aug. 29	II, Feb. 6	I, Oct. 24	II, Apr. 3
VLDL								
Control I[3] and II	55	2	20	49	50	143	94	28
DMBA I and II	52	42	19	55	24	109	105	15
LDL								
Control I and II	29	36	65	38	38	51	39	64
DMBA I and II	20	34	32=(S$_f$6-15)=23		14=(S$_f$6-15)=16		16=(S$_f$6-15)=31	
			34=(S$_f$0- 6)=15		25=(S$_f$0- 6)=30		17=(S$_f$0- 6)=30	
HDL$_2$(S$_f$0-4)								
Control I and II	85	129	168	152	148	138	191	163
DMBA I and II	101	127	157	119	110	80	166	119
HDL$_3$								
Control I and II	90	115	124	157	104	122	89	121
DMBA I and II	100	100	121	140	80	131	85	103

1 DMBA: 9,10-dimethyl-1,2-benz(a)anthracene. 25 female Sprague-Dawley rats, 50 days old, were given a single dose of 15 mg DMBA in 1.0 ml sesame oil. At days stated, blood was drawn from the retroorbital venous plexus after a 12-hour fast.

2 Lipoproteins: VLDL = total fraction with D <1.006 g/ml; LDL (S$_f$0-10) = major component in fraction with D <1.0635 g/ml; HDL$_2$ (S$_f$0-4) = major component in fraction with D <1.125 g/ml; HDL$_3$ = major component in fraction with D <1.210 g/ml.

3 25 female Sprague-Dawley rats, 50 days old, were given a single oral dose of 1.0 ml sesame oil. Blood was drawn at times stated and in manner as experimental animals, see footnote 1 above.

Table VIII. Lipoprotein values in serum from control rats and those in whom the Walker carcinosarcoma 256 had been implanted

Lipo-proteins[1]	Experiments and dates							
	I, March 29		II, April 21		III, June 22		IV, Jan. 4	
	control[2]	W256[2]	control	W256	control	W256	control	W256
LDL (S_f0–10)	58	100	61	76	52	35	40	65
HDL$_2$ (S_f0–4)	72	52	83	63	56	50	73	53
HDL$_3$	61	42	61	83	45	54	40	40

1 LDL (S_f0–10) = Major component in fraction with D <1.0635 g/ml; HDL$_2$ (S_f0–4) = major component in fraction with D <1.125 g/ml; HDL$_3$ = major component in fraction with D <1.210 g/ml. All values are in mg/100 ml serum.
2. In each experiment at the date listed, 50 CFN Wistar male rats, 50 days old, were used; 25 rats were implanted by trocar with 0.25 cm^3 of standardized fragments of the W256 tumor; 25 rats were reserved as controls. 14 days after implantation the rats were deprived of food for 12 h, anesthetized with Nembutal, and exsanguinated by cardiac puncture. Serum was obtained and the lipoprotein analyses were started immediately.

whom two additional components are often present [6, 13]. Table VII shows that values for serum HDL$_2$ increased in control female rats as they aged. This is not surprising since these rats were passing into their period of sexual maturity (8–10 weeks of age). The values for HDL$_2$ in both experiments were lower in rats with tumors than in control rats at every time period after receiving the DMBA. In the rats who received DMBA but had no overt evidence of tumors, the levels of HDL$_2$ were identical to those in the animals with palpable tumors.

In rats with implanted W256 tumor, serum HDL$_2$ values were also lower than in controls (table VIII). In another experiment involving rats implanted with the W256 tumor, blood was drawn from three groups of rats: controls (two pools, six analytical determinations); rats which rejected the tumor implants (two pools, four analytical determinations); and rats in which the tumor grew (two pools, four determinations). Each pool involved eight or ten animals.

The magnitudes of the values for the different lipoproteins in serum of these control rats resembled those from rats of the same age and strain studied previously (table VIII), although the levels of $S_f 0-10$ (LDL) were slightly lower. The rats in which the W256 tumor did not grow had higher values for both HDL_2 and HDL_3 than either the controls or the rats in which the tumor grew. These lipoproteins were the lowest in serum of rats in which the tumor grew; the comparative values for HDL_2 resembled those seen in table VIII also dealing with animals implanted with the W256 tumor. It is possible that the higher levels for HDL in serum had a protective effect in rats which rejected the implants.

NARAYAN [46] and NARAYAN and MORRIS [50] described experiments in which polyacrylamide gel (disc) electrophoresis and usually preparative ultracentrifugal separations as well, were utilized to measure the levels of serum HDL in rats fed N-2-fluorenylacetamide. This carcinogen was fed at different levels to young male Holtzman rats. Tracings of the discs showed increases in the intensity of the stain in the area(s) to which the HDL would migrate in the percentage of polyacrylamide gel employed. This was especially pronounced when the greater amount (0.03 % of purified diet) of the carcinogen was fed.

In a continuation of these experiments, using the same carcinogen and an essential fatty acid-deficient diet over a 52-week period, NARAYAN [47] showed that increases occurred in total HDL in the preneoplastic stage. There were also increases in the LDL as the effects of the carcinogen became more manifest. The great increase in LDL was attributed to a lipoprotein component with electrophoretic and ultracentrifugal characteristics similar to HDL_1 ($S_f 0-3$, 1.0635, in the LDL, fig. 1).

During the course of his experiments, NARAYAN observed that when tumors were present in organs other than the liver, the serum HDL levels decreased markedly. In order to determine whether the effects upon serum lipoproteins previously observed resulted from direct toxicity (primarily in the liver) of the carcinogen or from the tumors in the livers of the rats, NARAYAN and MORRIS [50] investigated other types of tumors. Serum from rats with the Morris hepatoma 7777 appeared to have wider discs in the HDL area, but serum from rats with Ehrlich ascites [18] and from those with regenerating liver [49] showed decreased HDL and increased LDL and VLDL.

The authors concluded [48, 50] that when the liver was involved directly (as in certain carcinogen-induced tumors) the levels of lipoproteins in serum increased. They suggest that there may be enhanced synthesis of all lipoproteins by the liver tumor; that normal regulation of lipoprotein synthesis and/or feed-back mechanisms are absent. In experiments with the Morris hepatoma 7777, NARAYAN [47, 48] and NARAYAN and MORRIS [50] observed a greatly increased incorporation of leucine-U-C^{14} by this tumor, suggesting uncontrolled synthesis of proteins. Since these tumors are, for the most part 'minimally deviated' from normal liver, and even secrete bile, it is also feasible that they may synthesize lipoproteins. The authors do not report metastatic lesions in the liver proper.

NISHIOKA et al. [52] reported that when serum from patients with primary liver carcinoma was analyzed by means of polyacrylamide gel, they saw what they described as a 'unique additional band situated in the albumin and the α_1-lipoprotein region'. This unique band proved to be α-feto-protein when immunochemical procedures were applied to the polyacrylamide gel segments containing the band. This band was not present in serum from normal subjects. The more pronounced band attributed to HDL by NARAYAN [46, 48] and NARAYAN and MORRIS [50] could be associated with α-feto-protein.

In most diseases of the liver, other than cancer, there is impaired synthesis of most proteins and lipoproteins [3]. Most of the published data concerned with the relations of serum lipoproteins to cancer in man have not dealt with liver cancer directly, perhaps because the cellular changes in a liver under stress would mask the effects of the tumor. As reported by NARAYAN et al. [49], the stress of partial hepatectomy resulted in substantially reduced concentrations of HDL$_2$ in serum of the rats. Thus, most investigators who wished to study the effects of cancer upon lipoproteins avoided subjects (or animals) with liver disease. Nevertheless, observations which discriminate between high levels of HDL in serum of rats with cancer plus liver involvement, and low levels when the liver is not involved should serve to add emphasis to a more meticulous selection of patients or types of tumors in animals for studies of these kinds. Even of greater importance are *any* clues as to why low levels of HDL$_2$ are related to cancer directly or indirectly.

VI. Summary

The presence of cancer in man and experimental animals influences the metabolism of lipoproteins. This has been convincingly demonstrated quantitatively by measurements of lipoproteins in serum or plasma by a variety of procedures. Some distinct patterns or relationships emerge. The most regular are: (1) elevation of VLDL (possibly contributing to the classical lipemia) concurrently with significantly low levels of α-lipoproteins and specifically HDL_2, and (2) unusually high values for HDL_1 (1.0635) together with significantly low values for other high-density lipoproteins. These patterns occur especially and strikingly in men with cancer, but also in some women who have cancer. There is a strong tendency for the first pattern to occur in men, clinically normal, but who have a high incidence of cancer in first degree relatives. Both men and women with this positive family history of cancer have significantly lower HDL_2 (density < 1.125 g/ml) than observed in their normal counterparts who have a minimum of cancer in their first degree relatives.

VII. Acknowledgments

It is a pleasure to acknowledge the cooperation of the Sisters of Mercy, Philadelphia, Pa. who provided some of the serial samples and filled in questionnaires originally provided by Dr. W. L. ELKINS whom we wish to thank for providing their serum samples. We wish also to thank Dr. J. B. ATKINSON for the subsequent physical examinations of these subjects, and especially Mr. EDWARD M. GREENE for negotiating and carrying out the 5- and 10-year surveys on these subjects. We are indebted to the late Dr. HAROLD W. DARGEON for selecting the children and providing their samples and to Dr. RICHARD J. KAUFMAN for selecting and providing the samples from the adult patients with cancer and many of the controls.

It is a pleasure to acknowledge the valuable advice and assistance of Dr. RALPH K. BARCLAY in the preparation of this chapter and especially to thank Mrs. DORMAN D. ISRAEL for her devoted and valuable literature search. We wish to thank Dr. C. CHESTER STOCK for constant encouragement and ELSIE BUCK MARTIN for inestimable inspiration.

This work was supported in part by research grant CA-08748 from the National Cancer Institute, US Public Health Service, and the Elsa U. Pardee Foundation.

VIII. References

1 ALTMAN, P. L.: Ultracentrifugal analysis of serum lipoproteins and proteins: man; in DITTMER Blood and other body fluids, pp. 64–72 (Federation of Amer. Soc. for Experimental Biology, Washington 1961).
2 ALVAREZ, R. R. DE and GOODELL, B. W.: Serum lipid pirtitions and fatty acid composition (using gas chromatography) in gynecological cancer. Amer. J. Obstet. Gynec. *88:* 1039–1060 (1964).
3 BARCLAY, M.: Lipoprotein class distribution in normal and diseased states; in

NELSON Blood lipids and lipoproteins, quantitation, composition and metabolism, pp. 585–704 (Wiley & Sons, Chichester 1972).
4   BARCLAY, M.; BARCLAY, R. K., and SKIPSKI, V. P.: High-density lipoprotein concentrations in men and women. Nature, Lond. *200:* 362–363 (1963).
5   BARCLAY, M.; BARCLAY, R. K.; SKIPSKI, V. P.; TEREBUS-KEKISH, O.; MUELLER, C. H.; SHAH, E., and ELKINS, W. L.: Fluctuations in human serum lipoproteins during the normal menstrual cycle. Biochem. J. *96:* 205–209 (1965).
6   BARCLAY, M.; BARCLAY, R. K.; TEREBUS-KEKISH, O.; SHAH, E. B., and SKIPSKI, V. P.: Disclosure and characterization of new high-density lipoproteins in human serum. Clin. chim. Acta *8:* 721–726 (1963).
7   BARCLAY, M.; CALATHES, D. N.; DiLORENZO, J. C.; HELPER, A., and KAUFMAN, R. J.: The relation between plasma lipoproteins and breast carcinoma: effect of degrees of breast disease on plasma lipoproteins and the possible role of lipid metabolic aberration. Cancer, Philad. *12:* 1163–1170 (1959).
8   BARCLAY, M.; COGIN, G. E.; ESCHER, G. C.; KAUFMAN, R. J.; KIDDER, E. D., and PETERMANN, M. L.: Human plasma lipoproteins. I. In normal women and in women with advanced carcinoma of the breast. Cancer, Philad. *8:* 253–260 (1955).
9   BARCLAY, M.; DARGEON, H. W.; GREENE, E. M.; TEREBUS-KEKISH, O., and SKIPSKI, V. P.: Serum lipoproteins in children with cancer. Clin. chim. Acta *24:* 225–231 (1969).
10 BARCLAY, M.; ESCHER, G. C.; KAUFMAN, R. J.; TEREBUS-KEKISH, O.; GREENE, E. M., and SKIPSKI, V. P.: Serum lipoproteins and human neoplastic disease. Clin. chim. Acta *10:* 39–47 (1964).
11 BARCLAY, M.; SKIPSKI, V. P.; TEREBUS-KEKISH, O.; GREENE, E. M.; KAUFMAN, R. J., and STOCK, C. C.: Effects of cancer upon high-density and other lipoproteins. Cancer Res. *30:* 2420–2430 (1970).
12 BARCLAY, M.; SKIPSKI, V. P.; TEREBUS-KEKISH, O.; MERKER, P. L., and CAPPUCCINO, J. G.: Serum lipoproteins in rats with tumors induced by 9,10-dimethyl-1,2-benzanthracene and with transplanted Walker carcinosarcoma 256. Cancer Res. *27:* 1158–1167 (1967).
13 BARCLAY, M.; TEREBUS-KEKISH, O.; SKIPSKI, V. P., and BARCLAY, R. K.: Additional evidence for the existence of 'new' high density lipoproteins in human serum. Clin. chim. Acta *11:* 389–394 (1965).
14 BEGG, R. W. and DICKINSON, T. E.: Systemic effects of tumors in force-fed rats. Cancer Res. *11:* 409–412 (1951).
15 BERG, G.; SCHEIFFARTH, F., and MARWAN, G.: Clinical experiences with lipide (lipoprotein) electrophoresis. Klin. Wschr. *35:* 215–220 (1957).
16 BRADY, R. O.: The abnormal biochemistry of inherited disorders of lipid metabolism. Fed. Proc. *32:* 1660–1667 (1973).
17 COHN, E. J.; GURD, F. R. N.; SURGENOR, D. M.; BARNES, B. A.; BROWN, R. K.; DEROUAUX, G.; GILLESPIE, J. M.; KAHNT, F. W.; LEVER, W. F.; LIU, C. H.; MITTELMAN, D.; MOUTON, R. F.; SCHMID, K., and UROMA, E.: A system for the separation of the components of human blood: quantitative procedures for

the separation of the protein components of human plasma. J. amer. chem. Soc. 72: 465–474 (1950).
18 CREININ, H. L. and NARAYAN, K. A.: Effect of Ehrlich ascites tumor cells on mouse plasma lipoproteins. Z. Krebsforsch. 75: 93–98 (1971).
19 EISENBERG, S.; BILHEIMER, D. W., and LEVY, R. I.: The metabolism of very low density lipoprotein proteins. II. Studies on the transfer of apoproteins between plasma lipoproteins. Biochim. biophys. Acta 280: 94–104 (1972).
20 EULER, H. VON; HASSELQUIST, H. und LIMNELL, I.: Vergleichende Versuche an normalen und cancerösen Blutseren durch Dünnschicht-Elektrophorese. Z. Krebsforsch. 65: 404–408 (1963).
21 FIDGE, N. H. and FOXMAN, C. J.: *In vivo* transformation of rat plasma very low density lipoprotein into higher density lipoproteins. Austr. J. exp. Biol. med. Sci. 49: 581–593 (1971).
22 FREDERICK, G. L. and BEGG, R. W.: A study of hyperlipemia in the tumor-bearing rat. Cancer Res. 16: 548–552 (1956).
23 FREDRICKSON, D. S.; LEVY, R. I., and LEES, R. S.: Fat transport in lipoproteins: an integrated approach to mechanisms and disorders. New Engl. J. Med. 276: 34–43, 94–103, 148–155, 215–224, 273–281 (1967).
24 GUPTA, R. M.: Serum immunoelectrophoresis in patients with Ewing's sarcoma. Lancet ii: 1136–1139 (1969).
25 HATCH, F. T. and LEES, R. S.: Practical methods for plasma lipoprotein analysis. Adv. Lipid Res. 6: 1–68 (1968).
26 HAVEN, F. L.: The nature of the fatty acids of rats growing Walker carcinoma 256. Cancer Res. 9: 619–623 (1951).
27 HAVEN, F. L. and BLOOR, W. R.: Lipids in cancer. Adv. Cancer Res. 4: 237 (1956).
28 HIGAZI, A. M.; ATA, A. A.; ABDEL-RAHMAN, Y. M.; MALEK, A., and MANSOUR, K.: Electrophoretic pattern of serum proteins and lipids in leukaemias. J. egypt. med. Ass. 49: 679–689 (1966).
29 KALININA, E. V.: The fractional protein, lipo- and glucoprotein compositions of blood serum in cancer patients. Tr. Leningr. nauchn-issled. Inst. Ekspertizy trudosposobnosti i oraniz. Truda Invalidov 2: 316–327 (1960).
30 KAMENEVA, T. I.: Lipoproteins in serum in leukemia in children. Vopr. Okhrany Materinstva i Detstva 8: 16–20 (1963).
31 KAUFMAN, R. J.; BARCLAY, M.; KIDDER, E. D.; ESCHER, G. C., and PETERMANN, M. L.: Human plasma lipoproteins. II. The effect of osseous metastases in patients with advanced carcinoma of the breast. Cancer, Philad. 8: 888–889 (1955).
32 KELLEN, J.: The serum beta-lipoproteins in different human malignant diseases. Neoplasma 15: 139–143 (1968).
33 KUNKEL, H. G. and TRAUTMAN, R.: The α_2 lipoproteins of human serum. Correlation of ultracentrifugal and electrophoretic properties. J. clin. Invest. 35: 641–648 (1956).
34 LALLA, O. DE and GOFMAN, J. W.: Ultracentrifugal analysis of serum lipoproteins; in GLICK Methods of biochemical analysis, pp. 459–478 (Interscience, New York 1954).

35 LENNARD-JONES, J. E.: Myelomatosis with lipaemia and Xanthomata. Brit. med. J. i: 781–783 (1960).
36 LEVY, R. I.; LEES, R. S., and FREDRICKSON, D. S.: The nature of pre-beta (very low density) lipoproteins. J. clin. Invest. 45: 63–77 (1966).
37 LEWIS, L. A. and PAGE, I. H.: Serum proteins and lipoproteins in multiple myelomatosis. Amer. J. Med. 17: 670–673 (1954).
38 LINDGREN, F. T.; JENSEN, L. C., and HATCH, F. T.: The isolation and quantitative analysis of serum lipoproteins; in NELSON Blood lipids and lipoproteins, quantitation, composition and metabolism, pp. 182–274 (Wiley & Sons, Chichester 1972).
39 LOSTICKY, C.; REJNEK, J., and BEDNARIK, T.: Incidence of an abnormal immunoelectrophoretic pattern of α_1-lipoprotein. Casopis Lekaru Coskych 101: 1291–1294 (1962).
40 MANCINI, G.; CARBONARA, A. O., and HEREMANS, J. F.: Immunochemical quantitation of antigens by single radial immunodiffusion. Immunochemistry 2: 235–254 (1965).
41 MIETTINEN, M.: Serum lipides and lipoproteins in cancer, leukemia, and malignant lymphogranulomatosis. Ann. Med. intern. Fenn. 46: 103–108 (1957).
42 MILLER, B. J. and ERF, L.: The serum proteins and lipoproteins in patients with carcinoma and in subjects free of recurrence. Surg. Gynec. Obstet. 102: 487–491 (1956).
43 MUELLER, P. S. and WATKIN, D. M.: Plasma unesterified fatty acid concentrations in neoplastic disease. J. Lab. clin. Med. 57: 95–108 (1961).
44 NAKAGAWA, H.: Studies on α-lipoprotein in serum of patients with cancer. J. jap. Soc. intern. Med. V 55: 1–11 (1966).
45 NANAVA, I. G. and TSINTSADZE, T. M.: Changes in serum lipoproteins in patients with breast cancer. Tr. nauchn-issled. Inst. Onkol. Gruz. SSR 1: 139–145 (1961).
46 NARAYAN, K. A.: Serum lipoproteins during chemical carcinogenesis. Biochem. J. 103: 672–676 (1967).
47 NARAYAN, K. A.: Serum lipoproteins of rats fed an essential fatty acid-deficient diet and N-2-fluorenylacetamide. Cancer Res. 30: 1185–1191 (1970).
48 NARAYAN, K. A.: Rat serum lipoproteins during carcinogenesis of the liver in the preneoplastic and the neoplastic state. Int. J. Cancer 8: 61–70 (1971).
49 NARAYAN, K. A.; MARY, G. E. S., and KUMMEROW, F. A.: Rat serum lipoproteins after partial hepatectomy. Proc. Soc. exp. Biol. Med. 129: 6–12 (1968).
50 NARAYAN, K. A. and MORRIS, H. P.: Serum lipoproteins of rats bearing transplanted Morris hepatoma 7777. Int. J. Cancer 5: 410–414 (1970).
51 NEUFELD, A. H.; HALPENNY, G. W., and MORTON, H. S.: Beta-2-lipoprotein myelomatosis. Canad. J. Biochem. 42: 1499–1508 (1964).
52 NISHIOKA, M.; HIRONAGA, K., and FUJITA, T.: Polyacrylamide gel electrophoresis of sera from patients with primary liver carcinoma. Clin. chim. Acta 31: 439–446 (1971).
53 NYDEGGER, U. E. and BUTLER, R. E.: Serum lipoprotein levels in patients with cancer. Cancer Res. 32: 1756–1760 (1972).

54 OHYA, A.: Immunochemical analysis of human serum lipoprotein. Studies on the immunochemical behavior of human serums from patients with hypertension or cancer. Juzen Igakukai Zasshi *64:* 387–408 (1960).

55 POTOP, I.; NEACSU, C.; BINER, J., and MREANA, G.: Electrophoresis of proteins and lipoproteins in tumor-bearing rats under the influence of thymus fractions. Rev. sci. med., Akad. Rep. Populaire Romaine *5:* 83–87 (1960).

56 REJNEK, J.; BEDNAŘÍK, T.; ŘEŘÁBKOVÁ, E., and DOLEŽAL, A.: Investigations of the influence of serum from pregnant women on the growth of cell cultures in conjunction with the occurrence of abnormal α_1-lipoprotein. Clin. chim. Acta *8:* 108–115 (1963).

57 REJNEK, J.; BEDNAŘÍK, T.; ŘEŘÁBKOVÁ, E., and PEŠKOVÁ, D.: The effect of human serum fractions on cell growth and its relation to the state of serum alpha$_1$-lipoprotein. Exp. Cell Res. *37:* 65–78 (1965).

58 SAKAMOTO, Y.: Changes of the correlation between serum total cholesterol and α-lipoprotein caused by treatments and their clinical evaluation in malignant disease. Gan no Rinsho (Clinical Cancer) *15:* 49–55 (1969).

59 SCANU, A.: Serum high density lipoproteins; in TRIA and SCANU Structural and functional aspects of lipoproteins in living systems, pp. 425–445 (Academic Press, New York 1969).

60 SKIPSKI, V. P.: Lipid composition of lipoproteins in normal and diseased states; in NELSON Blood lipids and lipoproteins, quantitation, composition and metabolism, pp. 471–583 (Wiley & Sons, Chichester 1972).

61 SKIPSKI, V. P.; BARCLAY, M.; BARCLAY, R. K.; FETZER, V. A.; GOOD, J. J., and ARCHIBALD, F. M.: Lipid composition of human serum lipoproteins. Biochem. J. *104:* 340–352 (1967).

62 SPIKES, J. L., JR.; COHEN, L., and DJORDJEVICH, J.: The identification of a myeloma serum factor which alters serum beta lipoproteins. Clin. chim. Acta *20:* 413–421 (1968).

63 STEWART, A. G. and BEGG, R. W.: Systemic effects of tumors in force-fed rats. III. Effect on the composition of the carcass and liver and on the plasma lipids. Cancer Res. *13:* 560–565 (1953).

64 TOYODA, N.: Serum lipoprotein in cancer patients. I. The influence of the grade of cancer and its surgical treatment. Hirosaki med. J. *14/2:* 275–282 (1962).

65 UHR, J. W. and VITETTA, E. S.: Synthesis, biochemistry and dynamics of cell surface immunoglobulin on lymphocytes. Fed. Proc. *32:* 35–40 (1973).

66 WOLFFRAM, E. VON: Die serum-lipoproteine beim weiblichen genitalkarzinom. Zentr. Gynakol. *81:* 805–818 (1959).

Authors' address: MARION BARCLAY and VLADIMIR P. SKIPSKI, Memorial Sloan-Kettering Cancer Center, *New York, NY 10021* (USA)

Tumor Proteolipids

VLADIMIR P. SKIPSKI, MARION BARCLAY, FRANCIS M. ARCHIBALD and C. CHESTER STOCK

Memorial Sloan-Kettering Cancer Center, New York, N.Y.

Contents

I. Introduction 112
II. Presence of Neoproteolipids in Tumors 113
III. Neoproteolipid-W 115
   A. Chemical Composition 115
   B. Occurrence of Neoproteolipid-W in Tissues of Normal and Cancer-Bearing Rats 118
IV. Neoproteolipid-S 123
V. Neoproteolipids in Serum of Cancer Patients 124
VI. Discussion 126
VII. Summary 130
VIII. Acknowledgments 131
IX. References 131

I. Introduction

Thin-layer chromatography of lipids extracted from serum high-density lipoproteins of cancer patients and lipids extracted from rat Walker carcinosarcoma 256 revealed the presence of a lipid which we could not match with any known available standard lipid. Subsequently, this unidentified lipid was isolated from lipid extracts of Walker carcinosarcoma 256 and subjected to chemical and physicochemical studies. It contained proteins or polypeptides and its acid hydrolyzate contained fatty acids and sphingosines. Thus, the new lipid appeared to be a protein-

Abbreviations used: NPL-W = neoproteolipid-W; NPL-S = neoproteolipid-S.

lipid complex, soluble in chloroform-methanol mixtures and some other lipid solvents, and insoluble in aqueous media.

These properties of the isolated complex, in general, met the definition given by FOLCH and LEES [6] for proteolipids discovered by them: 'Proteolipids are lipoproteins having as constituents a lipid moiety and protein moiety but, while other known lipoproteins are soluble in water or salt solution, proteolipids are insoluble in water and soluble in chloroform-methanol mixtures; i.e. their solutibilities are akin to those of lipids.' Therefore, the protein-lipid complex isolated from Walker carcinosarcoma 256 in our laboratory was tentatively classified as a proteolipid. However, this complex differed in many respects in its chemical and physicochemical properties from the classical Folch-Lees proteolipid isolated from nervous tissues [7, 8, 25, 51] and hence the term *neoproteolipid* was introduced. Neoproteolipid isolated from Walker carcinosarcoma 256 is called *neoproteolipid-W* or *NPL-W*.

More recent observations reveal that lipids extracted from other animal tumors, e.g. sarcoma 180 (Crocker) and Morris hepatoma 5123, contain a protein-lipid complex which has substantially different Rfs on certain thin-layer chromatographic systems from NPL-W. This proteolipid is termed *neoproteolipid-S,* or *NPL-S*. The differences in chromatographic behavior of neoproteolipids on thin-layer silica gel chromatoplates and on silicic acid columns presently provided the basis for a classification of neoproteolipids. Subsequent isolation and chemical analysis of these two neoproteolipids substantiated the existence of these two types of neoproteolipids [41].

II. Presence of Neoproteolipids in Tumors

Different transplanted animal tumors, chemically induced tumors and spontaneously originated animal and human tumors, were studied for the presence of neoproteolipids and their types (table I). In all tumors presently tested we observed neoproteolipids in measurable quantities [41]. Table I shows the predominant type (90% or more) of neoproteolipid present in each tumor. In some tumors small quantities of the other type of neoproteolipid are also present. Thus, Walker carcinosarcoma 256 contains 90–95% of NPL-W and 5–10% of NPL-S. Similar results were obtained with Morris hepatoma 5123: about 95% is NPL-S and approximately 5% of total neoproteolipid is NPL-W.

Table I. Tumors investigated for the presence and types of neoproteolipid

Tumors	Species	Type of neoproteolipid present (exclusively or predominantly)
Transplanted tumors		
Walker carcinosarcoma 256	rat	W
Novikoff hepatoma	rat	W
Sarcoma 180 (Crocker)	mouse	S
Ehrlich solid carcinoma	mouse	S
T241 Lewis sarcoma	mouse	S
E0771 adenocarcinoma	mouse	S
B16 melanoma	mouse	S
Morris hepatoma 5123tc	rat	S
Taper liver tumor	mouse	S
Ridgway osteogenic sarcoma	mouse	S
Jensen sarcoma	rat	S+W
Chemically induced tumors		
By 3-hydroxyxanthine (fibroma or fibrosarcoma)	rat	W
Spontaneous tumors		
Mammary adenocarcinoma	mouse	S
Breast carcinoma, primary	human	S
Lung carcinoma, primary	human	S
Lung carcinoma, metastasis from testicular carcinoma	human	S
Lung carcinoma, metastasis from osteogenic sarcoma	human	S

However, this problem has not been investigated in detail. Rat Jensen sarcoma contains 80 % NPL-S and 20 % NPL-W. Apparently, the type of neoproteolipid present predominantly in any particular tumor is not species-specific. In different rat tumors either NPL-W or NPL-S or both may be present. The type of neoproteolipid is not determined by the origin of the tumor since NPL-S is present in both carcinomas and sarcomas. It is obvious that NPL-S is the more common type in the malignant tumors tested to date.

Table II. Chemical composition of neoproteolipid-W. Data adapted from SKIPSKI [40]

	Weight %
Polypeptides	8–11
Fatty acids[1]	25–31
Sphingosine bases	10–14
Carbohydrates (as total monosaccharides)[2]	40–45
Galactose	20.5
Glucose	7.7
Fucose	5.5
N-Acetylhexosamine	8.8
Cholesterol, total	1.0 – 1.5
Phosphorus (after digestion, total)	0.4 – 0.5

1 Average accepted molecular weight 327 daltons.
2 Values for individual monosaccharides are given for median value of carbohydrates (42.5%).

III. Neoproteolipid-W

A. Chemical Composition.

Neoproteolipid-W was isolated from lipid extracts of Walker carcinosarcoma 256 tumor tissue, utilizing silicic acid column chromatography, with step-wise gradient elution by chloroform-methanol mixtures [40, 44]. With these selected conditions NPL-W was eluted with chloroform: methanol, 1:3 (v/v) as a pure complex. Purities of the products obtained were checked by several thin-layer chromatographic systems [40].

The chemical composition of NPL-W is shown in table II. NPL-W contains 8–11% polypeptides as determined by the HESS and LEWIN [17] adaptation of the LOWRY et al. [23] procedure for the determination of total protein in proteolipids. Somewhat smaller values (8–9%) were obtained for amounts of polypeptides by calculation of the recovery of individual amino acid derivatives on gas-liquid chromatograms [40]. The discrepancy in the values for protein content in NPL-W probably resulted from some losses of amino acids during long hydrolysis of this neoproteolipid as well as during the preparation of N-trifluoroacetyl n-butyl esters of amino acids for gas-liquid chromatography [34].

NPL-W contains 25–31 % fatty acids. The quantities of fatty acids were determined by gas-liquid chromatography [50] and by semimicro-titration of fatty acids recovered from acid hydrolysis of NPL-W [43, 45]. Both methods gave similar values.

From 10 to 14 % of NPL-W are sphingosine bases. The content of sphingosine bases in NPL-W was determined spectrophotometrically according to the procedure of YAMAMOTO and ROUSER [59] with some modifications [40], as trinitrobenzenesulfonic acid derivatives, and by gas-liquid chromatography as trimethylsilyl ethers of N-acetyl sphingosine bases, according to VANCE and SWEELEY [54] and WINDELER and FELDMAN [57]. Both spectrophotometric and gas-liquid chromatographic methods gave essentially similar results.

Carbohydrates present in hydrolyzates of NPL-W were analyzed as monosaccharides by gas-liquid chromatography along with sphingosine bases as trimethylsilyl ethers of methyl glycosides [54, 57]. The presence or absence of neuraminic acid or its derivatives in NPL-W was investigated with the gas-liquid chromatographic procedure of YU and LEDEEN [62].

The acid hydrolyzate of NPL-W contains 40–45 % monosaccharides. The following monosaccharides are present in the carbohydrate fraction: galactose (in the highest quantity), glucose, fucose and hexosamine, which appears to be N-acetylgalactosamine. No neuraminic acid or its derivatives is detected in NPL-W. Results obtained by the gas-liquid chromatographic procedure are confirmed qualitatively (and semi-quantitatively) in several thin-layer chromatographic systems [40].

NPL-W also contains small quantities of cholesterol (both in free and esterified forms). The sulfuric acid digest of NPL-W contains 0.4–0.5 % inorganic phosphorus. At least part of this phosphorus originates from phospholipids present in the neoproteolipid, since partial delipidation of NPL-W and subsequent thin-layer chromatography of lipid fractions reveals the presence of significant amounts of phospholipids.

In table III is shown the fatty acid composition of NPL-W determined by gas-liquid chromatography before and after hydrogenation [40]. The pattern of fatty acids in this complex is characterized by a high content of long-chain fatty acids. The total content of fatty acids with chain lengths of 24 or more carbon atoms is about 58 weight % of the total fatty acids present in NPL-W. Over 60 weight % of total fatty acids present in NPL-W are saturated fatty acids. The number of fatty acids with

Table III. Fatty acid composition of neoproteolipid-W (results expressed as weight % of total fatty acids)

Chain length and unsaturation	Hydrogenated	Nonhydrogenated
14:0	3.7	1.5
14:1		1.7
16:0	14.4	12.2
16:1		1.9
18:0	17.1	7.2
18:1		8.0
18:2		tr.
18:3		tr.
20:0	3.7	3.9
20:4		tr.
22:0	4.0	3.4
22:1		1.5
23:0	1.3	1.5
24:0	52.8	35.1
24:1		20.6
25:0	1.6	0.0
26:0	1.9	1.9

two or more double bonds is low in NPL-W. NPL-W does not contain hydroxy fatty acids as shown by two thin-layer chromatographic systems [24, 49].

The data reported here were obtained on the whole NPL-W complex, since at present quantitative delipidation of NPL-W has not been accomplished. This may affect, to a certain degree, the precision of some of the values obtained. However, the results obtained permit the formulation of the main chemical features of NPL-W. The large amount of sphingosine bases, the patterns of monosaccharides and fatty acids and their quantities in hydrolyzates of NPL-W strongly suggest that the major part of NPL-W is neutral glycosphingolipid(s). Thus, NPL-W is essentially a polypeptido-neutral glycosphingolipid complex with some phospholipids and neutral lipids (cholesterol and cholesteryl esters) associated with it. The nature of the bonding between these constituents has not been elucidated, but it appears that it is sufficiently strong to withstand chromatography in relatively strong acidic or basic solvents on silicic acid chromatoplates.

We would like to add some speculation concerning the type of glycosphingolipid which apparently participates in the formation of NPL-W, assuming that all carbohydrates present in neoproteolipid are present as glycolipids and that only one type of glycosphingolipid is present in the complex. The molecular ratio of sphingosine:glucose: galactose:fucose:N-acetyl-galactosamine is 1.00:1.04:2.83:0.84:0.99, or in round numbers 1:1:3:1:1. Thus, it seems that in NPL-W there is a hexaglycosylceramide which contains one glucose molecule, three galactose molecules, one fucose and one N-acetylgalactosamine molecule.

B. Occurrence of Neoproteolipid-W in Tissues of Normal and Cancer-Bearing Rats

The possible occurrence of NPL-W in different tissues was investigated to determine whether or not NPL-W was specifically associated with malignancy as well as the degree of specificity. The occurrence of NPL-W in different rat tissues was studied primarily by thin-layer chromatography of lipids extracted from these tissues [40, 44]. Several essentially different thin-layer chromatographic systems were used. They separate NPL-W from neutral lipids, phospholipids and glycosphingolipids. Certain systems also separate NPL-W from NPL-S. Four systems are described here:

System A. Developing solvent: chloroform:methanol:acetic acid:water, 100:50:18:8 (v/v) [46, 47]. Thin layer chromatoplates (0.5 mm thickness) were prepared from silica gel slurry in 0.001 M Na_2CO_3 solution. This system of TLC does not discriminate between NPL-W and NPL-S. Both neoproteolipid spots will be located below lysophosphatidyl choline on chromatograms.

System B. Developing solvent: chloroform:methanol: 2 N ammonium hydroxide, 90:78:22 (v/v). Adsorbent: 'neutral' silica gel. There is discrimination between NPL-W and NPL-S. The NPL-W spot is located below lysophosphatidyl choline. NPL-S has a much higher Rf and its spot usually coincides with other lipids. With this system NPL-S is not detectable when lipid extracts from whole tissues are used. Glucose may have an Rf close to that of NPL-W in this system. Therefore, a control with glucose is necessary. This TLC system is very useful when it is necessary to discriminate between NPL-W and NPL-S. System B is a modification of the procedures by MÜLDNER *et al.* [27].

System C. Developing solvent: chloroform:methanol:water, 45:55:4 (or 5) (v/v). Adsorbent: 'neutral' silica gel. There is discrimination between NPL-W and NPL-S. Both neoproteolipids are located above phosphatidyl choline, but have different Rfs. This system is a modification of the procedure by WAGNER *et al.* [55].

System D. Developing solvent: chloroform:methanol:water, 60:35:7 (v/v). Thin-layer chromatoplates were prepared from silica gel G slurry in 1% $Na_2B_4O_7 \cdot 10\ H_2O$. There is discrimination between NPL-W and NPL-S. Both spots are located close to the origin of chromatograms (below lysophosphatidyl choline) on a rather dark background when sulfuric acid spray is used for detection of spots. Much better results are obtained after orcinol spray [48, 52].

The sensitivities of these systems (except system D) are in the order of 0.3–0.4 μg per sample applied on the thin-layer chromatoplate. Quantitative measurements of NPL-W on thin-layer chromatograms were performed densitometrically [40].

Table IV shows the distribution of NPL-W in different normal and cancer-bearing rat tissues and organs with the approximate amounts when NPL-W is present. Most of the tissues or organs tested from normal rats: serum, blood cells, liver, muscle, kidney, heart and brain do not contain NPL-W, or if it is present, the concentration in lipid extracts is less than 0.03%. Substantial amounts of NPL-W are present in spleen and embryo. Trace amounts of NPL-W (0.04% of lipid extracts) are present in lung and small intestines. The highest quantities of NPL-W are present in tumor tissues, averaging 1.09% of total lipids extracted. The concentration of NPL-W in lipids extracted from spleen of tumor-bearing rats is threefold higher than in lipids extracted from spleen of control animals. An increase in the size of the spleen was also observed in Walker carcinosarcoma 256-bearing rats. Whereas the average weight of spleen of normal rats is 0.64 g, the average weight of spleen of cancer bearing rats is 1.23 g; the content of lipids per gram of wet tissues in both types of spleens remains approximately the same. Thus, the total amount of NPL-W in the spleen of a cancer-bearing rat is about six times higher than in the spleen of a normal rat.

The occurrence of NPL-W in blood plasma/serum and blood cells of normal rats and Walker carcinosarcoma 256-bearing rats was investigated by more sensitive tests [40, 42, 44]. Lipids extracted from these tissues were first chromatographed on silicic acid columns and certain eluates from the column, in which NPL-W is concentrated if it is present

Table IV. Distribution of neoproteolipid-W in rat tissues tested by thin-layer chromatography (TLC).[1] Data adapted from SKIPSKI [40]

Tissues	Concentration of neoproteolipid-W in tissue lipid extracts, weight %
Normal rats	
Serum	–
Blood cells	–
Liver	–
Muscle	–
Lung	0.04
Kidney	–
Heart	–
Brain	–
Spleen	0.23
Small intestine	0.04
Embryo (11-day-old)	0.10
Tumor-bearing rats	
Walker carcinosarcoma 256	1.09
Spleen	0.75
Blood cells	0.03[2]

1 Sensitivity of TLC test is 0.03%. Therefore, a dash means that neoproteolipid-W is absent, or present at a concentration of <0.03%.
2 See table V.

in original lipid extracts, were subsequently chromatographed on thin-layer chromatoplates [40]. When lipid extracts from blood plasma/serum are 'concentrated' or 'enriched' in eluates after analytical column chromatography, the sample tested corresponds to 18–20 mg of the original lipid extract. Whereas direct thin-layer chromatography of lipids extracted from blood plasma or serum (application of 1–2 mg lipid extract per chromatogram) did not show the presence of NPL-W in plasma/serum of normal or cancer-bearing rats, 'enriched' eluates from plasma/serum of tumor-bearing rats revealed the presence NPL-W.

The plasma and serum from rats in which the Walker carcinosarcoma 256 was 7 days (0.2–0.4 g weight of tumor) to 15 days old (up to 15 g weight of tumor) contained on the average 16 µg NPL-W per 100 mg of lipids. In plasma/serum from normal rats tested under identical conditions no NPL-W was observed or it may have been present in quantities below the sensitivity of the test (<2.5 μg NPL-W per 100 mg of lipids).

Direct thin-layer chromatography of lipids extracted from total blood cells of normal rats and Walker carcinosarcoma 256-bearing rats showed the presence of NPL-W in the latter (table IV). However, when the lipids extracted from blood cells were first enriched by column chromatography,

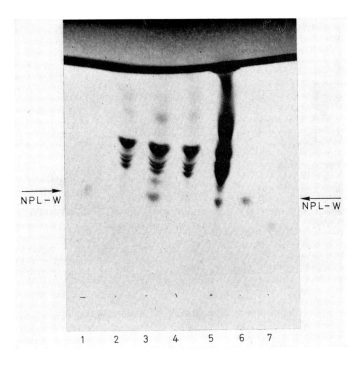

Fig. 1. Thin-layer chromatogram (system B, see text) of lipids extracted from blood cells. Prior to thin-layer chromatography, extracts were subjected to silicic acid column chromatography for enrichment of NPL-W. Lane 1: NPL-W, 3 μg. Lane 2: eluate from chromatographic column of lipids extracted from blood cells of normal rats; eluate applied is equivalent to 3.6 mg original lipid extract. Lane 3: eluate from chromatographic column of lipids extracted from blood cells of rats with Walker carcinosarcoma 256 (15 days after tumor transplantation); eluate applied is equivalent to 3.6 mg original lipid extract. Lane 4: eluate from chromatographic column of lipids extracted from blood cells of rats which rejected the transplanted tumor; eluate applied is equivalent to 3.6 mg original lipid extract. Lane 5: lipids extracted from Walker carcinosarcoma 256 tissue, 1 mg applied. Lane 6: NPL-W 5 μg. Lane 7: glucose, 15 μg. Detection of spots: sulfuric acid spray. Reproduced from SKIPSKI *et al.* [42].

Table V. Effect of tumor age upon the amount of neoproteolipid-W in lipid extracts from blood cells. Reproduced from SKIPSKI et al. [42]

Age of tumor, days	Number of determinations	Neoproteolipid-W average amount, in μg/100 mg lipid extracts
Rats with Walker carcinosarcoma 256		
7	4	31
10	2	75
11	6	94
12	6	106
13	2	139
15	19	147
Normal rats		
0	29	14

thin-layer chromatograms showed the presence of NPL-W in both normal and cancer-bearing rats (fig. 1). Examination of thin-layer chromatograms revealed that only small quantities of NPL-W were present in lipids extracted from blood cells of normal rats (fig. 1, lane 2) and also from rats which rejected transplanted tumors (fig. 1, lane 4), but a substantial amount of NPL-W was present in tumor-bearing rats (fig. 1, lane 3).

Lipids extracted from blood cells were tested in one thin-layer chromatographic system only, system B (see above), since other systems available at that time for the separation of NPL-W did not separate this neoproteolipid because of the presence of some substances in the lipid extracts from blood cells which have Rfs close to those of NPL-W. Therefore, for positive identification of NPL-W from blood cells, NPL-W was isolated on a large silicic acid chromatographic column [42] by the procedure used for the isolation of NPL-W from Walker carcinosarcoma 256 [40]. The isolated NPL-W from blood cells of rats bearing the Walker carcinosarcoma 256 by this method chromatographed identically with NPL-W isolated from tumor tissue in all thin-layer chromatographic systems used (systems A, B and C, see above). Thus, NPL-W isolated from blood cells is indeed a complex similar or identical with NPL-W isolated from Walker carcinosarcoma 256.

Table V shows changes in the amounts of NPL-W in lipids extracted from blood cells of Walker carcinosarcoma-bearing rats at different stages of tumor growth. The amount of NPL-W in blood cells from rats with the 7-day-old tumor is twice the normal value, and this increases to tenfold in rats with 15-day-old tumors [42].

Separation of erythrocytes from leukocytes by centrifugation [42] and subsequent analysis of lipids in these two cell fractions showed that NPL-W is present only in erythrocytes.

Results thus showed that blood plasma/serum from Walker carcinosarcoma 256-bearing rats contains NPL-W in measurable quantities, whereas this neoproteolipid is absent in blood plasma/serum of normal rats. When blood cells were tested there is a substantially greater amount of NPL-W in animals with tumors (two to tenfold higher, depending upon the age of the tumor).

Probably there is some relationship between the increase of NPL-W in erythrocytes of tumor-bearing rats and its accumulation in spleen of these rats. Erythrocytes undergo breakdown in the reticuloendothelial system. The spleen, apparently, plays an important role in this process, especially in the removal of slightly injured red cells [19, 58]. Erythrocytes carrying NPL-W may very well be 'injured on a molecular level' by the adsorption of NPL-W on the surface or by the incorporation of neoproteolipid into the plasma membrane. The removal of such injured erythrocytes from circulating blood by the spleen probably occurs preferentially and results in accumulation of NPL-W in this organ especially if the destruction of the erythrocytes is more rapid than the metabolism of NPL-W.

IV. Neoproteolipid-S

Although we were previously aware of the existence in some tumors of a type of neoproteolipid other than NPL-W, only recently have we started systematic studies of neoproteolipid-S or NPL-S [41]. It appears, at least among the tumors so far investigated, that NPL-S is more often present than NPL-W (table I). However, in most cases, our conclusions are based on chromatographic behavior of neoproteolipids rather than on actual isolation and chemical characterization of these complexes. At present NPL-S has been isolated only from mouse sarcoma 180 and rat Morris hepatoma 5123tc.

Neoproteolipid-S was isolated by silicic acid column chromatography using chloroform:methanol mixtures for stepwise gradient elution. The purity of isolated NPL-S was checked by thin-layer chromatography in systems described above. The NPL-S isolated either from sarcoma 180 or Morris hepatoma contains 7–10 weight % of polypeptides, as determined by the procedure of HESS and LEWIN [17]. It also contains sphingosine bases as determined by gas-liquid chromatography [54, 57] and by the method of YAMAMOTO and ROUSER [59]. Acid hydrolyzates of NPL-S from both tumors also contain somewhat higher quantities of fatty acids than those in NPL-W as shown by semi-microtitration [43, 45] and gas-liquid chromatography [50].

More detailed studies of the chemical composition of NPL-S were performed only on the NPL-S isolated from the Morris hepatoma. The acid hydrolyzate of this NPL-S contains 38–45 weight % of monosaccharides, similar to NPL-W. The determination of monomeric carbohydrates was performed by gas-liquid chromatography [54, 57]. The following monomeric carbohydrates were detected in the NPL-S hydrolyzate: galactose, glucose, galactosamine and N-acetylneuraminic acid. No fucose was detected in NPL-S complexes.

Thus, our preliminary results [41] show that the main difference between NPL-S and NPL-W resides in the carbohydrate fractions: NPL-S contains N-acetylneuraminic acid, but no fucose, whereas NPL-W contains no N-acetylneuraminic acid but does contain fucose. These differences indicate that different glycosphingolipids participate in the formation of NPL-S and NPL-W. Biological studies on the occurrence of NPL-S, so far, show that there is a substantially greater amount of this neoproteolipid in serum from the rats with the Morris hepatoma (12–15 μg/100 mg lipid extract) than in normal rat serum (0–2.5 μg NPL-S).

V. Neoproteolipids in Serum of Cancer Patients

Blood serum of patients with cancer and normal control subjects was analyzed for the presence of neoproteolipids. In most cases however, lipids extracted from high-density lipoproteins with density range 1.0635–1.210 g/ml were tested for the presence of neoproteolipids rather than lipid extracts from whole serum. Actually, the neoproteolipids were first observed in cancer patients in such lipid extracts. High-density lipoproteins were isolated from serum as described by BARCLAY et al. [1, 2], lipids extracted and subjected to thin-layer chromatography with system A (see above) which does not discriminate between NPL-W and

Table VI. Neoproteolipid in high-density lipoproteins from sera or in whole sera of patients with cancer and from normal subjects. Data adapted from SKIPSKI *et al.* [44]

Sources of Sera	Number of Cases	Sex	Presence of neoproteolipid
Normal subjects			
Individual samples	16	F	–
Pooled sample	1	M and F	–
Cancer patients			
Series I (no treatment before blood samples)			
Carcinoma of breast	8	F	+
Ovarian carcinoma	2	F	+
Bronchiogenic carcinoma	1	F	+
Cancer of the colon	1	M	+
Chronic myelocytic leukemia	1	M	±
Series II (all patients under treatment)			
Melanoma	1	M	+
Epidermoid bronchiogenic carcinoma[1]	1	M	+
Neuroblastoma	1	F	+
Leukemia, child, 4 years old	1	F	+
Carcinoma of uterus[1], metastases	1	F	+
Carcinoma of breast[1], bone metastases	1	F	+
Carcinoma of breast[1] (1 week post-mastectomy)	1	F	±
Carcinoma of breast[1] (minimal recurrence 2 years post-mastectomy)	1	F	–
Carcinoma of breast, with metastases	1	F	+ (NPL-S)
Leiomyosarcoma	1	F	+ (NPL-S)

[1] Primary tumor was removed.

NPL-S. As a matter of fact, at the time when most of these experiments were performed we were not aware of the existence of two types of neoproteolipids. Other lipoprotein fractions: very low-density lipoproteins, low-density lipoproteins and ultracentrifugal residue (density >1.210 g/ml) did not contain any detectable amounts of neoproteolipids, either in normal subjects or in cancer patients.

In the last several experiments with cancer patients and normal control subjects, whole blood serum was subjected to lipid extraction and these lipids were chromatographed on silicic acid columns and certain fractions of the eluate chromatographed on silica gel plates. In other words, the same procedure was used as with the serum from normal and cancer-bearing rats. Both these procedures: silicic acid column chromatography and isolation of high-density lipoproteins served the same purpose: enrichment of the lipid extracts with neoproteolipids when they are present. Table VI shows that lipids extracted from serum high-density lipoproteins or whole serum from normal subjects do not contain measurable quantities of neoproteolipids (17 cases). 23 patients with different forms and stages of cancer were studied also.

In series I lipids were extracted from serum high-density lipoproteins. All patients had advanced forms of malignancy, mostly inoperable. At the time when blood samples were taken from these patients, they had not received medication. Neoproteolipid was detected in lipid extracts from high-density lipoprotein in 12 cases out of 13 tested. Results were inconclusive in one patient who was in remission from chronic myelocytic leukemia.

In series II all patients were under some form of medical treatment. In some patients primary tumors had been removed by surgery, but there was evidence of metastases. Neoproteolipids were present in lipid extracts from high-density lipoproteins or whole serum in 8 out of 10 patients tested. The result was inclusive in one patient studied one week after radical mastectomy. In another patient, the presence of neoproteolipid was not detected in a sample drawn at a time of minimal recurrence of the disease 2 years after radical mastectomy.

In two cases: carcinoma of the breast and leiomyosarcoma, working with whole serum lipid extracts, the type of neoproteolipid was determined. In both cases NPL-S was present. This was anticipated for the patient with breast cancer since the presence of NPL-S in breast tumor tissue was previously demonstrated (table I).

VI. Discussion

The possible appearance and/or accumulation in malignant tissues of specific protein-lipid or polypeptide-lipid complexes soluble in certain organic solvents was suggested by WALLACH *et al.* [56] more than ten years ago. These authors observed a higher molar ratio of nitrogen to phosphorus in lipid extracted from Ehrlich ascites cells, than one would expect from their phospholipid composition. WALLACH *et al.* [56] believed that some amino acids or polypeptides were bound to phospholipids. However, the authors did not succeed in isolating such complexes from lipid extracts of Ehrlich ascites cells.

PETERING et al. [29] attempted to isolate 'malignolipin' [21] from tumor tissues, normal tissues and blood of normal and cancer-bearing animals. The authors were able to isolate from Walker carcinosarcoma 256, on silicic acid columns, a rather heterogeneous fraction which contained lipids associated with three amines, one of which was identified as glutamic acid. However, this lipid complex apparently was not 'malignolipin' with the chemical formula attributed to it by KOSAKI et al. [21]. The lipid portion of the complex isolated by PETERING et al. [29] was characterized by a high content of long-chain fatty acids (with chain lengths of 20–24 carbons).

Relatively recently DYATLOVITSKAYA et al. [5] isolated a lipopeptide fraction from phospholipids of the Jensen sarcoma. This lipopeptide fraction appeared to be a heterogeneous mixture of peptidolipids [3]. Apparently, an identical fraction of peptidolipids was isolated by the same authors [5] from normal rat liver, indicating that these peptidolipids were not specific for tumor cells. The presence of similar or identical peptidolipids in lipid extracts from Walker carcinosarcoma 256 was confirmed in our laboratory [SKIPSKI et al., unpublished observation, 1970].

All the evidence (chromatographic behavior, chemical composition) strongly suggest that neoproteolipids are different from all protein-(peptide)-lipid complexes described above. The exception may be the 'lipid-amines' fraction from Walker carcinosarcoma 256 isolated by PETERING et al. [29]. The only indications of some relationship between these two complexes are: the similarity of their elution from silicic acid chromatographic columns, and the patterns of the fatty acid composition (high content of long-chain fatty acids in both of them). However, our experience with the silicic acid column chromatography of lipids extracted from Walker carcinosarcoma 256 tissues showed that some 'lipids' of an unknown nature preceded NPL-W in the process of elution [40]. It is not excluded that these lipids rather than NPL-W or a mixture of these lipids and NPL-W were actually isolated and studied by PETERING et al. [29].

The chemical composition of both NPL-W and NPL-S, namely, the high content (up to 70 %) of glycosphingolipids in these complexes, indicates some relationship between glycosphingolipids, both neutral and ganglioside types, and neoproteolipids.

Alterations in glycosphingolipid patterns and the metabolism of these lipids in malignant tissues have been reported in a number of recent publications, some of which are discussed below. According to HAKOMORI: 'alteration of membrane heteroglycans and translocation of membrane carbohydrate sites could be the most general phenomena of malignancy...' [11].

Changes in glycosphingolipids in tumors were first demonstrated by RAPPORT [30] and RAPPORT et al. [31–33], using in most cases, immunochemical techniques. Independently, HAKOMORI and co-workers [10–14, 60] presented evidence for the appearance of several new neutral glycosphingolipids containing fucose and other monomeric carbohydrates in human adenocarcinoma of the colon and pancreas and their metastatic lesions in the liver of the patient. Some of these glycosphingolipids were purified further and their structures have been determined. They are known as Le^a-glycolipids and X-hapten glycolipids [11, 60]. SEIFERT [37], SEIFERT and UHLENBRUCK [38], and UHLENBRUCK and GIELEN [53] observed an accumulation of hematoside or disialohematoside in different types of brain tumors. KOSTIĆ and BUCHHEIT [22] also studied gangliosides in brain tumors and came to the conclusion that the levels of polar gangliosides were decreased, whereas the amounts of less polar gangliosides were substantially increased.

SIDDIQUI and HAKOMORI [39] investigated the glycolipid patterns of three lines of Morris hepatomas (5123, 5123C, and 7800) along with neonatal and adult liver of rats. The glycolipid patterns of neonatal livers and normal adult livers were characterized by the presence of substantial amounts of trisialogangliosides and high quantities of hematosides and disialogangliosides. Adult livers also contained high levels of monosialogangliosides. The glycolipid patterns of the Morris hepatomas resembled each other, but were different from both neonatal and adult normal livers. The hepatoma patterns were characterized by very high levels of disialogangliosides, high levels of monosialogangliosides, small amounts of hematosides and the complete disappearance of trisialogangliosides, accompanied by a great increase in neutral glycosphingolipids and ceramides. KAWANAMI [18] observed that the level of neutral glycosphingolipids in mouse Nakahara-Fukuoka sarcoma was at least ten times higher than in normal liver.

Studies of the relationship between malignancy and glycosphingolipid composition of tissues have been extended to cultured normal cells and cells transformed *in vitro* [4, 9, 16, 20, 26, 35, 36, 61]. Although there is no complete agreement among investigators concerning alterations in the glycosphingolipid pattern in cells transformed *in vitro,* basically the results obtained confirmed those in human and animal tumors: that there are changes in the glycosphingolipid patterns with malignancy. More detailed reviews on relationship between malignancy and glycolipids are given in the literature [3, 11, 12, 15, 28].

The composition of neoproteolipids, whose basic components are neutral glycosphingolipids or gangliosides, strongly suggests that the appearance or accumulation of those complexes is probably directly related to alterations in the metabolism of glycosphingolipids. However, it is not excluded that the appearance or accumulation in malignant tumors of new lipophilic polypeptides or proteins which interact with glycosphingolipids, may be responsible for the formation of neoproteolipids. Until the precise structures of glycosphingolipids and polypeptides and the type of bonds between them are elucidated, further discussion concerning the relationship between alterations in glycosphingolipid patterns and the appearance/accumulation of neoproteolipids is untimely.

Elucidation of the formation and metabolism of neoproteolipids in tumors may significantly contribute to our understanding of biochemical processes characteristic of malignancy and the effects of the tumor on the host. In addition, neoproteolipids may eventually also have a practical importance. The appearance of NPL-W in measurable quantities in blood plasma/serum of Walker carcinosarcoma 256-bearing rats and substantial increase of NPL-S (minimum fivefold) in the plasma/serum of Morris hepatoma-bearing rats as compared with normal rats, and the appearance of neoproteolipids (in most cases the type of neoproteolipid was not identified) in blood serum of most cancer patients, provide a reasonable biochemical basis for an attempt to develop a blood test for cancer. In addition, the level of the neoproteolipids in blood plasma/serum of cancer patients may provide a biochemical approach for the evaluation of the efficacy of the chemotherapy of cancer.

The most critical problem in developing a cancer test is the question of how early in the course of malignancy the neoproteolipids will appear in blood plasma/serum in measurable quantities. At present, we have no answer to this question with respect to human tumors. However, with the Walker carcinosarcoma 256-bearing rats our experiments showed that it is possible to detect chemically NPL-W in plasma/serum and to observe a twofold increase in the content of NPL-W in erythrocytes when the transplanted tumor starts to grow (7-day-old tumors weighing 0.2–0.4 g). To what extent these observations can be projected to other types of animal tumors and especially to human tumors is impossible to predict.

VII. Summary

1. A protein-lipid complex soluble in certain organic solvents and insoluble in aqueous media was isolated from Walker carcinosarcoma 256. It has been classified as a proteolipid-type complex and called neoproteolipid-W. Neoproteolipid-W contains 8–11 % polypeptides or proteins. The acid hydrolyzates of neoproteolipid-W contain 25–31 % fatty acids, 40–45 % monomeric carbohydrates, 10–14 % sphingosine bases and a small amount of cholesterol. The monomeric carbohydrate fraction contains four different monosaccharides: glucose, galactose, fucose and N-acetylgalactosamine with the approximate ratios 1:3:1:1 in the order listed. It appears that neoproteolipid-W is a complex formed by hexaglycosyl-ceramides and polypeptides (or protein) with small amounts of neutral lipids and phospholipids attached to them.

2. Neoproteolipid-W is not specifically a cancer-associated proteolipid, but is present in highest quantities in tumor tissues. In tissues of normal rats, it is present in significant quantities only in spleen and embryo. In rats bearing Walker carcinosarcoma 256, neoproteolipid-W appears in blood plasma/serum in measurable quantities (16 μg/100 mg lipids). The amounts also increased up to tenfold in erythrocytes, depending upon the age of the tumors. The amount of neoproteolipid-W increased severalfold in spleen of cancer-bearing animals.

3. Systematic studies of different tumors revealed that most tumors have a type of neoproteolipid other than neoproteolipid-W. This new type is called neoproteolipid-S because it was isolated first from mouse sarcoma 180 and then from Morris hepatoma 5123. Neoproteolipid-S apparently contains amounts of proteins similar to neoproteolipid-W. The carbohydrate moiety of neoproteolipid-S has a different monosaccharide composition: it contains N-acetylneuraminic acid and does not have fucose. The quantity of neoproteolipid-S in blood serum from Morris hepatoma-bearing rats increases at least fivefold compared with serum from normal control rats.

4. Blood serum of normal human subjects and patients with cancer in different forms and stages of malignancy were tested for the presence of neoproteolipids. In most cases lipid extracts from high-density lipoproteins, with which neoproteolipids are associated, were tested rather than lipid extracts from whole serum. In most experiments the presence or absence of neoproteolipids was determined without characterization of the type of neoproteolipid. In 17 control samples from normal subjects

no neoproteolipids were detected. Neoproteolipids were detected in 20 of 23 patients; in two cases the results were inconclusive and in one case the result was negative.

VIII. Acknowledgements

The research project reviewed was supported in part by the Damon Runyon Memorial Fund for Cancer Research, grants DRG 839 and 1115, grant from the Elsa U. Pardee Foundation and grant CA 08748 from the National Cancer Institute.

IX. References

1 BARCLAY, M.; ESCHER, G. C.; KAUFMAN, R. J.; TEREBUS-KEKISH, O.; GREENE, E. M., and SKIPSKI, V. P.: Serum lipoproteins and human neoplastic disease. Clin. chim. Acta *10:* 39–47 (1964).
2 BARCLAY, M.; SKIPSKI, V. P.; TEREBUS-KEKISH, O.; GREENE, E. M.; KAUFMAN, R. J., and STOCK, C. C.: Effects of cancer upon high-density and other lipoproteins. Cancer Res. *30:* 2420–2430 (1970).
3 BERGELSON, L. D.: Tumor lipids. Progr. chem. fat and other lipids, vol. 13, part 1, pp. 1–59 (Pergamon Press, Oxford 1972).
4 BRADY, R. O. and MORA, P. T.: Alteration in ganglioside pattern and synthesis in SV_{40}- and polyoma virus-transformed mouse cell lines. Biochim. biophys. Acta *218:* 308–319 (1970).
5 DYATLOVITSKAYA, E. V.; TORKHOVSKAYA, T. I., and BERGELSON, L. D.: Tumor lipids. An investigation of the phospholipids of the Jensen sarcoma. Biokhimiya *34:* 177–182 (1969).
6 FOLCH, J. and LEES, M.: Proteolipids, a new type of tissue lipoproteins. J. biol. Chem. *191:* 807–817 (1951).
7 FOLCH-PI, J.: Brain proteolipids; in FOLCH-PI and BAUER Brain lipids and lipoproteins, and the leucodystrophies, pp. 18–30 (Elsevier, Amsterdam 1963).
8 FOLCH-PI, J. and STOFFYN, P. J.: Proteolipids from membrane systems. Ann. N. Y. Acad. Sci. *195:* 86–107 (1972).
9 HAKOMORI, S.: Physiological variation of glycolipid pattern of cultured cells and its changes in the transformed cells. Proc. nat. Acad. Sci. Wash. *67:* 1741–1747 (1970).
10 HAKOMORI, S.: Glycosphinogolipids having blood-group ABH and Lewis specificities. Chem. Phys. Lipids *5:* 96–115 (1970).
11 HAKOMORI, S.: Glycolipid changes associated with malignant transformation; in WALLACH and FISCHER The dynamic structure of cell membranes. 22. Coll. der Gesellschaft biologische Chemie, Mosbach/Baden 1971, pp. 65–96 (Springer, Berlin 1971).
12 HAKOMORI, S.: Fucolipids and blood group glycolipids in normal and tumor tissue. Progr. biochem. Pharmacol., vol. 10, pp. 167–196 (Karger, Basel 1975).

13 HAKOMORI, S. and ANDREWS, H. D.: Sphingoglycolipids with Le^b activity, and co-presence of Le^a-, Le^b-glycolipids in human tumor tissue. Biochim. biophys. Acta 202: 225–228 (1970).
14 HAKOMORI, S. and JEANLOZ, R. W.: Isolation of a glycolipid containing fucose, galactose, glucose, and glucosamine from human cancerous tissue. J. biol. Chem 239: PC3606–3607 (1964).
15 HAKOMORI, S.; KIJIMOTO, S., and SIDDIQUI, B.: Glycolipids of normal and transformed cells. A difference in structure and dynamic behavior; in Fox Membrane research. 1st ICN-UCLA Symp. Molecular Biol., pp. 253–277 (Academic Press, New York 1972).
16 HAKOMORI, S. and MURAKAMI, W. T.: Glycolipids of hamster fibroblasts and derived malignant-transformed cell lines. Proc. nat. Acad. Sci., Wash. 59: 254–261 (1968).
17 HESS, H. H. and LEWIN, E.: Microassay of biochemical structural components in nervous tissues. II. Methods for cerebrosides, proteolipid proteins and residue proteins. J. Neurochem. 12: 205–211 (1965).
18 KAWANAMI, J.: Lipid of cancer tissues. II. Neural glycolipids of Nakahara-Fukuoka sarcoma tissue. J. Biochem., Tokyo 62: 105–117 (1967).
19 KEENE, W. R. and JANDL, J. H.: Studies of the reticulo-endothelial mass and sequestering function of rat bone marrow. Blood 26: 157–175 (1965).
20 KIJIMOTO, S. and HAKOMORI, S.: Enhanced glycolipid: α-galactosyltransferase activity in contact-inhibited hamster cells, and loss of this response in polyoma transformants. Biochim. biophys. Res. Commun. 44: 557–563 (1971).
21 KOSAKI, T.; IKODA, T.; KOTANI, Y.; NAKAGAWA, S., and SAKA, T.: A new phospholipid, malignolipin, in human malignant tumors. Science 127: 1176–1177 (1958).
22 KOSTIĆ, D. and BUCHHEIT, F.: Gangliosides in human brain tumors. Life Sci. 9: part II, pp. 589–596 (1970).
23 LOWRY, O. H.; ROSEBROUGH, N. J.; FARR, A. L., and RANDALL, R. J.: Protein measurement with Folin phenol reagent. J. biol. Chem. 193: 265–275 (1951).
24 MALINS, D. C. and MANGOLD, H. K.: Analysis of complex lipid mixtures by thin-layer chromatography and complementary methods. J. amer. Oil Chem. Soc. 37: 576–578 (1960).
25 MATSUMOTO, M.; MATSUMOTO, R., and FOLCH-PI, J.: The chromatographic fractionation of brain white matter proteolipids. J. Neurochem. 11: 829–838 (1964).
26 MORA, P. T.; BRADY, R. O.; BRADLEY, R. M., and MCFARLAND, V. W.: Gangliosides in DNA virus-transformed and spontaneously transformed tumorigenic mouse cell lines. Proc. nat. Acad. Sci., Wash. 63: 1290–1296 (1969).
27 MÜLDNER, H. G.; WHERRETT, J. R., and CUMINGS, J. N.: Thin-layer chromatography in the study of cerebral lipids. J. Neurochem. 9: 607–611 (1962).
28 NIGAM, V. N. and CANTERO, A.: Polysaccharides in cancer: glycoproteins and glycolipids. Adv. Cancer Res. 17: 1–80 (1973).
29 PETERING, H. G.; GIESSEN, G. J. VAN; BUSKIRK, H. H.; CRIM, J. A.; EVANS, J. S., and MUSSER, E. A.: The isolation, characterization and antigenicity of malignolipin-like material from normal and tumor tissue. Cancer Res. 27: 7–14 (1967).

30 RAPPORT, M. M.: Immunological properties of lipids and their relation to the tumor cell. Ann. N.Y. Acad. Sci. *159:* 446–450 (1969).
31 RAPPORT, M. M.; GRAF, L.; SKIPSKI, V. P., and ALONZO, N. F.: Cytolipin H, a pure lipid hapten isolated from human carcinomas. Nature, Lond. *181:* 1803–1804 (1958).
32 RAPPORT, M. M.; GRAF, L.; SKIPSKI, V. P., and ALONZO, N. F.: Immunochemical studies of organ and tumor lipids. VI. Isolation and properties of cytolipin H. Cancer, Philad. *12:* 438–445 (1959).
33 RAPPORT, M. M.; SKIPSKI, V. P., and SWEELEY, C. C.: The lipid residue in cytolipin H. J. Lipid Res. *2:* 148–151 (1961).
34 ROACH, D. and GEHRKE, C. W.: Direct esterification of the protein amino acids; gas-liquid chromatography of N-TFA n-butyl esters. J. Chromat. *44:* 269–278 (1969).
35 SAKIYAMA, H. and ROBBINS, P. W.: Glycolipid synthesis and tumorigenicity of clone isolated from Nil 2 line of hamster embryo fibroblasts. Fed. Proc. *32:* 86–90 (1973).
36 SAKIYAMA, H. and ROBBINS, P. W.: The effect of dibutyryl adenosine 3′:5′-cyclic monophosphate on the synthesis of glycolipids by normal and transformed Nil cells. Arch. Biochem. *154:* 407–414 (1973).
37 SEIFERT, H.: Über ein weiteres hirntumorcharakteristisches Gangliosid. Klin. Wschr. *44:* 469–470 (1966).
38 SEIFERT, H. und UHLENBRUCK, G.: Über Ganglioside in Hirntumoren. Naturwissenschaften *52:* 190 (1965).
39 SIDDIQUI, B. and HAKOMORI, S.: Change of glycolipid pattern in Morris Hepatomas 5123 and 7800. Cancer Res. *30:* 2930–2936 (1970).
40 SKIPSKI, V. P.: Proteolipids associated with malignancy; in WOOD Tumor lipids. Biochemistry and metabolism, pp. 225–243 (Amer. Oil Chem. Soc. Press, Champaign, Ill. 1973).
41 SKIPSKI, V. P.; ARCHIBALD, F. M.; ADLER, C. R.; BARCLAY, M.; TARNOWSKI, G. S., and STOCK, C. C.: Neoproteolipids in malignant tumors. Proc. amer. Ass. Cancer Res. *14:* 43 (1973).
42 SKIPSKI, V. P.; ARCHIBALD, F. M.; ADLER, R. C.; FINE, A. W.; BARCLAY, M., and STOCK, C. C.: Neoproteolipid-W in blood cells of normal rats and rats with cancer. Res. Commun. chem. Pathol. Pharmacol. *3:* 165–176 (1972).
43 SKIPSKI, V. P. and BARCLAY, M.: Thin-layer chromatography of lipids; in LOWENSTEIN Methods in enzymology, vol. 14, pp. 530–598 (Academic Press, New York 1969).
44 SKIPSKI, V. P.; BARCLAY, M.; ARCHIBALD, F. M.; LYNCH, T. P., jr., and STOCK, C. C.: A new proteolipid apparently associated with cancer. Proc. Soc. exp. Biol. Med. *136:* 1261–1264 (1971).
45 SKIPSKI, V. P.; GOOD, J. J.; BARCLAY, M., and REGGIO, R. B.: Quantitative analysis of simple lipid classes by thin-layer chromatography. Biochim. biophys. Acta *152:* 10–19 (1968).
46 SKIPSKI, V. P.; PETERSON, R. F., and BARCLAY, M.: Quantitative analysis of phospholipids by thin-layer chromatography. Biochem. J. *90:* 374–378 (1964).

47 SKIPSKI, V. P.; PETERSON, R. F.; SANDERS, J., and BARCLAY, M.: Thin-layer chromatography of phospholipids using silica gel without calcium sulfate binder. J. Lipid Res. *4:* 227–228 (1963).
48 SKIPSKI, V. P.; SMOLOWE, A. F., and BARCLAY, M.: Separation of neutral glycosphingolipids and sulfatides by thin-layer chromatography. J. Lipid Res. *8:* 295–299 (1967).
49 SKIPSKI, V. P.; SMOLOWE, A. F.; SULLIVAN, R. C., and BARCLAY, M.: Separation of lipid classes by thin-layer chromatography. Biochim. biophys. Acta *106:* 386–396 (1965).
50 STOFFEL, W.; CHU, F., and AHRENS, E. H., JR.: Analysis of long-chain fatty acids by gas-liquid chromatography. Analyt. Chem. *31:* 307–308 (1959).
51 STOFFYN, P. and FOLCH-PI, J.: On the type of linkage binding fatty acids present in brain white matter proteolipid apoprotein. Biochem. biophys. Res. Commun. *44:* 157–161 (1971).
52 SVENNERHOLM, L.: The quantitative estimation of cerebrosides in nervous tissue. J. Neurochem. *1:* 42–53 (1956).
53 UHLENBRUCK, G. und GIELEN, W.: Hirntumorcharakteristische Glykolipoide. Med. Welt, Stg. *20:* 332–335 (1969).
54 VANCE, D. E. and SWEELEY, C. C.: Quantitative determination of neutral glycosyl ceramides in human blood. J. Lipid Res. *8:* 621–630 (1967).
55 WAGNER, H.; HÖRHAMMER, L. und WOLFF, P.: Dünnschichtchromatographie von Phosphatiden und Glykolipiden. Biochem. Z. *334:* 175–184 (1961).
56 WALLACH, D. F. H.; SODERBERG, J. and BRICKER, L.: The phospholipides of Ehrlich ascites carcinoma cells composition and intracellular distribution. Cancer Res. *20:* 397–402 (1960).
57 WINDELER, A. S. and FELDMAN, G. L.: The isolation and partial structural characterization of some ocular gangliosides. Biochim. biophys. Acta *202:* 361–366 (1970).
58 WINTROBE, M. M.: Clinical hematology (Lea & Febiger, Philadelphia 1967).
59 YAMAMOTO, A. and ROUSER, G.: Spectrophotometric determination of molar amounts of glycosphingolipids and ceramide by hydrolysis and reaction with trinitobenzenesulfonic acid. Lipids *5:* 442–444 (1970).
60 YANG, H.-J. and HAKOMORI, S.: A sphingolipid having a novel type of ceramide and lacto-N-fucopentaose III. J. biol. Chem. *246:* 1192–1200 (1971).
61 YOGEESWARAN, G.; SHEININ, R.; WHERRETT, J. R., and MURRAY, R. K.: Studies on the glycosphingolipids of normal and virally transformed 3T3 mouse fibroblasts. J. biol. Chem. *247:* 5146–5158 (1972).
62 YU, R. K. and LEDEEN, R. W.: Gas-liquid chromatographic assay of lipid-bound sialic acids: measurement of gangliosides in brain of several species. J. Lipid Res. *11:* 506–516 (1970).

Authors' address: VLADIMIR P. SKIPSKI, MARION BARCLAY, FRANCIS M. ARCHIBALD and C. CHESTER STOCK, Sloan-Kettering Institute for Cancer Research, Walker Laboratory, 145 Boston Post Road, *Rye, NY 10580* (USA)

Lipids in Normal and Tumor Cells in Culture[1]

BARBARA V. HOWARD and WILLIAM J. HOWARD

Department of Biochemistry, School of Dental Medicine, and Department of Medicine, School of Medicine, University of Pennsylvania, and Clinical Research Center, Philadelphia General Hospital, Philadelphia, Pa.

Contents

```
  I. Introduction .................................................... 135
 II. Nutrition ....................................................... 138
III. Lipid Composition ............................................... 140
     A. Major Lipid Classes .......................................... 140
     B. Fatty Acids .................................................. 141
     C. Lipid Ethers ................................................. 144
 IV. Lipid Metabolism ................................................ 144
     A. Lipid Biosynthesis ........................................... 144
     B. Regulation of Lipid Metabolism ............................... 147
  V. Membranes ....................................................... 151
     A. Evidence for Alteration after Transformation ................. 151
     B. Lipid Composition of Membranes ............................... 152
     C. Phospholipid Metabolism ...................................... 152
     D. Glycolipid Metabolism ........................................ 153
     E. Lipid Transport .............................................. 154
 VI. Summary ......................................................... 156
VII. References ...................................................... 157
```

I. Introduction

When applied to cells in culture, the use of the terms 'normal' and 'tumor' can be ambiguous and must be carefully qualified. To this end, a

---

1 This review was aided by grants RR-107 and AM-14526 from the National Institutes of Health.

brief description of the nature of cells in culture will be presented and the distinction between normal and tumor cells will be considered. When a piece of tissue from almost any fetal or adult organ is explanted into a compatible culture environment, the usual observation is that cells of fibroblast-like morphology migrate out from the explant and remain viable *in vitro*. This state, first observed by CARREL [25] and LEWIS and LEWIS [78] is called a primary culture[2]. If these resultant cells are then serially subcultivated, a diploid cell line[2] is formed. These diploid fibroblast lines have been thoroughly characterized from a number of species, including human [66, 134], mammalian [140], and avian [65], and they have been found to retain a number of traits common to cells *in vivo*. They have a stable euploid karyotype characteristic of their tissue of origin. They cannot be cultivated indefinitely, that is they have a definite and limited life span. Moreover, they have certain characteristic growth properties, such as low saturation density, minimal ability to form colonies on feeder layers or soft agar, and high requirement for serum growth stimulating factors.

Through a phenomenon referred to as transformation, a diploid cell line can be converted to an established, or mixoploid cell line[2]. This can be induced by irradiation [19] or carcinogens [14], but most often through the action of an oncogenic virus [132, 138]. In this latter case, the phenomenon is also referred to as conversion. The mechanism of the viral conversion is under intensive investigation and for a more detailed discussion of this subject, the reader is referred to recent reviews by BUTEL *et al.* [23], SAMBROOK [119] and TEMIN [136]. The main feature of transformed cells that distinguishes them from the nontransformed is their insensitivity to the controls that regulate cell multiplication [119]. They have an indefinite potential for proliferation, are heteroploid in chromosome development and are morphologically pleiomorphic. They attain a higher cell density at confluency, and have minimum requirement for serum growth promoting factors. They have high plating efficiency, and can form colonies on feeder layers and in soft agar. Although the relationship has never been clearly defined, it is generally considered that transformation *in vitro* is analogous to the malignant process *in vivo*, since established cell lines can arise directly when explants of tumors are

---

[2] Terminology according to recommendation of Committee on Nomenclature, Tissue Culture Association. The terms 'mixoploid' and 'heteroploid' are synonymous and are used interchangeably in this review.

cultured *in vitro,* and since they can often form tumors when transplanted into a compatible host.

The seemingly simplistic comparison must be immediately qualified. In the first place, exceptions to each of the listed characteristics can be found, and variants can be selected from among a transformed cell population which have reverted from any of the above-listed properties [21, 23, 100, 101]. Secondly, other types of cell cultures have been described which do not fit into the above classification. These include lymphoblastoid cell cultures [75] which are derived from normal leukocytes. They maintain diploid chromosome composition but have unlimited life *in vitro,* probably due to an episome of viral origin [85].

Also there are diploid cell cultures recently established from normal tissues which are not fibroblastic [18, 29], but can undergo viral transformation [17]. Finally, diploid cultures from certain species consistently transform spontaneously. Several resultant established lines, especially the 3T3 from mouse [141] and the BHK21 [88] from hamster, have been selected to maintain characteristics of normal diploid cells, such as fibroblastic morphology, low saturation density, reduced ability to form colonies, and increased requirement for serum factors. These cells, although of heteroploid karyotype and unlimited lifespan, can then be transformed by oncogenic viruses; their growth characteristics then become similar to those described above for transformed cells.

Since the transformation process, either from diploid fibroblast cells or from the hamster or mouse established fibroblast lines, is considered to be analogous to the neoplastic process *in vivo,* the use of normal and transformed or tumor cell lines affords a system for the study of the comparative biochemistry of the tumor process. It has the advantage of precise control of the environment in which the cells grow, so that meaningful quantitative experiments can be done. Also, one can compare directly the transformed cell to its immediate progenitor and thus eliminate the variables presented by multidifferentiated cell types in tissues *in vivo*. In view of the above description of cell types, confusion arises when earlier cell culture studies are considered because many of these studies compared lines derived from tumor cells to cell lines originally derived from normal tissues which had already been transformed *in vitro*. Consideration will be limited to comparative lipid studies of diploid fibroblasts and their transformed derivatives, or studies of the specially selected mouse and hamster established fibroblast lines and their virus transformed derivatives. An attempt will be made to compile data of studies of individual diploid or tumor cell lines. For complete consideration of lipids in cell cultures, the reader is referred to reviews by ROTHBLAT [108, 109] and SPECTOR [126].

II. Nutrition

Almost all cells in culture require serum in the medium for optimum growth, although the untransformed cells generally have a higher serum requirement than do transformed cells [85, 88, 131]. When growthpromoting factors in serum have been isolated, they are usually not lipids. In most cases, they are proteins, usually a component of the α-globulins [137].

There have been few unequivocal demonstrations of requirements of cells in culture for lipids. Both diploid and established cell lines can synthesize lipids *de novo* (sect. IV. A). Moreover, uptake of fatty acids, cholesterol and glycerides has been demonstrated in a number of normal and tumor cells studied (sect. V. E). It has been reported by HOLMES *et al.* [67] that human diploid fibroblasts derived from cartilage and skin require cholesterol for optimum growth. A requirement for cholesterol has never been observed in any transformed cell lines, and it has also been clearly demonstrated by ROTHBLAT *et al.* [110] that the diploid fibroblast line WI-38 grows as well as its transformed derivative in a medium supplemented only with delipidized serum protein. Thus, a requirement for cholesterol is not a distinguishing characteristic of diploid cells when compared to transformed or established cell lines. It is possible that the requirement for cholesterol in some diploid fibroblasts may be related to the more complete inhibition of cholesterol biosynthesis observed in these cells (sect. IV. B).

The only other lipids that have been demonstrated to be required in certain instances are the essential fatty acids. Essential fatty acids have been reported to be required by a few established cell lines for either growth or mitochondrial function [48, 60]. However, the majority of cell lines appear to have no significant requirement for polyunsaturated fatty acids of any type (see review by BAILEY and DUNBAR [8]). The ability to biosynthesize polyunsaturated fatty acids seems to be a biochemical characteristic lost by most cells after prolonged periods in culture [61] (sect. III. B). These cells usually maintain adequate growth in lipid-free medium although cellular polyunsaturated fatty acid levels drop. Thus, requirement for essential fatty acid is not a distinguishing characteristic between the two cell types.

Table I. Lipid content of some diploid and heteroploid lines

	Human conn. tissue fibro-blasts[1] [16]	Human fibro-blasts[1] [55]	MAF – human fibro-blasts[1] [11]	BHK-21[2] [104]	Chang[2] liver [26]	L-929[2] [147]	L-M[2] [3]	Ehrlich[2] ascites [150]
Total lipid, mg/100 mg dry weight	15.0	NR[3]	NR	21	13.5	NR	7.9	12.5
Neutral lipid, % total lipid	22	NR	62	31	31	19.4	17	53
Cholesterol ester, % total lipid	NR	NR	8.7	NR	NR	0.4	0.4	8.8
Triglyceride, % total lipid	9.0	NR	15	NR	18.7	6.5	1.8	5.6
Free fatty acid, % total lipid	NR	NR	<2	7.1	NR	1.0	1.6	22
Cholesterol, % total lipid	12.8	NR	27	13	12.1	9.5	12.7	13
Phospholipid, % total lipid	78	NR	38	69	69	80.5	83	48
Phosphatidyl choline, % total phospholipid	52	47.4	25	52	38.6	43.7	47	46
Phosphatidyl ethanolamine, % total phospholipid	16	25.2	38	22	16.8	23.2	28	10
Phosphatidyl serine and inositol[4], % total phospholipid	5.9	13.0	20	5	5.1	10.7	6.2	NR
Sphingomyelin, % total phospholipid	5.6	11.5	6.6	12	22.9	9.1	9.2	26
Lysolecithin, % total phospholipid	NR	NR	8.5	NR	NR	2.0	NR	NR

1 Diploid cell lines.
2 Heteroploid cell lines.
3 NR = None reported.
4 Phosphatidyl serine and phosphatidyl inositol are often measured together; therefore, data for these two compounds are combined in all cases.

III. Lipid Composition

A. Major Lipid Classes

Although there have been many published reports on the lipid content of cells in culture, most studies have not been direct comparisons between diploid and heteroploid lines, or untransformed and transformed counterparts. Table I contains data compiled from individual studies of diploid fibroblast lines and several established mixoploid lines. Although it is difficult to compare data compiled in different laboratories using different culture conditions and methods for isolation and quantitation, it is clear that no striking or consistent differences occur in the nature or distribution of the major lipid classes. In fact, table I shows that variation in cell lipid induced by culture in different sera (i. e. human connective tissue fibroblasts and the MAF human fibroblasts) are far greater than those observed between cell types (i. e. human connective tissue fibroblasts and Chang liver cells).

There have been some reports of direct comparisons of the lipids of normal and tumor cells in culture. The first study was reported by GOTTFRIED [53], who found that leukemic lymphocytes, in comparison with normal lymphocytes or polymorphonuclear leukemic cells, contain relatively more phosphatidyl choline, less sphingomyelin, and less cholesterol. These differences were observed in short-term cultures, however, and a more recent study from the same laboratory found that leukocytes maintained in long-term culture had altered lipid patterns similar to the changes found in leukemic leukocytes [54]. These findings suggest that the observed changes are a reflection of cell immaturity rather than a characteristic peculiar to the leukemic state. TSAO and CORNATZER [142] have compared the phospholipid composition of subcellular particles of cultured cells derived from tumor (HeLa and KB) and normal tissues (heart and liver). The four different lines contained the major phospholipids which are found in normal mammalian tissues, and there were no substantial differences in distribution between the cell types. It must be pointed out, however, that the two cell lines derived from normal tissue were established cell lines and, therefore, presumably already transformed.

The first direct comparison of a diploid fibroblast line with its transformed derivative was made by HOWARD and KRITCHEVSKY [71] using the human fibroblast WI-38 and the line WI-38VA13A derived by transformation with the oncogenic virus SV_{40}. It was found that the total lipid content in these two cell types was similar (table II). There was, however, a statistically significant decrease in the amount of phospholipid in the transformed cells. There were no significant differences in the distribution of neutral lipid and phospholipid subclasses between these two cell types. QUIGLEY et al. [99] have performed similar studies of lipids of chick embryo fibroblasts before and after transformation with Rous sarcoma virus (RSV). They reported no differences in cholesterol/phospholipid molar ratios, or in the distribution of phospholipids of the two cell types. This study correlates well with the report of FIGARD and LEVINE [44] who found no major differences in cholesterol and phospholipid composition between the chorioallantoic membrane and tumors induced in the membranes by RSV.

CHAO et al. [27] have recently reported studies comparing hamster embryo fibroblasts with their transformed lines which were derived spontaneously or through the action of polyoma virus. Both transformed derivatives were similar to the normal cultures in cholesterol content and phospholipid distribution, but had a somewhat decreased phospholipid content.

The data obtained from the above three comparative studies (human, chick and hamster) are presented in table II. It is evident that the direct comparative studies corroborate the data presented in table I that the nature and amounts of major lipids present in normal and transformed or tumor cells in culture are similar and resemble the lipid content of tissue *in vivo* [34]. The only interesting observation is the decrease in phospholipid content noted in the SV_{40} transformed human line and in both the transformed hamster lines. The significance of this decrease in cell phospholipid is not clear at this time, but it is possible that it might be related to other changes in membranes of tumor cells, as discussed in section V.

B. Fatty Acids

Data compiled from individual studies of fatty acid composition of lipids of normal and transformed cell cultures are shown in table III. One again finds no striking differences in the content or distribution of fatty acids between the two cell types. The fatty acid composition of the diploid line WI-38 and SV_{40} transformed WI-38VA13A were directly compared by HOWARD and KRITCHEVSKY [71]. The same fatty acids were present

Table II. Direct comparison of lipids in normal and transformed cell cultures

	Human [71]		Chick [99]		Hamster [27]		
	WI-38	SV$_{40}$-transformed	chick embryo fibroblasts	RSV-transformed	hamster embryo fibroblasts	spontaneously transformed	polyoma-transformed
Total lipid, mg/100 mg dry weight	21	18					
Neutral lipid, mg/100 mg dry weight	6.0	7.4					
Cholesterol ester[1]	5.0	4.0					
Triglyceride[1]	27	25					
Free fatty acid[1]	21	26					
Cholesterol[1]	35	35			7.6[2]	5.3[2]	6.4[2]
Phospholipid, mg/100 mg dry weight	15	11			23.8[2]	17.4[2]	19.1[2]
Phosphatidyl ethanolamine[3]	12	13	31	33–36	23	26	16
Phosphatidyl inositol[3]	13	10	12	12–13	2.6	5.3	2.7
Phosphatidyl serine[3]	12	4.2	12	12–13	14	12	17
Phosphatidyl choline[3]	57	57	46	42	46	40	42
Sphingomyelin[3]	7.4	13	10	8–9	15	17	23
Lysolecithin[3]	2.9	2.0					

1 % Total neutral lipid.
2 μM/g.
3 % Total phosphlipid.

in the glycerolipids of both cells, but there was a significant decrease in the amount of arachidonic acid present in both the neutral lipid and phospholipids of the transformed cells. This decrease is compensated for by a parallel increase in the unsaturated oleic acid, reflecting the tendency described by WHITE *et al.* [148], of cells to maintain a constant ratio of unsaturated to saturated fatty acids. YAU and WEBER [151] have recently studied the acyl group composition of phospholipids from chicken embryonic fibroblasts after transformation by Rous sarcoma virus. They also observed a decrease in arachidonic acid and an increase in oleic acid in the cells after transformation. No similar change was observed during

Table III. Distribution of fatty acids in some diploid and heteroploid lines (% total)

		WI-38[1] [71]	BHK21[2] [104]	L-929[2] [147]	L-M[2] [3]	HeLa[2] [50]	Ehrlich ascites[2] [150]
Palmitic	16:0	22	21	20	14	20	20
Palmitoleic	16:1	<1	NR	4.4	6.2	4	trace
Stearic	18:0	19	8	23	16	12	12
Oleic	18:1	31	48	31	31	33	22
Linoleic	18:2	7.5	12	2.9	31	19	2
Linolenic	18:3	1.0	NR	5.9	trace	1	NR
Arachidonic	20:4	18	NR	6.8	trace	<1	7

1 Diploid cell line.
2 Heteroploid cell line.

lytic infection. One possible explanation for the decreased arachidonate content is that the transformed cells are unable to biosynthesize arachidonic acid from serum linoleate by elongation and desaturation. WI-38 cells have been observed to be capable of the desaturation of exogenous 18:2, whereas the WI-38VA13A are not [39]. Thus, these cells would not be able to convert serum linoleic acid to arachidonic acid. TAKAOKA and KATSUTA [135] found that a number of established cell lines adapted to serum-free medium lacked desaturating ability. However, some transformed and tumor lines have been shown to be able to biosynthesize polyunsaturates [39]. Therefore, it is most likely that loss of the ability to desaturate is not a property of tumor cells, but simply a phenomenon that occurs as a result of prolonged passage in culture, as suggested by HARARY et al. [61].

It must also be pointed out that the above comparative studies of tumor and normal cell fatty acids were all conducted on cells cultured in serum-supplemented medium. When cells are cultured in the presence of adequate amounts of serum lipid, the majority of the cell fatty acid is derived from serum fatty acid [70, 81]. The only differences noted in comparative studies of fatty acids were among the polyunsaturated fatty acids which are present in small amounts in serum. A comparison of the requirements for and the ability to synthesize other fatty acids should be conducted on cells cultured in lipid-free medium.

C. Lipid Ethers

There has been great interest recently in the lipid ethers, which are present in relatively small amounts in most cells. Elevated levels of both O-alkyl and O-alk-1-enyl ethers have been observed by SNYDER and WOOD [123, 124] in tumors as compared to normal tissues, and this topic is reviewed elsewhere in this volume. These investigators have also observed a high level of glyceryl ethers in L-M cells, a line derived from L cells which has a high incidence of tumors when transplanted into suitable hosts [3]. However, when glyceryl ethers were measured in the diploid line WI-38 and a number of established lines from normal and tumor origin [68], the levels were relatively high in all the cells, corresponding to those found in rapidly growing hepatomas and other tumors. The lipid ether content of the WI-38 diploid cell line was as high as in most of the heteroploid lines. In this study the levels of glyceryl ethers were found to correlate with decreased levels of the enzyme α-glycerolphosphate dehydrogenase in both the cells in culture and in tumor homogenates. This enzyme can influence the concentration of dihydroxyacetone phosphate, the acylated derivative of which is the precursor of lipid ethers [2]. It is proposed that the elevated lipid ethers and decreased α-glycerolphosphate dehydrogenase might be more closely related to the dedifferentiation and adaptation associated with increased growth rather than to some unique characteristic of the neoplastic process.

IV. Lipid Metabolism

A. Lipid Biosynthesis

Most cells are routinely grown in serum-supplemented medium. When there is an adequate supply of serum lipid, cells obtain their lipid from the exogenous lipid. This has been demonstrated both for established cell lines [7, 49], and for diploid fibroblasts [70]. In most cases the bulk of the cell lipid is derived from serum-free fatty acid [70, 81, 126]. This observation indicates that both diploid and heteroploid lines are able to synthesize the full complement of glycerides and phospholipids.

On the other hand, both diploid [73, 110] and transformed lines [7, 43] have been cultured in lipid-free medium. Thus, both cell types are able to biosynthesize sufficient amounts of long-chain fatty acids and

Table IV. Ability of normal and transformed cell lines to synthesize lipid

Lipid class	Cell type	References
Fatty acids	Diploid fibroblasts	JACOBS et al. [73]
	L-929	BAILEY [7]
	HeLa	HARARY et al. [61]
	Ehrlich ascites	SPECTOR and STEINBERG [129]
Cholesterol	Diploid fibroblasts	BERLINER et al. [13]
	WI-38 (diploid)	HOWARD and KRITCHEVSKY [70]
		ROTHBLAT et al. [110]
	L-929	BAILEY [7]
	L-5178Y (mouse leukemia)	ROTHBLAT et al. [111]
	HTC (hepatoma)	WATSON [144]
	HeLa	AVIGAN et al. [5]
Phospholipids	Primary bone cells	DIRKSEN et al. [36]
	Chick embryo fibroblasts	BLAIR and BRENNAN [15]
	Primary brain cultures	MENKES [86]
	BHK21	BEN-PORAT and KAPLAN [12]
	Ehrlich ascites	DONISCH and ROSSITER [38]
	L-929	BAILEY [7]
Glycerides	Primary bone cells	DIRKSEN et al. [36]
	Primary brain cultures	MENKES [86]
	HEK (transformed)	MCINTOSH et al. [84]

cholesterol to satisfy growth and metabolic needs. Fatty acid biosynthesis has been demonstrated directly in established cell lines [7, 61, 129] and in diploid fibroblasts [73]. Cholesterol biosynthesis has also been demonstrated in diploid fibroblasts [13, 71, 110], as well as in established cell lines [7, 113]. It is possible, however, that certain fibroblast lines are not able to biosynthesize enough cholesterol to satisfy cell requirements [67]. Data on studies of lipid biosynthetic abilities in individual diploid or transformed cell lines are compiled in table IV. It seems that ability to biosynthesize major lipid classes is not lacking in either cell type, although it must be emphasized that the majority of the studies were conducted in the presence of exogenous serum lipid.

There have been very few direct comparisons of lipid metabolism between diploid or untransformed and transformed or established cell lines. FIGARD and LEVINE [44] have measured the incorporation of labeled glucose into the lipids of the chorioallantoic membrane and tumors induced in the membrane by Rous sarcoma virus. They found, interestingly, that the primary lipid product was triglyceride in both cells, but that the incorporation rate was six times higher in the tumors. It must be emphasized that this system is more analogous to whole tissues than to isolated cell cultures. When the phospholipid metabolism of chick embryo fibroblasts and their RSV-transformed derivatives were compared in an *in vitro* system [99], ^{32}P was incorporated into all phospholipid subclasses in approximately the same relative proportion as these phospholipid subclasses occur quantitatively in the cells; there were no differences between the two cell types. These workers also evaluated the turnover of individual phospholipid classes by assaying ^{32}P incorporation following 15 min pulses and found no differences in the pattern of label incorporated into phosphatidyl choline, sphingomyelin, phosphatidyl ethanolamine or phosphatidyl serine (plus inositol) between the normal and transformed cells.

EICHBERG *et al.* [42] studied phosphoinositide metabolism in normal and neoplastic astrocytes and found that the two cell types contained similar amounts of phosphatidyl inositols. However, the relative turnover rate, as measured by ^{32}P incorporation expressed per mg inositide or compared to total cell phospholipid, was higher in both the di- and triphosphoinositide fractions of the astrocytomas. This observation might be more clearly elucidated by studying changes in phospholipid metabolism which occur as a consequence of the mitogenic stimulation of lymphocytes. LUCAS *et al.* [79] found a stimulation of ^{32}P incorporation into phosphoinositides which correlated with stimulation of DNA synthesis in these cells. RESCH and FERBER [106] monitored ^{14}C-choline, ^{14}C-oleate and ^{14}C-acetate incorporation into lipids in the same system. They observed, after treatment with three different stimulants of mitosis, that there was a two to sixfold generalized stimulation of lipid biosynthesis as measured by incorporation of acetate into fatty acids and oleate into neutral lipids. In addition, there was a 33-fold stimulation of acetate incorporation into phospholipid. It is possible, therefore, that the increase in turnover of phospholipids observed by Eichberg in the astrocytomas, reflects increased mitosis in these cells, as compared to the relatively slow-growing astrocytes. No difference would thus be expected in the above comparisons of diploid cultures such as chick embryo fibroblasts with their transformed derivatives which, although they have unlimited potential for division, do not usually have lower generation times [28]. Another possibility is that the observed stimulation of phospholipid synthesis in transformed cells described above is related to the generalized stimulation of phospholipid turnover often observed upon cell perturbation, such as reported for cultured cells infected by virus [12, 15, 84].

There have been some suggestions that tumor cells in culture might utilize alternative pathways for lipid biosynthesis as compared to normal cells. AGRANOFF and HAJRA [2] have proposed that tumor cells, since they generally lack dehydrogenase shuttle enzymes, might have an increased activity of the acyl-dihydroxyacetone phosphate pathway for glycerolipid synthesis, and they have observed an increase in this pathway in Ehrlich ascites cells. However, the observation that α-glycerolphosphate dehydrogenase is low and ether lipid synthesis elevated in both diploid and heteroploid lines [68] might suggest that this path is more active in all cultured cells because of their adaptation to rapid growth. RYTTER and CORNATZER [115] have recently reported that the methylation pathway of phosphatidyl choline biosynthesis is low in tumor lines such as HeLa, KB and Ehrlich ascites cells, and DIRINGER et al. [35] have presented evidence for a new biosynthetic pathway for sphingomyelin synthesis in SV_{40} transformed mouse cells, via a transfer of choline phosphate from lecithin to sphingomyelin. However, it must be emphasized that in neither of the latter two studies were untransformed or normal cells assayed; it is very likely that the alternative pathways are common to a number of cells in culture as a result of the dedifferentiation and adaptation to rapid proliferation *in vitro*. A similar conclusion was reached by PASTERNAK and co-workers [92, 93] who found that phospholipids in neoplastic mast cells could be divided into two pools, one of which is susceptible to continual enzymatic degradation and resynthesis, and the other of which is metabolically stable. They found that a similar two-pool system of phospholipid turnover is a common feature of viable cells and a prerequisite for cell division.

B. Regulation of Lipid Metabolism

Since there is an obvious lack of, or release from control of cell division in neoplastic cells, it is logical to pose the question of whether tumor cells have alterations in other areas of regulation of cell metabolism. Cell cultures provide an especially good system for the study of this problem since regulation of cell metabolism can be separated from the complexities of superimposed hormonal and other physiological processes.

Because a lack of dietary feedback control of cholesterol biosynthesis has been observed in several tumor systems [52, 117, 122], much attention has been focused on the regulation of cholesterol biosynthesis. SIPERSTEIN and FAGAN [121] proposed that the lack of observed feedback in tumors is due to a defect in the regulation of the enzyme β-hydroxy-β-methylglutaryl (HMG)-CoA reductase, but recently it has been alternatively proposed that the lack of feedback might be due to failure of the tumors to take up or retain adequate cholesterol [62, 116].

When regulation of cholesterol biosynthesis was assayed in cells in culture, it was demonstrated that a number of transformed cell lines demonstrated an efficient control of cholesterol biosynthesis; i.e. when exogenous lipid is supplied, cholesterol biosynthesis from both acetate and glucose is inhibited. This regulation was first observed by BAILEY [7] in L cells, which were originally derived by treating mouse embryo cells with carcinogens [41]. Regulation of cholesterol biosynthesis has also been demonstrated in HTC cells [144], which originated from a hepatoma, and in sarcoma 180 cells [77]. WATSON has further explored the mechanism of regulation in the HTC hepatoma cell cultures. He found that the removal of serum lipoprotein cholesterol resulted in a 3- to 6-fold stimulation of cellular sterol synthesis from a variety of precursors within 4–6 h. The data indicated that there is a protein synthesis-dependent regulatory mechanism at the level of the enzyme HMG-CoA reductase in the tumor cells in culture [145, 146]. This would suggest that the lack of feedback inhibition observed in tumors *in vivo* is not due to lack of control of this enzyme.

When the regulation of cholesterol biosynthesis was directly compared in the diploid line WI-38 and its transformed derivative WI-38VA13A, some differences were observed between the two cell types. The first indication of this occurred when the incorporation of acetate into cholesterol was compared using cells grown in serum supplemented medium [71]. The incorporation of acetate into cholesterol in the virus-transformed line was found to be greater than that of the normal cells, although in both cells the rate was less than required to satisfy cell sterol requirements. The data suggested that cholesterol biosynthesis in serum-supplemented medium was regulated in both cell types but there was a greater degree of inhibition in the WI-38. This difference was thoroughly examined by ROTHBLAT *et al.* [110] who cultured both cell lines in medium supplemented with delipidized serum protein. Under those conditions, both cell lines biosynthesized cholesterol at approximately the same rate, and intracellular cholesterol levels were similar in both cell types. As increased levels of unesterified cholesterol were added to the medium, both cell lines showed a decrease in *de novo* sterol biosynthesis, with the diploid line exhibiting a greater reduction than that observed in the transformed cells (fig. 1).

In addition to differing in extent of regulation of cholesterol biosynthesis, normal and transformed cells may also differ in the occurence of an additional regulatory point beyond mevalonic acid. This was suggested by a comparison of incorporation of ^{14}C-mevalonic acid into cholesterol in a number of normal and transformed lines, in the presence of exogenous cholesterol, which revealed a lower rate in the diploid fibroblasts as compared to the established lines [69]. A direct comparison of these parameters in the WI-38 cell compared to its transformed deriva-

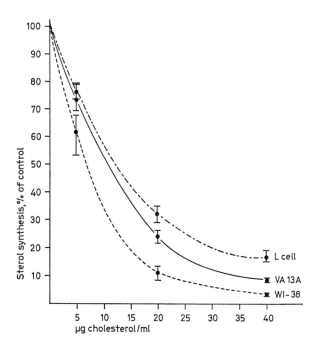

Fig. 1 Effect of increasing cholesterol concentration in the medium on cellular sterol biosynthesis in L cells, WI-38VA13A cells and WI-38 cells (from ROTHBLAT [109], p. 316). Growth medium contained 5 mg/ml delipidized calf serum protein, 20 µg/ml lecithin and the indicated amounts of cholesterol. Sterol biosynthesis was assayed after 24 h by incorporation of ^{14}C acetate into digitonin-precipitable material.

Table V. Stimulation of incorporation of labeled acetate into nonsaponifiable lipids by lipid-extracted serum and its reversal by whole serum[1]

	Relative radioactivity in nonsaponifiable lipids[2]	
	cells treated with 10% lipid-extracted serum	cells treated with 10% lipid-extracted serum followed by 10% whole serum
Human Fibroblasts	166.5	1.5
Hepatoma 7288C	6.3	1.1
HeLa	15.6	4.4
L-2071	2.9	1.7
BRL-62[3]	1.7	0.8

1 These data are taken from the work of AVIGAN *et al.* [5].
2 Ratio of radioactivity in nonsaponifiable lipids of cells, treated as indicated, to the radioactivity in identical cells treated with standard medium containing 10% fetal bovine serum.
3 Established cell line from rat liver.

tive WI-38VA13A indicated that there was only a twofold difference in mevalonate incorporation between the two cells [24]. Thus, there may be a control point beyond mevalonate that is less efficient in the transformed cells, but the difference is not large enough to account for the difference in magnitude of feedback inhibition shown in figure 1.

AVIGAN *et al.* [5] and WILLIAMS and AVIGAN [149] have studied regulation of cholesterol biosynthesis in human skin fibroblast cultures as compared to established cell lines. They also found that there was a lower rate of incorporation of acetate into sterol in the fibroblasts as compared to the transformed cell lines when both were cultured in the presence of serum lipid. In addition, they found that stimulation of cholesterol biosynthesis from both ^{14}C-acetate and tritiated water in lipid-free medium required the presence of solvent-extracted serum protein; under those conditions stimulation in the human fibroblasts was up to 166-fold as opposed to 2- to 15-fold for the established lines (table V). These workers suggested that this stimulation by solvent-extracted serum is a direct effect and not simply due to its ability to promote cell lipid efflux. This would imply a very basic difference in the regulatory mechanism between the normal fibroblasts and established lines in culture, which might be relevant to the differences in dietary feedback observed in tumors *in vivo*. However, a thorough investigation of control of cholesterol biosynthesis in cell cultures requires a controlled comparative study of cholesterol uptake and excretion in the normal and transformed cultured cells (sect. V. E).

Studies are scanty on the regulation of the biosynthesis of other lipid classes. There has been a report of lack of dietary feedback inhibition of fatty acid biosynthesis in hepatomas [82]. In this study, the lack of feedback was demonstrated to not be due to lack of regulation of acetyl-CoA carboxylase synthesis. Efficient regulation of fatty acid biosynthesis has been demonstrated in both diploid fibroblasts [73], as well as in a number of transformed cell lines [77, 149]. The stimulatory effect of delipidized serum proteins described by AVIGAN *et al.* [5] for sterol biosynthesis in fibroblasts was not as pronounced for fatty acids. No thorough comparison of the regulation of fatty acid biosynthesis in normal and transformed cells has been reported.

V. Membranes

A. Evidence for Alteration after Transformation

It has been suggested that tumor cells may contain modified surface membranes which are responsible for their altered growth behavior [91]. Cell adhesiveness may be a primary determinant of cell interactions [76], and an alteration of the surface of malignant cells could result in cells with altered surface properties and thus altered cell behavior. A number of experimental observations of changes in surface properties of tumor cells in culture support this hypothesis. They include loss of contact inhibition of motion [1], appearance of tumor antigens [33, 139], agglutination by plant lectins [22, 72, 120], alteration of membrane-bound enzymes [94, 153] and changes in cell transport [56, 63, 64]. As with other differentiating properties between normal and transformed cells, these surface differences are also not consistent in all cases [40], and the reader is referred to several extensive recent reviews [76, 83, 87]. Nevertheless, the above changes in surface properties strongly indicate that alterations in the cell surface are linked with the process of neoplasia; thus a consideration of studies of the lipids in membranes of normal and transformed cells is warranted.

B. Lipid Composition of Membranes

The most thorough study of lipid composition of membranes in cell cultures compared membranes isolated from uninfected and oncogenic RNA virus converted chick embryo fibroblasts, and membranes isolated from uninfected and oncogenic RNA virus converted parenchyma-like cells [95, 96]. The lipid composition of the cell homogenates were similar. When membranes were isolated into two fractions, A' and B', based on their migration in sucrose gradients, the plasma membrane in the A' fraction isolated from virus converted chick embryo fibroblasts had similar lipid content, increased levels of neutral sugar and decreased sialic acid content, as compared to the A' fractions of the untransformed cells [21]. These changes were not observed after infection with the virus. A similar comparison of membranes of parenchymal cells and their transformed derivatives showed no differences in lipid or sugar composition, but there were alterations in nucleotide phosphatases. Thus, these workers established that virus conversion, which results in a cell greatly altered in shape, cell association and adhesion, can be accomplished without any necessary changes in cell membrane lipid biochemistry. Although no other comparative studies have been reported, data in the literature of several varieties of cultured cells [104, 105, 147] and leukemic cells [37] indicate no consistent alteration in membrane composition associated with tumor cells in culture.

C. Phospholipid Metabolism

There has been one attempt to correlate observed changes in cell transport after transformation to alterations in phospholipid metabolism. CUNNINGHAM [31], and CUNNINGHAM and PARDEE [32] have examined phospholipid patterns in growing, density inhibited and serum-stimulated 3T3 cells. When CUNNINGHAM [31] studied phospholipid turnover after confluency, he found that there was no change in ^{32}P incorporation into phospholipids. However, there was a twofold increase in the rate of incorporation into phosphatidyl choline, a sevenfold decrease in phosphatidyl ethanolamine turnover and a twofold decrease in turnover of phosphatidyl serine after cells reached confluency. Upon stimulation of cell division, there were rapid increases in turnover of all phospholipids which paralleled a stimulation of transport of phosphate. No similar changes were observed in polyma-transformed 3T3 cells (table VI). He

Table VI. Incorporation of [^{32}P] phosphate into individual phospholipids during a 2-hour labeling period[1]

Phospholipid	Percentage of total labeled phospholipids			
	non-confluent 3T3	confluent	non-confluent Py3T3[2]	confluent Py3T3[2]
Phosphatidyl choline	27 ± 3	63 ± 6	15 ± 3	18 ± 2
Phosphatidyl ethanolamine	24 ± 4	4.8 ± 0.4	27 ± 3	18 ± 3
Phosphatidyl serine	42 ± 6	29 ± 3	45 ± 7	57 ± 7
Phosphatidyl inositol	3.6 ± 0.6	1.7 ± 0.4	6 ± 1	3 ± 1
Phosphatidic acid plus cardiolipin	2.4 ± 1	0.6 ± 0.2	5.3 ± 2	2.3 ± 1
Total	99.0	99.1	98.3	98.3

1 These data are taken from the work of CUNNINGHAM [31].
2 Polyoma-transformed derivative of 3T3.

concluded that the changes in phospholipid turnover and transport were related to the serum and density-dependent growth controls which are altered after transformation. Although this type of study is still preliminary, it initiates a promising approach to studies of surface changes in malignancy and indicates that *in vitro* studies on cells in culture can be very useful for these studies.

D. Glycolipid Metabolism

A number of biochemical comparisons have recently been made concerning the glycolipids of normal and transformed cell lines (this topic is reviewed elsewhere in this volume). Interest in glycosphingolipids was stimulated by RAPPORT *et al.* [102] who found accumulations of these compounds in tumors. HAKOMORI and co-workers [57–59] and ROBBINS and co-workers [107, 118] studied glycolipids of cells of hamster and chick origin and their transformed derivatives. These cells have monosialyllactosylceramide (G_{M3})[3] as the predominant glycolipid and the results indicated that the nonmalignant, nontransformed cells consistently contained glycolipids of longer carbohydrate chain length than did the various counterpart transformed or established lines. Moreover, these

3 Nomenclature is that of SVENNERHOLM [133].

changes could be related to density dependent changes in glycolipid content in these cells [57]; i.e. as the cell density of the contact-inhibited cells increases, there is an increase in the more complex neutral glycolipids, especially ceramide trihexoside (CTH).[3] There is a reduction in these glycolipids in transformed cells and no cell density-dependent effects are observed.

A similar pattern of decreases in the complexity of ganglioside patterns has been observed in mouse cells and their transformed derivatives [46, 89]. These cells are able to biosynthesize gangliosides, and BRADY and MORA [20] reported strikingly decreased amounts of gangliosides with longer carbohydrate chains (G_{M1} and G_{D1A})[3] in cells transformed by DNA viruses. These alterations have been linked to deficiency of a key glucosyl transferase in the transformed cells (N-acetylgalactosaminyl transferase) [30, 47]. Ganglioside changes in mouse cells appear not to be cell density dependent [45, 152] and appear to be more closely associated with viral transformation.

Studies complementary to the glycolipid studies have also been reported for glycoproteins [76, 143] and they again fall into the general pattern of decreased complexity of carbohydrate side chains as a result of defective glycosylation in the transformed cells. These studies provide one of the best documented correlations of a chemical change associated with altered surface properties, and thus yield a powerful tool for use in explorations of the mechanisms behind these changes and their relation to the malignant process.

E. Lipid Transport

A number of the above sections indicated the importance of considering lipid uptake and excretion in normal and transformed cells. No direct comparison has been made of lipid transport in the two cell types, but a considerable body of information has been accumulated on lipid uptake by cell cultures. The most extensive studies have been on free fatty acids, which are bound to albumin in serum. Although they account for only less than 10 % of the total lipid of serum, they can, if cell density is not too great, be the sole source of nonsterol lipid for the cells. This is the case both in diploid fibroblasts [70] and in established cell lines [81].

---

3   Nomenclature is that of SVENNERHOLM [133].

The uptake of fatty acids is very rapid, and the mechanism of this uptake has been extensively studied by SPECTOR [125, 126]. There does not seem to be a cellular mechanism to limit free fatty acid accumulation. When the medium contains elevated levels of free fatty acids, intracytoplasmic triglyceride droplets accumulate in the cells. This has been observed in diploid and primary lines [4, 74] and in established lines [49, 81, 90]. Free fatty acids can also be excreted from cells, if fatty acid-poor albumin is present in the medium [128].

Glycerides can also be taken up by cells. Monoglyceride uptake is quite rapid [10], and their uptake can either be mediated by serum lipases or monoglycerides can enter the cell intact [80]. Triglyceride uptake has also been observed in conjunctiva cells and HeLa cells [51], in MBIII cells [9], in L cells [10] and in Ehrlich ascites [127]. Uptake of triglyceride is relatively slow [10]. It may be mediated by hydrolysis if serum enzymes are present in sufficient quantities, but triglycerides can also enter intact [10, 127].

Very few studies have been performed on phospholipid utilization, but experimental evidence indicates that phospholipids can also enter intact into both diploid fibroblasts and established lines. REED [103] has observed transfer of sphingomyelin and lecithin from serum lipoproteins to erythrocytes. PETERSON and RUBIN [97, 98] observed exchange between serum lipoproteins and chick embryo fibroblasts, and BAILEY et al. [9] reported utilization of serum phospholipids by MBIII cells.

Sterol flux has been studied extensively in cell cultures (see review by ROTHBLAT [109]). Influx and efflux of free sterol seems to be a function of the serum lipoproteins present in the growth medium, and exchange between serum and cell lipoproteins seems to be a primary step in the transport of sterol. The cell's primary response to changes in sterol flux is at the level of sterol synthesis. Sterol esters are also taken up, but rate of incorporation is probably slower than with free cholesterol. Most cells then hydrolyze the ester to free sterol and free fatty acid.

Table VII summarizes the studies on lipid uptake reported so far in normal and tumor cells in culture. The primary conclusion of most studies of lipid uptake by cultured cells is that the rate and amount of lipid flux are dependent on the concentration of protein and other lipid in the media, and the density and metabolic state of the cells. Therefore, although the data so far do not point to any striking differences between the two cell types in the area of lipid flux, direct comparative studies under controlled conditions must be conducted.

Table VII. Lipid uptake by various cell lines

Lipid class	Cell type	Reference
Fatty acid	WI-38 (diploid)	HOWARD and KRITCHEVSKY [70]
	Cornea fibroblasts	KLINTWORTH and HIJMANS [74]
	Ehrlich ascites	SPECTOR et al. [130]
	HeLa	GEYER [49]
	L-929	MACKENZIE et al. [81]
Cholesterol	L-5178 (mouse leukemia)	ROTHBLAT et al. [112]
	L-929	BAILEY [6]
	MBIII (lymphoblasts)	BAILEY [6]
	Human aorta fibroblasts	RUTSTEIN et al. [114]
Phospholipids	Chick embryo fibroblasts	PETERSON and RUBIN [97]
	MBIII (lymphoblasts)	BAILEY et al. [9]
Glycerides	HeLa	GEYER and NEIMARK [51]
	MBIII (lymphoblasts)	BAILEY et al. [9]
	L-929	BAILEY et al. [10]
	Ehrlich ascites	SPECTOR and BRENNEMAN [127]

VI. Summary

Many aspects of lipid composition and metabolism in normal and transformed or tumor cells in culture have been considered in this chapter. There seem to be no differences in the composition of the major lipids of the two cell types, although a decrease in phospholipid content has been observed in direct comparisons of normal and transformed cells. Decreases in polyunsaturated fatty acids are noted in most tumor cells in culture, but this seems to be due to a loss of the capability for biosynthesis after prolonged culture. Some diploid fibroblasts require cholesterol for growth but this is not a unique characteristic for all diploid lines. Both cell types seem to synthesize and take up all of the major lipids, and thorough comparative studies have not been done in these areas.

The most important objective is not to find differences between the lipids of normal and tumor cells, but to assess which changes are directly relevant to the behavior of malignant cells *in vivo*, and to determine whether these characteristics are intimately linked to the neoplastic process. Although the relationship between the transformation process *in vitro* and malignancy *in vivo* is not unequivocally established, the cell culture system seems to be particularly useful in approaching these problems. This is evident in the consideration of the lipid ether content of cells, and the turnover of phospholipids. In both cases, changes in these characteristics appear to be related to dedifferentiation and adaptation to rapid growth, rather than to be essential to tumor cells.

One of the most promising approaches in the study of lipids in tumor cells lies in the studies of membrane lipids. Although the phospholipid and sterol composition of membranes of tumor cells in culture seems unchanged, there are definite alterations in glycolipids, and interesting changes in turnover of phospholipids. These might be linked to changes in membrane characteristics that occur in malignant cells.

The other interesting area concerns the control of metabolism in the two cell types. Definite differences seem to exist in the regulation of cholesterol biosynthesis in transformed cells. These differences must be explored more fully to link them directly to the malignant process *in vivo*, and particularly to eliminate the possibility that they occur simply because established cell lines can be maintained in culture for more generations. Moreover, the link between surface properties, transport rates, and control of metabolism must be thoroughly explored, because changes in one can obviously affect observations in the others.

One tool that might be employed in these studies is that of cell genetics; methods for induction and selection of mutants in animal cell lines have recently been developed which can be applied to the study of lipid metabolism in these cells and to the comparison of lipids in the two cell types. Finally, more emphasis must be placed on the characteristics of differentiated cells. Although most prior cell culture studies were performed on relatively differentiated cells, cell lines are now being established which retain differentiated functions in culture. These will surely be useful in the examination of changes which occur during neoplasia, which is most certainly a phenomenon related to differentiation.

VII. References

1 ABERCROMBIE, H. N. and HEAYSMAN, J. E. M.: Observations on the social behavior of cells in tissue culture. II. 'Monolayering' of fibroblasts. Exp. Cell Res. *6:* 293–306 (1954).
2 AGRANOFF, B. W. and HAJRA, A. K.: The acyl-dihydroxyacetone phosphate pathway for glycerolipid biosynthesis in mouse liver and Ehrlich ascites tumor cells. Proc. nat. Acad. Sci., Wash. *68:* 411–415 (1971).

3 ANDERSON, R. E.; CUMMING, R. B.; WALTON, M., and SNYDER, F.: Lipid metabolism in cells grown in tissue culture: O-alkyl, O-alk-1-enyl, and acyl moieties of L-M cells. Biochim. biophys. Acta *176:* 491–501 (1969).

4 APFFEL, C. A. and BARKER, J. R.: Lipid droplets in the cytoplasm of malignant cells. Cancer *17:* 176–184 (1964).

5 AVIGAN, J.; WILLIAMS, C. D., and BLASS, J. P.: Regulation of sterol synthesis in human skin fibroblast cultures. Biochim. biophys. Acta *218:* 381–384 (1970).

6 BAILEY, J. M.: Lipid metabolism in cultured cells. I. Factors affecting cholesterol uptake. Proc. Soc. exp. Biol. Med. *107:* 30–35 (1961).

7 BAILEY, J. M.: Cellular lipid nutrition and lipid transport; in ROTHBLAT and KRITCHEVSKY Lipid metabolism in tissue culture cells. Wistar Inst. Symp. Monogr., No. 6, pp. 85–109 (Wistar Institute Press, Philadelphia 1967).

8 BAILEY, J. M. and DUNBAR, L. M.: Essential fatty acid requirement of cells in tissue culture. Exp. molec. Path. *18:* 142–161 (1973).

9 BAILEY, J. M.; GEY, G. O., and GEY, M. K.: Utilization of serum lipids by cultured mammalian cells. Proc. Soc. exp. Biol. Med. *100:* 686–691 (1959).

10 BAILEY, J. M.; HOWARD, B. V., and TILLMAN, S. F.: Lipid metabolism in cultured cells. XI. Utilization of serum triglycerides. J. biol. Chem. *248:* 1240–1247 (1973).

11 BAILEY, P. J. and KELLER, D.: The deposition of lipids from serum into cells cultured *in vitro*. Atherosclerosis *13:* 333–343 (1971).

12 BEN-PORAT, T. and KAPLAN, A. S.: Phospholipid metabolism of Herpes virus-infected and uninfected rabbit kidney cells. Virology *215:* 252–264 (1971).

13 BERLINER, D. L.; SWIM, H. E., and DOUGHERTY, T. F.: Synthesis of cholesterol by a strain of human uterine fibroblasts propagated *in vitro*. Proc. Soc. exp. Biol. Med. *99:* 51–53 (1958).

14 BERWALD, Y. and SACHS, L.: *In vitro* cell transformation with chemical carcinogens. Nature, Lond. *200:* 1182–1184 (1963).

15 BLAIR, C. D. and BRENNAN, P. J.: Effect of Sendai virus infection on lipid metabolism in chick embryo fibroblasts. J. Virol. *9:* 813–822 (1972).

16 BOLE, G. G. and CASTOR, C. W.: Characterization of lipid constituents of human 'fibroblasts' cultivated *in vitro*. Proc. Soc. exp. Biol. Med. *115:* 174–179 (1964).

17 BOREK, C.: Neoplastic transformation *in vitro* of a clone of adult liver epithelial cells into hepatoma-like cells under conditions of nutritional stress. Proc. nat. Acad. Sci., Wash. *69:* 956–959 (1972).

18 BOREK, C.; HIGASHINO, S., and LOWENSTEIN, W. R.: Intercellular communication and tissue growth. IV. Conductance of membrane junctions of normal and cancerous cells in culture. J. membr. Biol. *1:* 274–293 (1969).

19 BOREK, C. and SACHS, L.: *In vitro* cell transformation by irradiation. Nature, Lond. *210:* 276–278 (1966).

20 BRADY, R. O. and MORA, P. T.: Alteration in ganglioside pattern and synthesis in SV_{40} and polyoma virus-transformed mouse cell lines. Biochim. biophys. Acta *218:* 308–319 (1970).

21 BRAUN, A. C.: On the origin of cancer cells. Amer. Sci. *58:* 307–320 (1970).

22 BURGER, M. M.: Forssman antigen exposed on surface membrane after viral transformation. Nature, Lond. 231: 125–126 (1971).
23 BUTEL, J. S.; TEVETHIA, S. S., and MELNICK, J. L.: Oncogenicity and cell transformation by papovavirus SV_{40} – the role of the viral genome. Adv. Cancer Res. 15: 1–55 (1972).
24 BUTLER, J. D. and BAILEY, J. M.: Personal commun. (1972).
25 CARREL, A.: The new cytology. Science 73: 297–303 (1931).
26 CHANG, R. S.: Biochemistry of human cells of normal and tumor origin. Nat. Cancer Inst. Monogr. 7: 249–259 (1962).
27 CHAO, F.; ENG, L. F., and GRIFFIN, A.: Compositional differences of lipid in transformed hamster fibroblasts. Biochim. biophys. Acta 260: 197–202 (1972).
28 CRISTOFALO, V. J.: Personal commun. (1964).
29 CULLING, C. A.; REID, D. E.; TRUEMAN, L. S., and DUNN, W. L.: A simple method for the isolation of viable epithelial cells of the gastrointestinal tract. Proc. Soc. exp. Biol. Med. 142: 434–438 (1973).
30 CUMAR, F. A.; BRADY, R. O.; KOLODNY, E. H.; McFARLAND, V. W., and MORA, P. T.: Enzymatic block in the synthesis of gangliosides in DNA virus-transformed tumorigenic mouse cell lines. Proc. nat. Acad. Sci., Wash. 67: 757–764 (1970).
31 CUNNINGHAM, D. D.: Changes in phospholipid turnover following growth of 3T3 mouse cells to confluency. J. biol. Chem. 247: 2464–2470 (1972).
32 CUNNINGHAM, D. D. and PARDEE, A. B.: Transport changes rapidly initiated by serum addition to 'contact-inhibited' 3T3 cells. Proc. nat. Acad. Sci., Wash. 64: 1049–1056 (1969).
33 DEFENDI, V.: Effect of SV_{40} immunization on growth of transplantable SV_{40} and polyoma virus tumors in hamsters. Proc. Soc. exp. Biol. Med. 113: 12–16 (1962).
34 DEUEL, H. J. jr.: The lipids. Biochemistry (vol. II), p. 521 (Lippincott, Philadelphia 1951).
35 DIRINGER, H.; MARGGRAF, W. D.; KOCH, M. A., and ANDERER, F. A.: Evidence for a new biosynthetic pathway for sphingomyelin in SV_{40} transformed mouse cells. Biochem. biophys. Res. Commun. 47: 1345–1352 (1972).
36 DIRKSEN, T. R.; MARINETTI, G. V., and PECK, W. A.: Lipid metabolism in bone and bone cells. I. The in vitro incorporation of ^{14}C glycerol and ^{14}C glucose into lipids of bone and bone cell cultures. Biochim. biophys. Acta 202: 67–79 (1970).
37 DODS, R. F.; ESSNER, E., and BARCLAY, M.: Isolation and characterization of plasma membranes from an L-asparaginase-sensitive strain of leukemia cells. Biochem. biophys. Res. Commun. 46: 1074–1081 (1972).
38 DONISCH, V. and ROSSITER, R. J.: Metabolism of phospholipids in Ehrlich ascites tumor. Cancer Res. 25: 1463–1467 (1965).
39 DUNBAR, L. M. and BAILEY, J. M.: Personal commun. (1972).
40 EAGLE, H.; FOLEY, G. E.; KOPROWSKI, H.; LAZARUS, H.; LEVINE, E. M., and ADAMS, R. A.: Growth characteristics of virus-transformed cells. J. exp. Med. 131: 863–879 (1970).
41 EARLE, W. R.: Production of malignancy in vitro. IV. The mouse fibroblast

cultures and changes seen in the living cells. J. nat. Cancer Inst. *4:* 165–212 (1943).
42 EICHBERG, J.; HAUSER, G., and SHEIN, H. M.: Polyphosphoinositides in normal and neoplastic rodent astrocytes. Biochem. biophys. Res. Commun *45:* 43–50 (1971).
43 EVANS, V. J.; BRYANT, J. C.; KERR, H. A., and SCHELLING, E. L.: Chemically defined media for cultivation of long-term cell strains from four mammalian species. Exp. Cell Res. *36:* 439–474 (1965).
44 FIGARD, P. H. and LEVINE, A. S.: Incorporation of labelled precursors into lipids of tumors induced by Rous sarcoma virus. Biochim. biophys. Acta *125:* 428–434 (1966).
45 FISHMAN, P. H.; BASSIN, R., and MCFARLAND, V.: Altered ganglioside biosynthesis in mouse cells during transformation by murine sarcoma virus. Fed. Proc. *32:* 1348, abstr. (1973).
46 FISHMAN, P. H.; BRADY, R. O., and MORA, P. T.: Altered glycolipid metabolism related to viral transformation of established mouse cell lines; in WOOD Tumor lipids, biochemistry and metabolism, pp. 250–268 (Amer. Oil Chem. Soc. Press, Champaign, Ill. 1973).
47 FISHMAN, P. H.; MCFARLAND, V. W.; MORA, P. T., and BRADY, R. O.: Ganglioside biosynthesis in mouse cells. Glucosyl transferase activities in normal and virally transformed lines. Biochem. biophys. Res. Commun. *48:* 48–57 (1972).
48 GERSCHENSON, L. E.; MEAD, J. G.; HARARY, H., and HAGGERTY, D. F.: Studies on the effects of essential fatty acids on growth rate, fatty acid composition, oxidative phosphorylation and respiratory control of HeLa cells in culture. Biochim. biophys. Acta *131:* 42–49 (1967).
49 GEYER, R. P.: Uptake and retention of fatty acids by tissue culture cells; in ROTHBLAT and KRITCHEVSKY Lipid metabolism in tissue culture cells. Wistar Inst. Symp. Monogr., No. 6, pp. 33–44 (Wistar Institute Press, Philadelphia 1967).
50 GEYER, R. P.; BENNETT, A., and ROHR, A.: Fatty acids of the triglycerides and phospholipids of HeLa cells and strain L fibroblasts. J. Lipid Res. *3:* 80–83 (1962).
51 GEYER, R. P. and NEIMARK, J. M.: Triglyceride utilization by human HeLa and conjunctiva cells in tissue culture. Amer. J. clin. Nutr. *7:* 86–90 (1959).
52 GOLDFARB, S. and PITOT, H. C.: The regulation of β-hydroxy-β-methylglutaryl coenzyme A reductase in Morris hepatomas 5123C, 7800 and 9618A. Cancer Res. *31:* 1879–1882 (1971).
53 GOTTFRIED, E. L.: Lipids of human leukocytes; relation to cell type. J. Lipid Res. *8:* 321–327 (1967).
54 GOTTFRIED, E. L.: Lipid patterns in human leukocytes maintained in long-term culture. J. Lipid Res. *12:* 531–537 (1971).
55 GOTTFRIED, E. L.; CEDERQVIST, L. L., and DANES, B. S.: Lipids of fresh and cultured cells derived from amniotic fluid. Biochim. biophys. Acta *231:* 250–253 (1971).
56 GRIFFITHS, J. B.: The effect of cell population density on nutrient uptake and

cell metabolism: a comparative study of human diploid and heteroploid cell lines. J. Cell Sci. 10: 515–524 (1972).

57 HAKOMORI, S.: Cell density dependent changes of glycolipid concentrations in fibroblasts, and loss of this response in virus transformed cells. Proc. nat. Acad. Sci., Wash. 67: 1741–1747 (1970).

58 HAKOMORI, S. and MURAKAMI, W. T.: Glycolipids of hamster fibroblasts and derived malignant transformed cell lines. Proc. nat. Acad. Sci., Wash. 59: 254–261 (1968).

59 HAKOMORI, S.; SAITO, T., and VOGT, P. K.: Transformation by Rous sarcoma virus. Effects on cellular glycolipids. Virology 44: 609–621 (1971).

60 HAM, R. G.: Albumin replacement by fatty acids in clonal growth of mammalian cells. Science 140: 802–803 (1963).

61 HARARY, I.; GERSCHENSON, L. E.; HAGGERTY, D. F.; DESMOND, W., and MEAD, J. F.: Fatty acid metabolism and function of cultured heart and HeLa cells; in ROTHBLAT and KRITCHEVSKY Lipid metabolism in tissue culture cells. Wistar Inst. Symp. Monogr., No. 6, pp. 17–31 (Wistar Institute Press, Philadelphia 1967).

62 HARRY, D. S.; MORRIS, H. P., and MCINTYRE, N.: Cholesterol biosynthesis in transplantable hepatomas: evidence for impairment of uptake and storage of dietary cholesterol. J. Lipid Res. 12: 313–317 (1971).

63 HATANKA, M. and HANAFUSA, H.: Analysis of a functional change in membrane in the process of cell transformation by Rous sarcoma virus; alteration in the characteristics of sugar transport. Virology 41: 647–652 (1970).

64 HATANKA, M.; HEUBNER, R. T., and GILDEN, R. V.: Alterations in the characteristics of sugar uptake by mouse cells transformed by murine sarcoma viruses. J. nat. Cancer Inst. 43: 1091–1096 (1969).

65 HAY, R. J. and STREHLER, B. L.: The limited growth span of cell stains isolated from the chick embryo. Exp. Geront. 2: 123–135 (1967).

66 HAYFLICK, L. and MOORHEAD, P. S.: The serial cultivation of human diploid cell strains. Exp. Cell Res. 25: 585–621 (1961).

67 HOLMES, R.; HELMS, J., and MERCER, G.: Cholesterol requirement of primary diploid human fibroblasts. J. Cell Biol. 42: 262–271 (1969).

68 HOWARD, B. V. and BAILEY, J. M.: Ether lipids, α-glycerol phosphate dehydrogenase, and growth rate in tumors and cultured cells. Cancer Res. 32: 1533–1538 (1972).

69 HOWARD, B. V.; BUTLER, J. D., and BAILEY, J. M.: Lipid metabolism in normal and tumor cells in culture; in WOOD Tumor lipids, biochemistry and metabolism, pp. 200–214 (Amer. Oil Chem. Soc. Press, Champaign, Ill. 1973).

70 HOWARD, B. V. and KRITCHEVSKY, D.: The source of cellular lipid in the human diploid cell stain WI-38. Biochim. biophys. Acta 187: 293–301 (1969).

71 HOWARD, B. V. and KRITCHEVSKY, D.: The lipids of normal diploid (WI-38) and SV_{40}-transformed human cells. Int. J. Cancer 4: 393–402 (1969).

72 INBAR, M. and SACHS, L.: Interaction of the carbohydrate-binding protein concanavalin A with normal and transformed cells. Proc. nat. Acad. Sci., Wash. 63: 1418–1425 (1969).

73 JACOBS, R. A.; SLY, W. S., and MAJERUS, P. W.: The regulation of fatty acid biosynthesis in human skin fibroblasts. J. biol. Chem. 248: 1268–1276 (1973).
74 KLINTWORTH, G. K. and HIJMANS, J. C.: The induction by serum of lipid storage in cells of the cornea grown in culture. Amer. J. Path. 58: 403–418 (1970).
75 KOHN, G.; DIEHL, V.; MELLMAN, W. J.; HENLE, W., and HENLE, G.: C-group chromosome marker in long term leukocyte cultures. J. nat. Cancer Inst. 41: 795–804 (1968).
76 KRAEMER, P. M.: Complex carbohydrates of mammalian cells in culture; in ROTHBLAT and CRISTOFALO Growth nutrition and metabolism of cells in culture, vol. 1, pp. 371–426 (Academic Press, New York 1972).
77 LENGLE, E. and SMITH, J. L.: The effect of culturing Sarcoma 180 cells with lipid free serum. Fed. Proc. 28: 688, abstr. (1969).
78 LEWIS, M. R. and LEWIS, W. H.: Cultivation of tissue from chick embryos in solutions of NaCl, $CaCl_2$, KCl and $NaHCO_3$. Anat. Rec. 5: 277–293 (1911).
79 LUCAS, D. O.; SHOHET, S. B., and MERLER, E.: Changes in phospholipid metabolism which occur as a consequence of mitogenic stimulation of lymphocytes. J. Immunol. 106: 768–772 (1971).
80 LYNCH, R. D. and GEYER, R. P.: Uptake of rac-glycerol-1-oleate and its utilization for glycolipid synthesis by strain L fibroblasts. Biochim. biophys. Acta 260: 547–557 (1972).
81 MACKENZIE, C. G.; MACKENZIE, J. B., and REISS, O. K.: Regulation of cell lipid metabolism and accumulation. V. Quantitative and structural aspects of triglyceride accumulation caused by lipogenic substances; in ROTHBLAT and KRITCHEVSKY Lipid metabolism in tissue culture cells. Wistar Inst. Symp. Monogr., No. 6, pp. 63–81 (Wistar Institute Press, Philadelphia 1967).
82 MAJERUS, P. W.; JACOBS, R.; SMITH, M. B., and MORRIS, H. P.: The regulation of fatty acid biosynthesis in rat hepatomas. J. biol. Chem. 243: 3588–3595 (1968).
83 MARTINEZ-POLOMO, A.: The surface coat of animal cells. Int. Rev. Cytol. 29: 29–75 (1970).
84 MCINTOSH, R.; PAYNE, S., and RUSSELL, W. C.: Studies on lipid metabolism in cells infected with adenovirus. J. gen. Virol. 10: 251–265 (1971).
85 MELLMAN, W. J. and CRISTOFALO, V. J.: Human diploid cell cultures: their usefulness in the study of genetic variations in metabolism; in ROTHBLAT and CRISTOFALO Growth nutrition and metabolism of cells in culture, vol. 1, pp. 327–369 (Academic Press, New York 1972).
86 MENKES, J. H.: Lipid metabolism of brain tissue in culture. Lipids 7: 135–141 (1972).
87 MEYER, G.: Viral genome and oncogenic transformation: nuclear and plasma membrane events. Adv. Cancer Res. 4: 71–153 (1971).
88 MONTAGNIER, L.: Factors controlling the multiplication of untransformed and transformed BHK21 cells under various environmental conditions; in WOLSTENHOLME and KNIGHT Growth control in cultured cells, pp. 33–41 (Churchill, Livingstone, London, 1971).
89 MORA, P. T.; BRADY, R. O.; BRADLEY, R. M., and MCFARLAND, V. W.: Gang-

liosides in DNA virus-transformed and spontaneously transformed tumorigenic mouse cell lines. Proc. nat. Acad. Sci., Wash. *63:* 1290–1296 (1969).

90 MOSKOWITZ, M. S.: Fatty acid-induced steatosis in monolayer cell cultures; in ROTHBLAT and KRITCHEVSKY Lipid metabolism in tissue culture cells. Wistar Inst. Symp. Monogr., No. 6, pp. 49–59 (Wistar Institute Press, Philadelphia 1967).

91 PARDEE, A. B.: Cell division and a hypothesis of cancer. Nat. Cancer Inst. Monogr. *14:* 7–18 (1964).

92 PASTERNAK, R. A. and BERGERON, J. J. M.: Turnover of mammalian phospholipids – stable and unstable components in neoplastic mast cells. Biochem. J. *119:* 473–480 (1970).

93 PASTERNAK, G. A. and FRIEDRICKS, B.: Turnover of mammalian phospholipids – rates of turnover and metabolic heterogeneity in cultured human lymphocytes and in tissues of healthy, starved and vitamin A-deficient rats. Biochem. J. *119:* 481–488 (1970).

94 PEERY, C. V.; JOHNSON, G. S., and PASTAN, I.: Adenyl cyclases in normal and transformed fibroblasts in tissue culture. J. biol. Chem. *246:* 5785–5790 (1971).

95 PERDUE, J. F.; KLETZIEN, R., and MILLER, K.: The isolation and characterization of plasma membrane from cultured cells. I. The chemical composition of membrane isolated from uninfected and oncogenic RNA virus-converted chick embryo fibroblasts. Biochim. biophys. Acta *249:* 419–434 (1971).

96 PERDUE, J. F.; KLETZIEN, R.; MILLER, K.; PREDMORE, G., and WRAY, V. L.: The isolation and characterization of plasma membranes from cultured cells. II. The chemical composition of membrane isolated from uninfected and oncogenic RNA virus converted parenchyma-like cells. Biochim. biophys. Acta *249:* 435–453 (1971).

97 PETERSON, J. A. and RUBIN, H.: The exchange of phospholipids between cultured chick embryo fibroblasts and their growth medium. Exp. Cell Res. *58:* 365–378 (1969).

98 PETERSON, J. A. and RUBIN, H.: The exchange of phospholipids between cultured chick embryo fibroblasts as observed by autoradiography. Exp. Cell Res. *60:* 383–392 (1970).

99 QUIGLEY, J. P.; RIFKIN, D. P., and REICH, E.: Phospholipid composition of Rous sarcoma virus, host cell membranes and other enveloped RNA viruses. Virology *46:* 106–116 (1971).

100 RABINOWITZ, Z. and SACHS, L.: Reversion properties in cells transformed by polyoma virus. Nature, Lond. *220:* 1203–1206 (1968).

101 RABINOWITZ, Z. and SACHS, L.: The formation of variants with a reversion of properties of transformed cells. VIII. *In vitro* limited life span of variants isolated from tumors. Int. J. Cancer *10:* 607–612 (1972).

102 RAPPORT, M. M.; GRAF, L.; SKIPSKI, V. P., and ALONZO, N. F.: Immunochemical studies of organ and tumor lipids. VI. Isolation and properties of cytolipin H. Cancer *12:* 438–445 (1959).

103 REED, F.: Phospholipid exchange between plasma and erythrocytes in man and the dog. J. clin. Invest. *47:* 749–760 (1968).

104 RENKONEN, O.; GAHMBERG, C. G.; SIMONS, K., and KAARIAINEN, L.: The lipids

of the plasma membranes and endoplasmic reticulum from cultured baby hamster kidney cells. Biochim. biophys. Acta 255: 66–78 (1972).

105 RENKONEN, O.; KAARIAINEN, L.; SIMONS, K., and GAHMBERG, C. G.: The lipid class composition of Semliki forest virus and plasma membranes of the host cell. Virology 46: 318–326 (1971).

106 RESCH, K. and FERBER, E.: Phospholipid metabolism of stimulated lymphocytes. Effects of phytohemagglutinin, concanavalin A and anti-immunoglobulin serum. Europ. J. Biochem. 27: 153–161 (1972).

107 ROBBINS, P. W. and MACPHERSON, I.: Control of glycolipid synthesis in a cultured hamster cell line. Nature, Lond. 229: 569–570 (1971).

108 ROTHBLAT, G. H.: Lipid metabolism in tissue culture cells. Adv. Lipid Res. 7: 135–163 (1969).

109 ROTHBLAT, G. H.: Cellular sterol metabolism; in ROTHBLAT and CRISTOFALO Growth, nutrition and metabolism of cells in culture, vol. 1, pp. 297–326 Academic Press, New York 1972).

110 ROTHBLAT, G. H.; BOYD, R., and DEAL, C.: Cholesterol biosynthesis in WI-38 and WI-38 VA13A tissue culture cells. Exp. Cell Res. 67: 436–440 (1971).

111 ROTHBLAT, G. H.; BUCHKO, M. K., and KRITCHEVSKY, D.: Cholesterol uptake by L-5178Y tissue culture cells: studies with delipidized serum. Biochim. biophys. Acta 164: 327–338 (1968).

112 ROTHBLAT, G. H.; HARTZELL, R. W.; MAILHE, H., and KRITCHEVSKY, D.: The uptake of cholesterol by L-5178Y tissue culture cells: studies with free sterol. Biochim. biophys. Acta 116: 133–145 (1966).

113 ROTHBLAT, G. H.; HARTZELL, R.; MAILHE, H., and KRITCHEVSKY, D.: Cholesterol metabolism in tissue culture cells; in ROTHBLAT and KRITCHEVSKY Lipid metabolism in tissue culture cells. Wistar Inst. Symp. Monogr., No. 6, pp. 129–149 (Wistar Institute Press, Philadelphia 1967).

114 RUTSTEIN, D. D.; INGENITO, E. F.; CRAIG, J. M., and MARTINELL, M.: Effects of linoleic and stearic acids on cholesterol-induced lipoid deposition in human aortic cells in tissue culture. Lancet i: 545–552 (1958).

115 RYTTER, D. J. and CORNATZER, W. E.: Phospholipid metabolism in cells in culture. Lipids 7: 142–145 (1972).

116 SABINE, J. R.: Defective control of cholesterol metabolism and the development of liver cancer; in WOOD Tumor lipids, biochemistry and metabolism, pp. 21–33 (Amer. Oil Chem. Soc. Press, Champaign, Ill. 1973).

117 SABINE, J. R.; ABRAHAM, S., and CHAIKOFF, I. L.: Control of lipid metabolism in hepatomas: insensitivity of rate of fatty acid and cholesterol synthesis by mouse hepatoma BW 7756 to fasting and feedback control. Cancer Res. 27: 793–799 (1967).

118 SAKIYAMA, H.; GROSS, S. K., and ROBBINS, P. W.: Glycolipid synthesis in normal and virus-transformed hamster cell lines. Proc. nat. Acad. Sci., Wash. 69: 872–876 (1972).

119 SAMBROOK, J.: Transformation by Polyoma virus and Simian virus 40. Adv. Cancer Res. 16: 141–172 (1972).

120 SHOHAM, J. and SACHS, L.: Differences in the binding of fluorescent concana-

valin A to the surface membranes of normal and transformed cells. Proc. nat. Acad. Sci., Wash. *69:* 2479–2482 (1972).
121 SIPERSTEIN, M. D. and FAGAN, V. M.: Feedback control of mevalonate synthesis by dietary cholesterol. J. biol. Chem. *241:* 602–609 (1966).
122 SIPERSTEIN, M. D.; FAGAN, V. M., and MORRIS, H. P.: Further studies on the deletion of the cholesterol feedback system in hepatomas. Cancer Res. *26:* 7–11 (1966).
123 SNYDER, F. and WOOD, R.: The occurrence and metabolism of alkyl and alk-1-enyl ethers of glycerol in transplantable rat and mouse tumors. Cancer Res. *28:* 972–978 (1968).
124 SNYDER, F. and WOOD, R.: Alkyl and alk-1-enyl ethers of glycerol in lipids from normal and neoplastic human tissues. Cancer Res. *29:* 251–255 (1969).
125 SPECTOR, A. A.: Metabolism of free fatty acids. Progr. biochem. Pharmacol., vol. 6, pp. 130–176 (Karger, Basel 1971).
126 SPECTOR, A. A.: Fatty acid, glyceride and phospholipid metabolism; in ROTHBLAT and CRISTOFALO Growth, nutrition and metabolism of cells in culture, vol. 1, pp. 257–296 (Academic Press, New York 1972).
127 SPECTOR, A. A. and BRENNEMAN, D. E.: Uptake of very low density lipoprotein triglycerides by Ehrlich ascites tumor cells. Fed. Proc. *32:* 2587, abstr. (1973).
128 SPECTOR, A. A. and STEINBERG, D.: Release of free fatty acids from Ehrlich ascites tumor cells. J. Lipid Res. *7:* 649–656 (1966).
129 SPECTOR, A. A. and STEINBERG, D.: Turnover and utilization of esterified fatty acids in Ehrlich ascites tumor cells. J. biol. Chem. *242:* 3057–3062 (1967).
130 SPECTOR, A. A.; STEINBERG, D., and TANAKA, A.: Uptake of free fatty acids by Ehrlich ascites tumor cells. J. biol. Chem. *240:* 1032–1041 (1965).
131 STOKER, M. G. P.; HOLLEY, R. W., and KIERNAN, J. A.: Studies of serum factors required by 3T3 and SV3T3; in WOLSTENHOLME and KNIGHT Growth control in cell cultures, pp. 3–10 (Churchill, Livingstone, London 1971).
132 STOKER, M. and MCPHERSON, I.: Studies on transformation of hamster cells by polyoma virus *in vitro*. Virology *14:* 359–370 (1961).
133 SVENNERHOLM, L.: Chromatographic separation of human brain gangliosides. J. Neurochem. *10:* 613–623 (1963).
134 SWIM, H. E. and PARKER, R. F.: Culture characteristics of human fibroblasts propagated serially. Amer. J. Hyg. *66:* 235–243 (1957).
135 TAKAOKA, T. and KATSUTA, H.: Long-term cultivation of mammalian cell strains in protein and lipid-free chemically-defined synthetic media. Exp. Cell Res. *67:* 295–300 (1971).
136 TEMIN, H. M.: Malignant transformation of cells by viruses. Perspect. biol. Med. *14:* 11–26 (1970).
137 TEMIN, H. M.; PIERSON, R. W. jr., and DULAK, A. C.: The role of serum in the multiplication of avian and mammalian cells in culture; in ROTHBLAT and CRISTOFALO Growth, nutrition and metabolism of cells in culture, vol. 1, pp. 50–82 (Academic Press, New York 1972).
138 TEMIN, H. M. and RUBIN, H.: Characteristics of an assay for Rous sarcoma virus and Rous sarcoma cells in tissue culture. Virology *6:* 669–688 (1958).
139 TEVETHIA, S. S.; KATZ, M., and RAPP, F.: New surface antigen in cells trans-

formed by Simian papova virus SV_{40}. Proc. Soc. exp. Biol. Med. *119:* 896–901 (1965).
140 TIJO, J. H. and PUCK, T. T.: Genetics of somatic mammalian cells. II. Chromosomal constitution of cells in tissue culture. J. exp. Med. *108:* 259–268 (1958).
141 TODARO, G. and GREEN, H.: Quantitative studies of the growth of mouse embryo cells in culture and their development into established lines. J. Cell Biol. *17:* 299–313 (1963).
142 TSAO, S. S. and CORNATZER, W. E.: Chemical composition of subcellular particles from cultured cells of human tissues. Lipids *2:* 41–46 (1967).
143 WARREN, L.; FUHRER, J. P., and BUCK, C. A.: Surface glycoproteins of cells before and after transformation by oncogenic viruses. Fed. Proc. *32:* 80–85 (1973).
144 WATSON, J. A.: Regulation of lipid metabolism in *in vitro* cultured minimal deviation hepatoma 7288C. Lipids. *7:* 146–155 (1972).
145 WATSON, J. A.: Regulation of HMG-CoA reductase activity in HTC cells. Fed. Proc. *32:* 480, abstr. (1973).
146 WATSON, J. A.: Regulation of cholesterol synthesis in HTC cells (minimal deviation hepatoma 7288C); in WOOD Tumor lipids, biochemistry and metabolism, pp. 34–53 (Amer. Oil Chem. Soc. Press, Champaign, Ill. 1973).
147 WEINSTEIN, D. B.: The lipid composition of the surface membrane of the L-cell; in MANSON Biological properties of mammalian surface membrane. Wistar Inst. Symp. Monogr., No. 8, pp. 17–21 (Wistar Institute Press, Philadelphia 1968).
148 WHITE, H. B.; GALLI, C., and PAOLETTI, R.: Brain recovery from essential fatty acid deficiency in developing rats. J. Neurochem. *18:* 869–882 (1971).
149 WILLIAMS, C. D. and AVIGAN, J.: *In vitro* effects of serum proteins and lipids on lipid synthesis in human skin fibroblasts and leukocytes grown in culture. Biochim. biophys. Acta *260:* 413–423 (1972).
150 YAMAKAWA, T.; UETA, N., and IRIE, R.: Biochemistry of lipids of neoplastic tissue. I. Lipid composition of ascites tumor cells of mice. Jap. J. exp. Med. *32:* 289–296 (1962).
151 YAU, T. M. and WEBER, M. J.: Changes in acyl group composition of phospholipids from chicken embryonic fibroblasts after transformation by Rous sarcoma virus. Biochem. biophys. Res. Commun. *49:* 114–120 (1972).
152 YOGEESWARAN, G.; SHEININ, R.; WHERRETT, J. R., and MURRAY, R. K.: Studies on the glycosphingolipids of normal and virally transformed 3T3 mouse fibroblasts. J. biol. Chem. *247:* 5146–5158 (1972).
153 YOSHIKAWA-FUKADA, M. and NOJIMA, T.: Biochemical characteristics of normal and virally transformed mouse cell lines. J. cell. Physiol. *80:* 421–430 (1972).

Authors' address: BARBARA V. HOWARD and WILLIAM J. HOWARD, Clinical Research Center, Philadelphia General Hospital, *Philadelphia, PA 19104* (USA)

Fucolipids and Blood Group Glycolipids in Normal and Tumor Tissue[1]

SEN-ITIROH HAKOMORI

Department of Pathobiology, School of Public Health and Department of Microbiology, School of Medicine, University of Washington, Seattle, Wash.

Contents

I. Introduction 167
II. Fucolipids as a New Glycolipid Class and Cell Surface Marker 169
III. Fucolipids as a Blood Group Hapten of Human Erythrocyte Membrane 170
IV. Fucolipids of Gastrointestinal Tract and Glandular Tissues 175
V. Fucolipids of Tumor Tissue 177
    A. Blood Group A and B Fucolipids and Glycoproteins in Tumors 180
        1. Deletion of A and B Determinants 180
        2. Blocked Synthesis of A and B Determinants in Tumors 182
    B. Lewis Fucolipids in Tumors 185
    C. Presence of Incompatible Blood Group Antigens in Human Tumors 186
    D. Fucolipid Changes in Transformants *in vitro* 187
    E. Relation of 'Carcinoembryonic' Antigen and Blood Group Substances 188
VI. Epilogue 188
VII. References 190

I. Introduction

Study of contrasting properties of normal and tumor cell membranes has become a major topic in biochemical oncology, as the property changes of cell surface membranes occurring during the process of malignant transformation are now regarded as the central mechanism for establishing 'malignancy'. Loss of 'contact inhibition', decreased inter-

---

[1] The author's studies cited in this article are supported by National Cancer Institute grants CA10909 and CA12710 and by American Cancer Society grant BC-9B:C.

cellular linkages and cellular adhesiveness, increased nutrient transport, presence of tumor-specific transplantation antigen, and other immunogenic alterations of cell surface have been regarded as the basis for the common properties of tumor cells, such as tumorigenicity, metastasis, and uncontrolled cell growth. There is increasing evidence that the plasma membranes of tumor cells are profoundly different from normal cells; incomplete synthesis of membrane glycolipids [3, 4, 9, 11, 12, 16, 17, 35, 43, 44, 47, 54, 61, 75, 86], enhanced agglutinability by various phytoagglutinins [7, 57, 68, 99] and by anti-glycolipid antisera [41], and the change of chemical composition in the glycoprotein [5–7, 10, 30, 100, 101, 104] have been particularly noticeable. The enzymic basis of these changes in terms of activities of glycosyltransferase and hydrolases has been elucidated [12, 16, 30, 31, 61]. Significantly, a cell contact-dependent enhancement of some glycosyltransferase is consistently lost when 'contact inhibition' of cell growth is lost on viral transformation [61].

Although glycolipids have been significantly altered in all the transformed cells thus far examined, the change of other lipid classes is limited to only some transformed cells or tumor tissue. Phospholipids and neutral lipids of 3T3sv cells are essentially the same as normal 3T3 cells [56, 66], whereas a decreased phospholipid was observed in human fibroblasts transformed with SV40 virus [66]. No significant change of individual components of phospholipids or of neutral glycolipids has been observed [24]. An increased proportion of oleic acid and decrease of arachidonic acid in some transformed cells have been noticed [113], which could be a basis for the alleged fluid dynamic change of the lipid bilayer of transformed cells [77]. A significant increase of alkenyl or alkyl ether glyceride has been reported [92] which will be reviewed in a separate chapter of this monograph. As glycolipids display the most significant changes associated with transformation, this topic has been reviewed repeatedly including the changes of gangliosides and neutral glycolipids [3, 35–37]. Little attention has been paid, however, to the changes of fucose-containing glycolipids ('fucolipid') associated with malignant transformation. Due to its potential importance in determination of blood group and other cell surface specificities of allogeneic and syngeneic antigens, the change of fucolipids associated with malignant transformation could be extremely important.

Unequivocal demonstration of fucolipid changes, associated with malignant transformation and their enzymatic basis, is now available [93, 94]. It may be appropriate at this time to give the first review of this relatively small topic which promises an immense future development.

II. Fucolipids as a New Glycolipid Class and Cell Surface Marker

Fucolipids are the glycosphingolipids containing fucose at the nonreducing terminal of the carbohydrate chain. The term is tentatively used in contrast to gangliosides which have sialic acid (N-acyl-neuraminic acid) at the nonreducing terminal of the carbohydrate chain. This has been known since DISCHE [18] and DISCHE *et al.* [19] pointed out that the terminals of carbohydrate chains in glycoproteins are often glycosylated with either sialic acid or with fucose, i.e. the terminal group with a fucosyl residue is complementary to those with a sialyl residue in many heterosaccharide chains of glycoproteins. In glycolipids, however, only relatively small populations have been found to be substituted with a fucosyl residue; indeed, fucolipids are the minority lipids as compared to gangliosides and other neutral glycolipids [34]. Their distribution among various cells and tissues has not been extensively studied. It is likely, however, that they are present in relatively large quantities in gastrointestinal mucosal epithelia and in glandular tissues and to a lesser degree in other parenchymatous organs and mesenchymal cells. Although they constitute an essential part of blood group antigens in erythrocytes, they are still a minor component. No trace of fucolipid was found in nervous tissue in which gangliosides and cerebrosides predominate, although Folch's 'upper layer' fraction of brain occasionally contains unknown glycopeptide containing fucose [unpublished observation]. The lipid-bound fucose in HeLa cells [85] and in secondary mouse fibroblasts [8] was barely detectable, whereas the presence of a number of fucolipids in various mouse and baboon cell lines was clearly demonstrated by STEINER *et al.* [93].

Fucolipids are, however, very potent antigens compared to other neutral glycolipids and gangliosides. The fucolipid population in human erythrocytes is extremely small (less than 1 % of total glycolipid), however, in contrast to globoside, which occupies over 70 % of the total glycolipid of erythrocytes. Fucolipids produce antisera with a high titer, i.e. anti-A, anti-B, and anti-H, whereas antibody production against globoside and other abundant glycolipids is rather weak. Anti-A or anti-B agglutinates A or B erythrocytes very obviously, whereas anti-globoside barely agglutinates intact human erythrocytes, although a greatly increased agglutination occurred after trypsin treatment [34].

GAHMBERG [26] observed that fucose is greatly enriched in surface membrane compared to other carbohydrates which are distributed also in

intracellular membranes. BENNETT and LEBLOND [1, 2] observed that
³H-fucose rapidly incorporated into cell surface 'glycocalyx' of intestinal
columnar cells, as evidenced by autoradiographic electron microscopy.
Recent studies of FORSTNER and WHERRETT [25] indicated that isolated
plasma membranes of intestinal epithelial cells, such as microvillus and
brush border membranes, have a much higher content of glycolipids and
fucolipids than intracellular membrane.

Fucolipids as the blood group hapten of erythrocyte membranes have
been studied by three independent groups, those of YAMAKAWA [105–111],
KOSCIELAK [62–64], and of this reviewer [34, 45, 46]. The presence of
fucolipids as Lewis antigens in serum has been investigated by MARCUS
and CASS [70]. Since an accumulation of fucolipid occurring in some
human adenocarcinoma was first observed by this reviewer in 1964 [40],
studies of its chemical and immunological characterization have been
processed [34, 38, 42, 112], although much more work needs to be done.
Fucolipids also accumulate in a rare hereditary disease called 'fucosidosis'
in which H-like or Le^a-like glycolipids accumulate [20, 80]. The progress
of research on fucolipids since the first unequivocal description in 1964
is reported in table I.

III. Fucolipids as a Blood Group Hapten of Human Erythrocyte Membrane

The haptens of Wasserman's syphilis antigen, of heterophile Forssman antigen, and of blood group A and B antigens have been well established to be of lipid nature. The Wasserman hapten was characterized by PANGBORN et al. [79] as 'cardiolipin'; Forssman antigen was recently characterized as a ceramide pentasaccharide with a structure GalNAcα(1→3)GalNAcβ(1→3)Galα(1→4)Galα(1→4)Glc→Ceramide [87]. In contrast, the chemical nature of blood group haptens, although they have been studied repeatedly, has not yet been fully characterized. The lipid nature of blood group A and B haptens in erythrocytes was claimed by LANDSTEINER and VAN DER SCHEER [67], who discovered the blood group system in the early part of this century. The lipid theory of blood group hapten was also supported by SCHIFF and ADELBERGER [84], WITEBSKY [103], and HAMASATO [50–52], although others claimed that the real hapten should be 'polysaccharide' in nature [21, 49, 65, 96], and the activity associated with the lipid fraction is due to a contamination of

Table I. The development of fucolipid research, 1964–73

Enriched fucose in purified glycolipids of human adenocarcinoma that showed Le^a and H activities	Hakomori and Jeanloz [39]; Hakomori et al. [42]
Isolation and characterization of fucose-rich glycolipid fraction with human blood group activities	Yamakawa et al. [110]; Hakomori and Strycharz [46]
Presence of fucoglycolipid in dog or porcine intestinal tract	McKibbin [74]; Suzuki et al. [97]
Accumulation of fucoglycolipid and deletion of blood group A and B active glycolipids in human adenocarcinoma	Hakomori et al. [42]; Hakomori and Andrews [38]; Yang and Hakomori [112]
Accumulation of fucolipid in liver in fucosidosis	Durand et al. [20]; Philippart [80]
Lewis-active fucolipid of serum uptaken into erythrocytes	Marcus and Cass [70]
Fucoglycolipid as a blood group glycolipid of gastric mucous membrane and of intestinal mucosa	Slomiany and Horowitz [89]; Hiramoto et al. [55]
Isolation, structures, and polymorphic forms (variants) of A-active glycolipids	Hakomori et al. [45]
Structures of H-active glycolipids and B-active glycolipids	Stellner et al. [95]; Wherrett and Hakomori [102]; Koscielak et al. [64]
Accumulation of B-active fucolipids in Fabry's disease	Wherrett and Hakomori [102]
Localization and enrichment of fucolipids in microvillus and brush border membranes	Forstner and Wherrett [25]
Fucolipid changes detectable in in vitro transformed cells	Steiner et al. [93]

'real hapten' present in the lipid fraction [96]. A neutral glycolipid fraction, isolated and termed 'globoside' by Yamakawa and Suzuki [111] in 1952, was found to be blood group active in subsequent studies [105].

The active fraction in 'globoside' was later separated by chromatography on cellulose and on silicic acid columns [82, 106]. In these early studies, the presence of fucose was not clearly demonstrated, or the presence of fucose was not easily determined by classical colorimetry (Dische-Schettles method) in the presence of a large excess of hexoses. A trace amount of fucose was readily destroyed during hydrolysis and often overlooked by paper chromatography.

In the preparation of YAMAKAWA and IRIE [106], who obtained the blood group active fraction by immunoprecipitin with anti-A sera followed by extraction of the glycolipid with organic solvents, they were able to determine methyl pentose as a trace component. Similarly, the preparations of HANDA [53] and of KOSCIELAK [62], independently published in 1963, contained a trace amount (less than 3%) of methylpentose besides the major component hexoses (glucose, galactose), galactosamine, glucosamine, and as much as 10% of sialic acid.

The first unequivocal demonstration of fucosphingolipid was reported in 1964 by isolating a sphingolipid containing fucose, glucose, glucosamine, and galactose in a molar ratio of 1:1:1:2 from human adenocarcinoma tissue [40]. The content of fucose in this glycolipid was about 10%. It was predicted that this glycolipid was related to blood group hapten, and later on H and Le^a activities were detected in this 'tumor glycolipid' [42]. These results obviously encouraged the idea that cellular blood group haptens could be fucolipids and spurred the work for further extensive purification to obtain pure fucolipid haptens. Fucose content was found to be higher in a preparation subsequently separated on thin-layer chromatography by YAMAKAWA et al. [110].

The presence of multiple blood group active fucolipids was first demonstrated by HAKOMORI and STRYCHARZ [46]; three A-active components were separated from A_1-blood group erythrocytes, while only two active components were separated from A_2-erythrocytes. The component with the simpler composition seemed to be a ceramide hexasaccharide with a backbone structure βGal(1→4or3)βGlcNAc(1→3)βGal(1→4)βGlc(1→1)Ceramide [46]. In the above-mentioned studies, the presence of H-active and Le^b-active components was also demonstrated, although the presence of H-active glycolipid in erythrocyte membranes has been questioned by KOSCIELAK [62] and KOSCIELAK et al. [63]. They were unable to demonstrate H-active glycolipid in the glycolipid fraction; instead, H-activity was invariably demonstrated in glycoproteins (27).

In KOSCIELAK's hypothesis, H activity and A and B activities of the secretors were carried by glycoproteins, while only A and B activities of the nonsecretors were carried by glycolipids [27, 63]. This hypothesis strongly contradicted the presence of A and B glycolipids unless these glycolipids were biosynthesized through unusual precursors or those particular A and B glycolipids had an unusual structure. In fact, an unusual A structure was once proposed, based on methylation studies, i.e. GalNAcα(1→3)Galβ(1→3){L-Fucα(1→4)}GlcNAcβ(1→3)Galβ(1→4)Glc →Ceramide [63].

Consequently, an extensive fractionation of blood group H-active glycolipids has been carried out [95]. Four active components have been isolated, and the structure of the simplest component elucidated: αL-Fuc(1→2)βGal(1→4)βGlcNAc(1→3)βGal(1→4)Glc→Ceramide for H_1 glycolipid. The structure of component II (H_2 glycolipid) was assumed to be a ceramide heptasaccharide: αL-Fuc(1→2)βGal(1→4)βGlcNAc(1→3) βGal(1→4)βGlcNAc(1→3)βGal(1→4)Glc→Ceramide. The third component could have a branching structure with two oligosaccharides with the terminal structure αL-Fuc(1→2)βGal [95]. The fourth component has not been isolated in pure form. In all these H-active glycolipids of erythrocytes, only the 'type II' chain {βGal(1→4)GlcNAc} has been found, and no type I chain was detected by methylation studies [95].

The previous claim of Koscielak [62] and Koscielak et al. [63] that H-active glycolipids were absent in erythrocyte membranes could be based on the fact that H activity of H_1-active glycolipids was masked by globoside; in fact, H activity of H_1 glycolipid was greatly diminished or suppressed by mixing with globoside [95]. Koscielak et al. [64] also isolated in their subsequent studies two H-active components, which agreed with the structures proposed by us. Along with the studies on H-active glycolipids, components of blood group A-active glycolipids were thoroughly fractionated and the properties of each component were studied [45]. Four variants, termed A^a, A^b, A^c, and A^d glycolipids, were separated on DEAE-cellulose chromatography. Separation patterns of A and H variants are illustrated in figure 1.

The structure of the simplest variant (A^a) was identified as αGalNAc (1→3){αL-Fuc(1→2)}βGal(1→4)βGlcNAc(1→3)βGal(1→4)βGlc(1→1) Ceramide. The second variant is a ceramide octasaccharide. These two variants each contain a straight carbohydrate chain carrying the same A-determinant group, but differ in chain lengths. The A^c variant is a mixture of ceramide deca- to hendecasaccharide characterized by the presence of two different carbohydrate chains conjoined with a branching structure at the galactosyl residue, which is attached to glycosylceramide. The variant A^d is a mixture of ceramide dodeca- to tettareskaidecasaccharide with branchings and a highly complex structure. All these variants

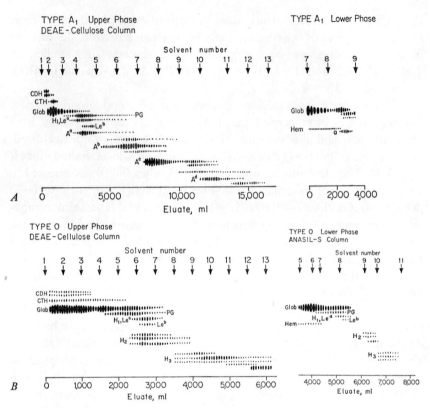

Fig. 1A. Separation pattern of long-chain neutral glycolipids by combination of chromatographies on DEAE-cellulose and thin-layer of silica gel H. Column size 2.5 × 30 cm. Solvent numbers indicated by chloroform-methanol-water mixture of the following proportions: (1) 90:20:0.2; (2) 87.5:12.5:0.2; (3) 85:15:0.2; (4) 80:20:0.2; (5) 77.5:22.5:0.2; (6) 75:25:0.2; (7) 70:30:0.2; (8) 67.5:32.5:0.4; (9) 65:35:0.4; (10) 60:40:0.5; (11) 57.5:42.5:0.5; (12) 55:45:0.5; (13) 0:100:0, solvent for thin-layer chromatography 60:35:8. CDH: lactosylceramide; CTH, Galα1→4Galβ1→4Glc→Cer; Glob: Globoside (GalNAcβ1→4Galα1→4Galβ1→4Glc→Cer); PG: paragloboside (Galβ1→4GlcNAcβ1→3Galβ1→4Glc→Cer) [SIDDIQUI and HAKOMORI, unpublished]; A^a, A^b, A^c, A^d: variants of A-active glycolipids. The majority of CDH, CTH, globosides and PG was found in the lower phase.

B. Separation pattern of glycolipids from lipid extract of human blood group O erythrocyte membrane: The pattern on the left (type O, upper phase) was obtained by combination of chromatographies on DEAE-cellulose and thin-layer silica gel H. Column size 2.5 × 30 cm. Solvent numbers (in parentheses) indicated by chloroform-methanol-water mixtures of the following proportions: (1) 9:1:0.02; (2) 8.75:1.25:0.02; (3) 8.5:1.5:0.02; (4) 8:2:0.02; (5) 7.5:2.5:0.02; (6) 7:3:0.02; (7) 6.75:3.25:0.04; (8) 6.5:3.5:0.04; (9) 6:4:0.05; (10) 5.75:4.25:0.05; (11) 5:4:0.05; (12) 5.25:4.75:0.05; (13) 5:5:0.05.

again contained exclusively type II chain, and no type I chain has been detected so far [45, 95].

Blood group B-fucolipids of erythrocytes have been studied by KOSCIELAK *et al.* [64]. Two variants, B-I and B-II glycolipids, were isolated and characterized; they were, respectively, ceramide hexasaccharide and octasaccharide without branching structure. It is anticipated that the third and fourth variants of B-active component could be present.

Lewis-active glycolipids of erythrocytes have been considered to be acquired from serum rather than synthesized *in situ*. An unequivocal demonstration of this theory is that Lewis-active glycosphingolipids are present in relatively large quantities in serum, and when an aqueous solution of Lewis glycolipids is incubated with Lewis-negative erythrocytes followed by washing, the glycolipid haptens are rapidly taken up onto cells which are converted to Lewis a- or b-positive erythrocytes [70]. Thus, Lewis-active factors of erythrocytes were fucolipids and not glycoproteins and were acquired by erythrocytes from serum. The origin of serum Lewis fucolipid is not known [70].

IV. Fucolipids of Gastrointestinal Tract and Glandular Tissues

Gastrointestinal mucosa contains relatively large quantities of fucolipids compared to other tissues. Isolation and partial characterization of fucolipids from dog intestine and from porcine intestine were carried out independently by MCKIBBIN [74] and by SUZUKI *et al.* [97]. MCKIBBIN's fucolipid fraction was isolated from total lipid extract by chromatography on silicic acid, eluting with 50 % methanol in chloroform ('S-50 fraction') followed by chromatography on columns of DEAE cellulose and Florisil. Two fucolipid fractions were separated: one fraction 'hexahexoside D1' and the other 'hexahexoside D2'. Both fractions were composed of galactose (2.6:2.4 mol), glucose (0.78:0.77 mol), fucose (1.4:1.7 mol), and glucosamine (1.2:1.3 mol). Further studies on separation and fractionation of fucolipids on thin-layer chromatography with new solvents (chloroform-methanol-H_2O-glacial acetic acid: 65/35/4/4 or 55/45/5/5; chloroform-methanol-conc. NH_4OH: 40/80/25) enabled the total fucolipids to separate completely from other lipid classes as a slow migrating spot on thin-layer chromatography [91]. The fraction showed, however, considerable variation in heterogeneity depending on the individual type of dog. Three types of dog intestinal fucolipids were distinguished: type I having galactosamine, type II lacking galactosamine, and type III having both type I and type II fucolipids. Fractionation of fucolipids was extremely difficult, as those with different numbers of fucose residues did not separate well, compared to other glycolipids and gangliosides which separate according to saccharide number. In a subsequent study by McKibbin's group, Hiramoto et al. [55] investigated immunological reactivities of fucolipids of dogs and human intestine. Blood group activities A and Le^a were detected in fucolipids of human intestine, although blood groups of the hosts were not given. Blood group H and Le^b activities were also detected in fucolipids of dog intestine [55].

Feline intestinal tract also contains a similar fucolipid; although chemical composition has not been determined, blood group A activity was detected [55].

Fucolipids of porcine intestine studied by Suzuki et al. were obtained in a fraction similar to McKibbin's fucolipids of dog intestine. They were eluted from silicic acid with 50–60 % methanol in chloroform and from DEAE-cellulose with 45 % methanol in chloroform. The fucolipid fraction was further separated by thin-layer chromatography on silica gel into fucolipids 1 and 2. Fraction 1 contained glucose, galactose, N-acetylhexosamine, fucose, and ceramide in the molar ratio of approximatively $1:2:1:1:1$, and fraction 2 contained glucose, galactose, N-acetylglucosamine, fucose, and ceramide in the molar ratio of $1:2:1:1:1$. Both N-acetylglucosamine and galactosamine were detectable in fucolipid 1, whereas only N-acetylglucosamine was detected in fucolipid 2. Both fractions showed essentially the same fatty acid profile. They contained 41–45 % of hydroxy fatty acids; C_{16}, C_{22}, and C_{24} were the major ones [97]. The analysis was made on pooled samples, and it is possible that these fractions were a mixture of heterogeneous glycolipids differing slightly from one group of individuals to another. In dog, human, and porcine, two types of fucolipids were demonstrated: one with fucose, glucosamine, galactose, and glucose; the other with additional galactosamine. As indicated by Smith et al. (cited by Hiramoto et al. [55] as 'in preparation'), three types of intestinal fucolipids could be distinguished, and their distribution was different, depending on the individual. Type 1 had N-acetylgalactosamide, type 2 lacked N-acetylgalactosamine, and type 3 had both type 1 and type 2 fucolipids.

Localization of fucolipids in rat intestine has been studied by Forstner and Wherrett [25]. They found two types of fucolipids (complex 2 and 3) exclusively present in plasma membrane preparations of rat small intestine that included brush border and microvillus membrane. The complex 2 fucolipid showed carbohydrate composition of fucose, N-acetylglucosamine, galactose, and glucose 0.3/0.6/2.1/1.0, and that of complex 3 showed 0.3/1.9/2.4/10. The 'complex 2' fucolipids were isolated from brush border membranes, and the 'complex 3' fucolipids were isolated from microvillus membrane.

The concentration of dog intestine fucolipids was estimated at 0.14 μmol/g (25–27 μg per gram of tissue). This value is about 30 times higher than the yield of total fucolipids from human erythrocyte membrane. If the fucolipid content of microvillus and brush border membrane is counted, fucolipid content of these intestinal membranes must be 200 times higher than the fucolipid content of plasma membrane of other cells, including erythrocyte membrane.

SLOMIANY and HOROWITZ [89] described isolation and structural determination of three kinds of blood group A-active fucolipids isolated from lipid extract of hog stomach mucosa. One component showed the same carbohydrate composition and structure as A^a glycolipids of human erythrocyte membrane. The second component contained two galactose per molecule as shown in table II. The third component had no N-acetylglucosamine but had 3 mol of galactose [90]. The structure was determined by enzymatic degradation and by methylation. No extensive studies have been made on fucolipids of other organs and tissues. It is probable, however, that all glandular epithelial tissues might contain fucolipids in much higher quantity than mesenchymal organs. Pancreas, for example, contains relatively large quantities of fucolipids showing blood group A, B, and H specificities [34]. H glycolipids were in fact first isolated from pancreas from which an oligosaccharide with the same migration rate as 'lacto-N-fucopentaose I' was identified. Recently, accumulation of blood group B fucolipid in pancreas of a patient with Fabry's disease was found, and its structure was identified as a ceramide hexasaccharide, shown in table II [102]. Pancreas fucolipid contains type 1 chain in striking contrast to the predominance of type 2 chain in erythrocyte fucolipid.

V. Fucolipids of Tumor Tissue

As described in the preceding sections, glandular epithelial tissue has fucolipids in appreciable quantities. Fucolipids also constitute an essential part of membrane-bound surface antigens with blood group ABH and

Table II. Structures of fucolipids

1. 'X Hapten' of human adenocarcinoma: YANG and HAKOMORI [112]
   Galβ(1→4)GlcNAcβ(1→3)Galβ(1→4)Glc→Ceramide
   $\uparrow^3_1\alpha$
   L-Fuc

2. Le^a Glycolipid of human adenocarcinoma: HAKOMORI and JEANLOZ [40]
   Galβ(1→3)GlcNAcβ(1→3)Galβ(1→4)Glc→Ceramide
   $\uparrow^4_1\alpha$
   L-Fuc

3. Le^b Glycolipid of human adenocarcinoma: HAKOMORI and ANDREWS [38]
   Galβ(1→3)GlcNAcβ(1→3)Galβ(1→4)Glc→Ceramide
   $\uparrow^2_1\alpha$ $\uparrow^4_1\alpha$
   L-Fuc   L-Fuc

4. Le^b Glycolipid of human adenocarcinoma: HAKOMORI and ANDREWS [38]
   Galβ(1→3)GlcNAcβ(1→3)Galβ(1→3 or 4)GlcNAcβ(1→3)Gal$\}*\beta$(1→4)
   $\uparrow^2_1\alpha$ $\uparrow^4_1\alpha$ Glc→Ceramide
   L-Fuc   L-Fuc   * sequence is tentative

5. H$_1$ Glycolipid: STELLNER *et al.* [95]; KOSCIELAK *et al.* [64]
   Galβ(1→4)GlcNAcβ(1→3)Galβ(1→4)Glc→Ceramide
   $\uparrow^2_1\alpha$
   Fuc

6. H$_2$ Glycolipid: STELLNER *et al.* [95]; KOSCIELAK *et al.* [64]
   Galβ(1→4)GlcNAcβ(1→3)Galβ(1→4)GlcNAcβ(1→3)Gal$\}*\beta$(1→4)
   $\uparrow^2_1\alpha$ Glc→Ceramide
   Fuc   * sequence is tentative

7a. B-Active glycolipid of pancreas accumulating in Fabry's disease: WHERRETT and HAKOMORI [102]
   Galα(1→3)Galβ(1→3*)GlcNAcβ(1→3)Galβ(1→4)Glc→Ceramide
   $\uparrow^2_1\alpha$
   L-Fuc   * 1→3 linkage 80%; 1→4 linkage 20%

Lewis specificities and with unknown specificities. Since the tumor-specific transplantation antigen is regarded as the altered isologous antigen, it is highly probable that the altered blood group antigen, as expressed in altered fucolipid synthesis, could be an important surface change associated with transformation of glandular epithelial cells.

Table II. Continuation

7b. B-I Glycolipid of erythrocytes: KOSCIELAK et al. [64]
Galα(1→3)Galβ(1→4)GlcNAcβ(1→3)Galβ(1→4)Glc→Ceramide
$\uparrow\,^2_1\alpha$
L-Fuc

8. B-II Glycolipid: KOSCIELAK et al. [64]
Galα(1→3)Galβ(1→4)GlcNAcβ(1→3)Galβ(1→4)GlcNAcβ(1→3)Galβ(1→4)
$\uparrow\,^2_1\alpha$ Glc→Ceramide
L-Fuc

9. A^a Glycolipid of human erythrocytes: HAKOMORI et al. [45]
GalNAcα(1→3)Galβ(1→4)GlcNAcβ(1→3)Galβ(1→4)Glc→Ceramide
$\uparrow\,^2_1\alpha$
L-Fuc

10. A^b Glycolipid: HAKOMORI et al. [45]
GalNAcα(1→3)Galβ(1→4)GlcNAcβ(1→3)Galβ(1→4)GlcNAc(1→3)Gal(1→4)
$\uparrow\,^2_1\alpha$ Glc→Ceramide
L-Fuc

11. A^c Glycolipid: HAKOMORI et al. [45]
GalNAcα(1→3)Galβ(1→4)GlcNAc
$\uparrow\,^2_1\alpha$ $^3_6$Gal(1→3)Galβ(1→4)GlcNAc(1→3)
L-Fuc   Gal(1→4)Glc→Ceramide
Galβ(1→4)GlcNAc
$\uparrow$
L-Fuc

12. A Glycolipid of hog stomach mucous membrane: SLOMIANY and HOROWITZ [89]
GalNAcα(1→3)Galβ1→GlcNAc→Gal→Gal→Glc→Ceramide
$\uparrow$
L-Fuc

13. A Glycolipid of hog stomach mucosa: SLOMIANY et al. [90]
GalNAcα(1→3)Galβ(1→3)Galβ(1→4)Galβ(1→4)Glc→Ceramide
$\uparrow$
L-Fuc

Glandular epithelial cells and erythrocytes have blood group determinants bound to glycoprotein as well [69]. It is unknown, however, which part of the determinants, that bound to lipid or that bound to protein, plays a major role in determining antigenicity of cells. The enzyme systems for synthesis and degradation of blood group determinants could be identical, regardless of whether they are glycolipid or glycoprotein. Fucolipid changes observed in tumors could be valid for glycoprotein as well [94]. In this section the change of blood group antigen associated with malignancy will be discussed, not exclusively with regard to fucolipid changes but also to glycoprotein changes.

For convenience of description, the five items shown below will be discussed: deletion of blood group A and B antigens due to impaired synthesis of these determinants; increase of Le^a and its positional isomer 'X-hapten glycolipid'; presence of a glycolipid hapten with an incompatible blood group specificity; fucolipid changes observed in *in vitro* transformed cells, and relationship of carcinoembryonic antigen to blood group antigen.

A. Blood Group A and B Fucolipids and Glycoproteins in Tumors

1. Deletion of A and B Determinants

Deletion of blood group A and B antigens in human gastric cancer was first recognized by OHUTI [78] in 1949, when his 'blood group polysaccharide' of tumor tissue failed to inhibit or only weakly inhibited A-specific hemagglutination. Optical rotation of these 'cancerous blood group substances' differed from that of normal blood group substances. Comparison of blood group polysaccharide of original gastric tumor, of normal gastric mucosa, and of metastatic lesion of liver confirmed the previous findings [73]. Differences in sialic acid content, consumption of periodate and hypoiodide were observed between normal and tumorous blood group polysaccharides [73]. Further studies were carried out by the same research group, avoiding drastic preparation procedures, and at least six glycoproteins were separated. Some were isolated in a homogeneous state on electrophoresis, and the fractions were noticed to carry blood group haptens [59, 71]. Blood group activities and chemical and physical properties were examined in greater detail. These properties are listed in table III.

Decreases of either A or B antigenic activities and corresponding chemical changes were noticed, i.e. either decrease of N-acetylgalactosamine or galactose in tumorous blood group glycoproteins, whereas 'H activity' determined by eel sera or by anti-human O saliva chicken sera, decreased to a lesser degree [71]. Unfortunately, these data were based on analysis of blood group polysaccharides isolated from pooled tumor

Table III. Comparison of blood group glycoproteins of normal gastric mucosa and of gastric cancer[1]

	Gastric cancer			Normal mucosa		
	A	B	O	A	B	O
Blood group activity[2]						
A-activity	6	200	200	0.01	200	200
B-activity	200	12.5	200	200	0.01	100
H-activity						
Eel serum	50	50	125	50	25	6
Anti-O saliva						
chicken serum	12.5	12.5	0.4	12.5	12.5	0.1
Sugar composition						
Hexosamine	31	33	32	34	32	34
GlcNAc/GalNAc ratio	3.1	3.2	3.2	2.3	3.5	3.1
Gal	28	30	31	28	30	29
Fuc	17	15	17	15	16	16
Sialic acid	2.3	4.2	4.3	0.9	1.1	1.4
$[\alpha]_D^{20}$	−38	−38	−41	−20	−28	−35
Molecular weight $\times 10^4$						
As rods	7.7			8.9		
As discs	11.1			12.8		

1 Compiled from MASAMUNE et al. [72] and KAWASAKI [59].
2 Minimum amount of substance in µg that inhibits respective hemagglutination.

tissues or pooled normal mucosa and, therefore, possible variation of individual tumor tissue was ignored. ISEKI et al. [58] isolated blood group substances from individual gastric tumor tissue and compared them with normal blood group substances isolated from individual normal gastric mucosa, prepared according to Kabat's pepsin digestion method. Significant variation in immunological activities was noticed; a polysaccharide appearing in some tumors had high H activity, whereas in others high Le[a] activity was shown; some tumors showed low H and Le[a] activities [58]. In addition to these altered blood group activities, a unique serological activity directed to aminosugar hapten (glucosamine, not N-acetylglucosamine) has been detected in anti-rabbit serum obtained by immunization of 'cancerous blood group substances'. A similar immunological specificity inhibitable by glucosamine rather than N-acetylglucamine was detected in the 'polysaccharide' fraction in urine of cancer-bearing individuals [58].

In addition to these immunochemical studies based on isolated blood group substances from tumors and normal tissues, extensive immuno-histological studies have been carried out. Using mixed hemagglutination technique, KAY and WALLACE [60] showed that A and B antigens were not detected in bladder carcinoma cells, but were detected in normal urinary tract cells. Similarly, NAIRN et al. [76] showed deletion of blood group A and B antigens in human adenocarcinoma cells. A study by TELLEM et al. [98] on mammary tumors showed, however, that blood group A and B antigens persisted in these tumor tissues, which contradicts the reports on other tissues.

Further extensive studies using immunohistological methods (immuno-fluorescence technique and mixed cell agglutination reaction on histologic specimens) by DAVIDSOHN et al. [14, 15] revealed that ABH isoantigens are detectable in normal tissues and are reduced or absent in the course of transformation to carcinoma. Such changes of isoantigens are obvious in adenocarcinoma of the gastrointestinal tract, ovary, and in squamous cell carcinoma of the skin, tongue, larynx, and urinary bladder [14]. Further histocyto-adherence reactions from 82 cases of benign and malignant cervical cancer indicated that isoantigens (ABH) were present in all benign lesions and were absent in all metastatic carcinomas. In 18 cases out of 21 early infiltrating carcinomas without clinical evidence of metastasis, the isoantigens were decreased or absent [15]. In these studies, isoantigens were always demonstrated in noncancerous metaplastic or dysplastic epithelium, whereas even in the early stages of infiltrative carcinoma, isoantigens were no longer demonstrated. Absence of ABH isoantigens in neoplastic oral epithelium was reported by PRENDERGAST et al. [81]. A study by DABELSTEEN and FULLING [13] has shown that A and B antigens are lost in premalignant dysplasia of oral epithelial tissues.

Deletion of H antigen in metastatic cervical cancer and gastric cancer was observed by DAVIDSON et al. [15] and by ISEKI et al. [58], whereas only a slight decrease of H activity was found to occur in the isolated blood group antigens from pooled gastric cancer tissues [59].

2. Blocked Synthesis of A and B Determinants in Tumors

Diminished or totally deleted A or B reactivities accompanying malignancy were detectable on histochemical sections at a very early stage of transformation, even in premalignant lesions ('dysplasia' or

Table IV. Conversion rate of H_1 into A^a glycolipid by enzymes from various sources. Yield of A^a glycolipid in CPM $\times$ 10^{-3} and percent conversion expressed by added UDP sugars

Enzymes of	Donor	Incorporation of ^{14}C-GalNAc into A^a glycolipid		
		blood group	cpm $\times$ 10^{-3}	added radioactivity, %
Serum	C. Ga.	A_1	8.33	9.2
	R. Je.	A_1	9.83	10.9
	N. Ga.	A_2	0.24	0.2
	J. Ca.	A_2	0.23	0.2
	G. Yo.	B	0.04	0.04
	K. St.	O	0.03	0.03
Tumor from mucosal epithelia	L. Ek.	A_1	14.34	15.5
	T. Wn.	AB	3,9; 2.1[1]	4.3; 2.2[1]
	N. Da.	A_2	4.2	4.6
	A. Wm.	A_1	5.1	5.6
Host's mucosal epithelia	L. Ek.	A_1	71.18	79.0
	T. Wn.	AB	28.17; 12.2[1]	31.3; 13.5[1]
	N. Da.	A_2	8.64	9.6
	A. Wm.	A_1	30.05	33.3

[1] H_1 to B conversion determined with ^{14}C-Gal under the same conditions as for H_1 to A conversion.

'atypia') [13, 15]. It is therefore essential to determine whether this sensitive change of A or B determinants is due to enhanced hydrolase or to blocked synthetase activity.

As seen in table IV and figure 2, a much higher rate of conversion took place when catalyzed by the enzyme of normal epithelial mucosa as compared to the same reaction catalyzed by the enzyme of adenocarcinoma tissue derived from epithelial mucosa of the same individual [94]. The difference was also clear when the yields of radioactive A^a glycolipid were compared by increasing the concentration of enzyme proteins or by increasing incubation time (fig. 2B). The hydrolase activity for A^a glycolipid of 'P2 + 3' fraction of normal mucosal tissue was nearly identical to that of adenocarcinoma (table IV). Higher A^a synthesis was observed from A_1 serum than from A_2 serum in agreement with SCHACHTER *et al.* [83].

Thus, the deletion of A and B reactivities in epithelial tumor is due

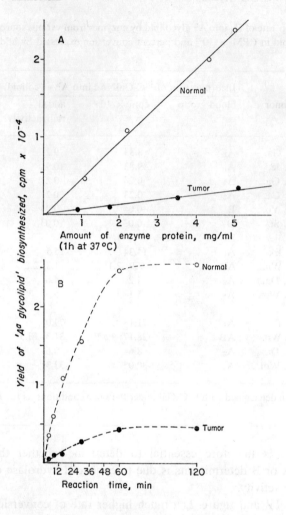

Fig. 2. Difference in A^a glycolipid synthesis from H_1 glycolipid between enzyme fractions of normal gastric mucosa and tumor: difference of reactivities depending on the amount of enzyme protein (A) and on reaction time (B). Enzyme protein in B: 5 mg/ml (300 µg/100 µl). Open circle: normal gastric mucosal epithelia; solid circle: tumor. Activities determined by the incorporation of UDR (^{14}C)-N-acetylgalactosamine into A^a glycolipid.

to a deficiency of glycosyltransferases for synthesis of A and B determinants but not to enhanced hydrolase activity. Blocked enzyme activities for synthesis of various glycolipids have been demonstrated in various transformed cells *in vitro* [12, 16, 61]. A similar enzyme block for glycoprotein synthesis was also demonstrated [30, 31]. The deficient A or B enzyme for glycolipid could be the same enzyme for synthesis of A or B glycoprotein as well, because blood group glycoproteins of tumor were also shown to be deficient [71, 72, 78].

B. Lewis Fucolipids in Tumors

Lewis antigen activities of blood group glycoprotein isolated from human gastric adenocarcinoma showed some inconsistency, *i. e.* enhanced Le[a] activity in some cases and enhanced H-Le[b] activity in other cases [58].

Accumulation of a fucose-containing glycosphingolipid in some human adenocarcinoma was noticed, having a carbohydrate composition fucose : glucose : galactose : GlcNAc 1 : 1 : 2 : 1 [40]. Subsequent studies showed that this glycolipid fraction inhibited Le[a] hemagglutination, although the fraction was heterogenous and inhibited cytoagglutination caused by wheat germ agglutinin [42]. By further fractionation of fucose-containing glycolipid as acetylated compounds on thin-layer chromatography, several components have been isolated. A major component besides Le[a]-active glycolipid was finally identified as Galβ(1→4) (L-Fucα (1→3) (GlcNAcβ(1→3) Galβ(1→4) Glcβ(1→1) Ceramide, i. e. 'lacto-N-fucopentaosyl-III-ceramide'. The ceramide moiety of this lipid was characterized as having 4-hydroxyoctadecasphinganine as the major fatty base component [112]. This compound is highly antigenic in rabbits, easily producing precipitating antibodies which cross-react with Le[a] glycolipid. It is assumed that normal blood group glycolipids of glandular tissue contain a 'type 1' carbohydrate chain, i. e. Galβ(1→3) GlcNAcβ(1→3) Gal→R, as the major component [102], whereas adenocarcinoma tissue contains a 'type 2' carbohydrate chain for the backbone structure in the majority of glycolipids [112].

In five cases of gastrointestinal adenocarcinoma, the co-presence of Le[a] and Le[b] glycolipids has been noticed. Two Le[b]-active glycolipids were separated, and their structures were determined [38]. The large amount of Le[a]-active glycolipids in tumors of Le[b] individuals can be easily understood, as blocked synthesis of Le[b] glycolipid resulted in accumulation of the precursor glycolipid, which is Le^a. However, the presence of Le^b-active glycolipids, though small in quantity, in the tumors of Le^a individuals cannot be easily interpreted at this time.

In many cases of human adenocarcinoma of gastric mucosa, pancreas, bronchogenic lung tumor, and cecal tumor, accumulation of two isomeric fucose-containing glycolipids were especially noticeable – Le^a and its isomer. This is probably due to the blocked synthesis of blood group glycolipids in general [42, 112].

The reactive site for anti-type XIV pneumococcal antisera is present in blood group ABH substances which increased on acid hydrolysis. The structure has been established as β-galactopyranosyl(1→4) (N-acetyl) glucosaminyl. The 'type 2' chain of H substance can be converted to this structure by hydrolysis of the terminal 1→2 structure. Such a structure is indeed greatly increased in 'blood group glycoprotein' of cancer tissue [WATANABE and HAKOMORI, unpublished observation]. The reactive site for wheat germ agglutinin is also present in blood group ABH substances, which is most probably an unfinished carbohydrate chain having N-acetyl-glucosaminosyl at the terminal [42]. Le^a glycoprotein or Le^a glycolipid can be converted to a highly active inhibitor for wheat germ agglutinin by Smith degradation [112].

The structure which can inhibit wheat germ phytoagglutinin also greatly increased in glycoproteins of human adenocarcinoma tissues [28, 42]. These data, together with deletions of blood ABH antigens, suggested strongly that synthesis of blood group glycoproteins is incomplete in cancer cells, which is closely related to the infiltrative properties of the tumor cell surface.

C. Presence of Incompatible Blood Group Antigens in Human Tumors

The presence of Le^a antigen in tumors of host Le^b has been described [38]. The antibodies directed against the glycolipid fraction and isolated from tumors of blood group O individuals showed anti-A properties [42]. HÄKKINEN [32] described the appearance of an A-like antigen demonstrated in tumors of host blood groups B and O. This is called 'neo-A antigen' and was distinguished from normal A antigen of normal tissues. The heterospecific A was distributed all over the surface of gastric mucosa which was invaded by tumor cells. The chemical nature of this antigen has not been determined; however, the possibility of Forssman antigen should be considered, as the appearance of Forssman antigen in human tumors was reported recently, although humans should be Forssman-negative animals. Sulfomucopolysaccharide isolated from human cancer showed a unique A-like antigen activity. It was reported that carcinomatous gastric juice contains a sulfated glycoprotein which differed immunologically from that of gastric juice [33]. In some cases of gastrointestinal tumor, appearance of blood group B antigen in tumors of host O or A has been reported. This, however, is due to contamination by blood group B-producing bacteria.

Association of the reactive site for wheat germ lectin with blood group A substance is of special interest. Blood group A substance can inhibit hemagglutination caused by wheat germ agglutinin stronger than blood group B or O substances [42, 48]. Synthesis of the wheat germ lectin reactive site is closely related genetically to the synthesis of A hapten [48]. The blood group substance isolated from human adenocarcinoma also showed high activity for inhibition of wheat germ agglutination [42]. A recent report by GOLD et al. [29] states that compound carcinoembryonic antigen can react with anti-A antibody, although the hapten structure of CEA is entirely different from blood group A hapten.

D. Fucolipid Changes in Transformants *in vitro*

Differences in chemical composition and immunologic activities in fucolipids have been observed between 'normal' and tumor tissues, but it has never been determined that the observed differences are really transformation-dependent, as the progenitor cells of spontaneous tumor have never been clear and the comparison has never been justified. A comparison of fucolipids of plasma membranes of gastrointestinal tumors with those of plasma membranes of gastrointestinal glandular cells would be interesting.

A recent paper by STEINER et al. [93], however, unequivocally demonstrated a clear change of fucolipid class in tissue culture cells upon transformation with oncornaviruses. Normal rat embryo cells (RE-2 cells) derived from Sprague-Dawley rats and their transformants with murine-sarcoma-murine-leukemic virus complex (MSV-MLV), normal rat embryo cells derived from Wistar rat cells (WIS cell line) infected with and producing murine leukemic virus (MLV), a Wistar rat cell line transformed with MSV-MLV, and RE cells transformed by RSV were compared.

A baboon cell line (SA571) transformed with feline sarcoma-leukemic virus (FSV-FLV) was also examined. Fucolipid profile was found dramatically altered in each case of transformed cells, but there was no change of MLV-infected but nontransformed cells. In normal cell lines, one or two closely associated fucolipids with low mobility were dominant, whereas in all the transformed cell lines, there was a sharp decrease in fucose incorporation into the larger fucosylglycolipids, and a corresponding rise in radioactivity in more mobile, apparently simple fucolipids has been observed.

E. Relation of 'Carcinoembryonic' Antigen and Blood Group Substances

Carcinoembryonic antigen (CEA), as first described by GOLD and FREEDMAN [28], is a specific antigen occuring in both embryonic gastrointestinal tract and tumor tissue derived from gastrointestinal tract, but is not present in adult gastrointestinal tract. The antigen is not precipitable by perchloric acid and is isolated by successive gel filtration with Sepharose and Sephadex chromatography followed by electrophoresis. Recently SIMMONS and PERLMANN [88] found that the CEA fraction is identical to the fraction of blood group substances. CEA cannot be distinguished from blood group substance by gel filtration and electrophoresis, and it showed identical solubility with similar carbohydrate and amino acid composition. Immunochemical studies support the view that CEA is an incomplete blood group substance of the ABO system. The deficient antigens have tumor-specific activity due to unmasking of a sequence that is cryptic in the normal structure. Two immunodominant groups were identified as N-acetylgalactosaminyl peptide and blood group I antigen complex, which are found in blood group substance [22, 23]. Incomplete synthesis of many carbohydrate chains associated with malignant transformation has been suggested by model experiments with the *in vitro* system [43] and in blood group antigens as well [48].

VI. Epilogue

Glycosphingolipids are, in general, components of plasma membrane and are present in much less quantity in intracellular membranes. Of a number of glycosphingolipids, a new class of compounds with terminal substitution of L-α-fucosyl residue has been found in relatively limited populations of cells. This class of glycosphingolipids, termed fucolipids, can be found in erythrocyte membrane, in plasma membrane of gastrointestinal tract, and in glandular epithelial cells. Up to now, a dozen such glycolipids have been isolated and their structures determined. Principally, they had glucosamine and galactose, as found in 'lacto-N-neotetraosyl' or 'lacto-N-tetraosyl' ceramide. Ten times higher concentration of fucolipids was found in gastrointestinal epithelial or glandular tissues compared with membranes of mesenchymal cells, including erythrocytes. A number of known, isolated fucolipids bear blood group ABH and Lewis specificities, although some bear unrelated unknown specificities. Fucolipids constitute, therefore, an essential part of membrane isoantigens, although they are minority lipids.

The changes of glycosphingolipids asociated with malignant transformation of cultured cells *in vitro* or *in vivo* have been studied extensively. In many transformed cell systems, including chick fibroblasts transformed with temperature-sensitive mutants of Rous sarcoma virus, incomplete synthesis of glycolipid with or without accompanying increase of precursor glycolipid has been observed [43].

The difference of fucolipid composition, observed between normal gastrointestinal mucosa and adenocarcinoma derived therefrom, suggests a probable presence of incomplete synthesis of blood group A or B fucolipid with simultaneous accumulation of H or Le^a glycolipid. This idea for incomplete synthesis of fucolipid in various *in vitro* transformed cells by oncornaviruses is indicated also from the results of labeling pattern of fucolipids *in vitro* with ^{14}C-fucose as precursor, which is the most clear demonstration of metabolic alteration of fucolipid synthesis by transformation. The exact relation of these fucolipid changes to the alteration of immunological properties of cell surfaces will be the most interesting topic awaiting elucidation.

Fucolipid changes associated with malignancy of epithelial cells may represent altered antigenicity as well as various functional changes of epithelial cell membrane, such as loss of growth control and weakened intercellular linkages, etc., just as the changes in gangliosides and neutral glycolipids are regarded as the basis of various transformation-dependent changes of membranes. Fucolipids, therefore, could be key molecules for understanding the unknown mechanism of malignant transformation in epithelial cells and glandular epithelial cells.

VII. References

1 BENNETT, G. and LEBLOND, C. P.: Formation of cell coat material for the whole surface of columnar cells in the rat small intestine, as visualized by radioautography with L-fucose-^{3}H. J. Cell Biol. *46:* 409–416 (1970).
2 BENNETT, G. and LEBLOND, C. P.: Passage of fucose-^{3}H label from the Golgi apparatus into dense and multivesicular bodies in the duodenal columnar cells and hepatocytes of the rat. J. Cell Biol. *51:* 875–881 (1971).
3 BRADY, R. O.; FISCHMAN, P., and MORA, P. T.: Membrane components and enzymes in virally transformed cells. Fed. Proc. *32:* 102–108 (1973).
4 BRADY, R. O. and MORA, P. T.: Alteration in ganglioside pattern and synthesis in SV40-transformed and polyoma virus-transformed mouse cell lines. Biochim. biophys. Acta *218:* 308–319 (1970).
5 BUCK, C.A.; GLICK, M. C., and WARREN, L.: A comparative study of glycoproteins from the surface control and Rous sarcoma virus transformed hamster cells. Biochemistry *9:* 4567–4576 (1970).
6 BUCK, C. A.; GLICK, M.C., and WARREN, L.: Glycopeptides from the surface of control and virus-transformed cells. Science *172:* 169–171 (1972).
7 BURGER, M. M.: Difference in the architecture of the surface membrane of normal and virally transformed cells. Proc. nat. Acad. Sci., Wash. *62:* 994–1001 (1969).
8 CHATTERJEE, S.: Studies on the metabolism of glycosphingolipids in cultured cells; Ph.D. thesis, Toronto (1973).
9 CHEEMA, P.; YOGEESWARAN, G.; MORRIS, H. P., and MURRAY, R. K.: Ganglioside patterns of three Morris minimal deviation hepatomas. FEBS Letters *11:* 181–184 (1970).
10 CHENG, S.; PIANTADOSI, C., and SNYDER, F.: Lipid droplets and glyceryl ether diesters in Ehrlich ascites cells grown in tissue culture. Lipids *2:* 193–194 (1967).
11 CRITCHLEY, D. R. and MACPHERSON, I.: Cell density-dependent glycolipids in NIL hamster cells, derived and transformed cell lines. Biochim. biophys. Acta *296:* 145–159 (1973).
12 CUMAR, F. A.; BRADY, R. O.; KOLODNY, E. H.; MCFARLAND, V. W., and MORA, P. T.: Enzymatic block in the synthesis of gangliosides in DNA-virus transformed tumorigenic mouse cell lines. Proc. nat. Acad. Sci., Wash. *67:* 757–764 (1970).
13 DABELSTEEN, E. and FULLING, J. H.: A preliminary study of blood group substances A and B in oral epithelium exhibiting atypia. Scand. J. dent. Res. *79:* 387–393 (1971).
14 DAVIDSOHN, I.; KOVARIK, S., and LEE, C. L.: A, B, and O substances in gastrointestinal carcinoma. Arch. Path. *81:* 381–390 (1966).
15 DAVIDSOHN, I.; KOVARIK, S., and NI, Y.: Isoantigens A, B, and H in benign and malignant lesions of the cervix. Arch. Path. *87:* 306–314 (1969).
16 DEN, H.; SCHULTZ, A. M.; BASU, M., and ROSEMAN, S.: Glycosyltransferase activities in normal and polyoma-transformed BHK cells. J. biol. Chem. *246:* 2721–2723 (1971).

17 DIRINGER, H.; STROBEL, G., and KOCH, M. A.: Glycolipids of mouse fibroblasts and virus transformed cell lines. Z. Physiol. Chem. *353:* 1769–1774 (1972).
18 DISCHE, Z.: Aminosugar-containing compounds in mucuses and in mucous membranes; in BALARZ and JEANLOZ The aminosugars, pp. 115–140 (Academic Press, New York 1965).
19 DISCHE, Z.; DISAIT'AGNESE, P.; PALLAVICINI, C., and YOULOS, J.: Composition of mucoid fractions from duodenal fluid of children and adults. Arch. Biochem. Biophys. *84:* 205–223 (1959).
20 DURAND, P.; PHILIPPART, M.; BORRONE, C.; DELLA, C. e BUGIANI, O.: Una nuova malattia da acumulo di glycolipidi. Minerva pediat. *19:* 2187–2196 (1967).
21 EILSER, M. und MORITSCH, P.: Untersuchungen über gruppenspezifische Reaktionen in menschlichen Blüte. Z. Immunitatsforsch. *57:* 421–454 (1928).
22 FEIZI, T.; KABAT, E. A.; VICAR, G.; ANDERSON, B., and MARSCH, W. L.: The I antigen complex: specificity differences among anti-I sera. J. Immunol. *106:* 1578–1592 (1971).
23 FEIZI, T.; KABAT, E. A.; VICAR, G.; ANDERSON, B., and MARSCH, W. L.: The I antigen complex precursors in A, B, H, Le[a] and Le[b] blood group systems. J. exp. Med. *133:* 39–52 (1971).
24 FIGARD, P. H. and LEVINE, A. S.: Incorporation of labeled precursors into lipids of tumors induced by Rous sarcoma virus. Biochim. biophys. Acta *125:* 428–434 (1966).
25 FORSTNER, G. G. and WHERRETT, J. R.: Plasma membrane and mucosal glycosphingolipids in the rat intestine. Biochim. biophys. Acta *306:* 446–459 (1973).
26 GAHMBERG, C. G.: Proteins and glycoproteins of hamster kidney fibroblasts. Biochim. biophys. Acta *249:* 81–95 (1971).
27 GARDAS, A. and KOSCIELAK, J.: A, B, and H blood group specificities and glycolipid fractions of human erythrocyte membrane. Vox Sang. *20:* 137–149 (1971).
28 GOLD, P. and FREEDMAN, S.: Specific carcinoembryonic antigens of the human digestive system. J. exp. Med. *122:* 467–481 (1965).
29 GOLD, P.; FREEDMAN, S., and GOLD, J.: Human anti-CEA antibodies detected by radioimmunoelectrophoresis. Nature new Biol. *239:* 60–62 (1972).
30 GRIMES, W. J.: Sialic acid transferases and sialic acid levels in normal and transformed cells. Biochemistry *9:* 5083–5092 (1970).
31 GRIMES, W. J.: Glycosyltransferase and sialic acid levels of normal and transformed cells. Biochemistry *12:* 990–996 (1973).
32 HÄKKINEN, I.: A-like blood group antigen in gastric cancer cells of patients in blood groups O or B. J. nat. Cancer Inst. *44:* 1183–1194 (1970).
33 HÄKKINEN, I.; JARVI, O., and GRONROOS, J.: Sulphoglycoprotein antigens in the human alimentary canal and gastric cancer. An immunohistological study. Int. J. Cancer *3:* 572–581 (1968).
34 HAKOMORI, S.: Glycosphingolipids having blood group ABH and Lewis specificities. Chem. Phys. Lipids *5:* 96–115 (1970).
35 HAKOMORI, S.: Cell density-dependent changes of glycolipid concentrations

in fibroblasts, and loss of this response in virus-transformed cells. Proc. nat. Acad. Sci., Wash. 67: 1741–1747 (1970).
36 HAKOMORI, S.: Glycolipid changes associated with malignant transformation. 22nd Coll. der Gesellschaft für Biologische Chemie, Mosbach/Baden 1971, pp. 65–96.
37 HAKOMORI, S.: Glycolipids of tumor cell membrane; in WEINHOUSE Advances in cancer research (in press).
38 HAKOMORI, S. and ANDREWS, H. D.: Sphingoglycolipids with Le^b activity and the co-presence of Le^a and Le^b glycolipids in human tumor tissue. Biochim. biophys. Acta 202: 225–228 (1970).
39 HAKOMORI, S. and JEANLOZ, R. W.: Isolation of a glycolipid containing fucose, galactose, glucose, and glucosamine from human cancerous tissue. J. biol. Chem. 239: 3606–3607 (1964).
40 HAKOMORI, S. and JEANLOZ, R. W.: Glycolipids as membrane antigens; in AMINOFF Blood and tissue antigen, pp. 149–162 (Academic Press, New York 1970).
41 HAKOMORI, S.; KIJIMOTO, S., and SIDDIQUI, B.: Glycolipids of normal and transformed cells – a difference in structure and dynamic behavior; in FOX Membrane research, pp. 253–257 (Academic Press, New York 1972).
42 HAKOMORI, S.; KOSCIELAK, J.; BLOCH, H., and JEANLOZ, R. W.: Studies on the immunological relation between the tumor glycolipids and blood group substances. J. Immunol. 98: 31–38 (1967).
43 HAKOMORI, S. and MURAKAMI, W. T.: Glycolipids of hamster fibroblasts and derived malignant transformed cell lines. Proc. nat. Acad. Sci., Wash. 59: 254–261 (1968).
44 HAKOMORI, S.; SAITO, T., and VOGT, P. K.: Transformation by Rous sarcoma virus: effects on cellular glycolipids. Virology 44: 609–621 (1971).
45 HAKOMORI, S.; STELLNER, K., and WATANABE, K.: Four antigenic variants of blood group A glycolipids: examples of highly complex, branched chain glycolipids of animal cell membrane. Biochem. biophys. Res. Commun. 49: 1061–1068 (1972).
46 HAKOMORI, S. and STRYCHARZ, G. D.: Investigations on cellular blood group substances. I. Isolation and chemical composition of blood group ABH and Le^b isoantigens of sphingoglycolipid nature. Biochemistry 7: 1285–1286 (1968).
47 HAKOMORI, S.; TEATHER, C., and ANDREWS, H. D.: Organizational difference of cell surface hematoside in normal and virally transformed cells. Biochem. biophys. Res. Commun. 33: 563–568.
48 HAKOMORI, S. and WATANABE, K.: Glycosphingolipid sharing reactivity with both wheat germ lectin and 'carcinoembryonic antisera (Gold)': partial identity of these reactive sites. FEBS Letters (in press).
49 HALLAUER, C.: Weitere Versuche zur Isolierung wasserlöslicher Gruppenstoffe aus menschlichen Erythrozyten. Z. Immunitatsforsch. 83: 114–123 (1934).
50 HAMASATO, Y.: The blood group substance in red blood cells. First report. Tohoku J. exp. Med. 52: 17–27 (1950).
51 HAMASATO, Y.: The blood group substance in red blood cells. Second report. Tohoku J. exp. Med. 52: 29–33 (1950).

52 HAMASATO, Y.: The blood group substance in red blood cells. Third report. Tohoku J. exp. Med. *52:* 35–38 (1950).
53 HANDA, S.: Blood group active glycolipid from human erythrocytes. Jap. J. exp. Med. *33:* 347–360 (1963).
54 HILDEBRAND, J.; STRYCHMANS, P. A., and VANHOUDE, J.: Gangliosides in leukemic and non-leukemic human leukocytes. Biochim. biophys. Acta *260:* 272–278 (1972).
55 HIRAMOTO, R.; SMITH, E. L.; GHANTA, V.; SHAW, J., and McKIBBIN, J. M.: Intestinal sphingoglycolipids with A and Le[a] activity from humans and A, H-like, and Le[b]-like activity from dogs. J. Immunol. *110:* 1037–1043 (1973).
56 HOWARD, B. V. and KRITCHEVSKY, D.: The lipids of normal diploid (WI-38) and SV40-transformed human cells. Int. J. Cancer *4:* 393–402 (1969).
57 INBAR, M. and SACHS, L.: Interaction of the carbohydrate-binding protein concanavalin A with normal and transformed cells. Proc. nat. Acad. Sci., Wash. *63:* 1418–1425 (1969).
58 ISEKI, S.; FURUKAWA, K., and ISHIHARA, K.: Immunochemical studies on polysaccharides from stomach cancer. Proc. Jap. Acad. Sci. *38:* 556–566 (1962).
59 KAWASAKI, H.: Molisch-positive mucopolysaccharides of gastric cancers as compared with the corresponding components of gastric mucosae. Second report. Tohoku J. exp. Med. *68:* 119–132 (1958).
60 KAY, H. E. H. and WALLACE, B. H.: A and B antigens of tumors arising from urinary epithelium. J. nat. Cancer Inst. *26:* 1349–1366 (1961).
61 KIJIMOTO, S. and HAKOMORI, S.: Enhanced glycolipid: α-galactosyl-transferase activity in contact inhibited hamster cells and loss of this response in polyoma transformants. Biochem. biophys. Res. Commun. *44:* 557–563 (1971).
62 KOSCIELAK, J.: Blood group A specific glycolipids from human erythrocytes. Biochim. biophys. Acta *78:* 313–328 (1963).
63 KOSCIELAK, J.; PIASEK, A., and GORNIAK, H.: Studies on the chemical structure of blood group A specific glycolipids from human erythrocytes; in AMINOFF Blood and tissue antigens, pp. 163–183 (Academic Press, New York 1970).
64 KOSCIELAK, J.; PIASEK, A.; GORNIAK, H., and GARDAS, A.: Structures of fucose-containing glycolipids with H and B blood group activity and of sialic acid and glucosamine-containing glycolipid of human erythrocyte membrane. Europ. J. Biochem. *37:* 214–225 (1973).
65 KOSSJAKOW, P. N.: Polysaccharide – Träger der Gruppeneigenschaften des Menschen. Z. ImmunForsch. *99:* 221–231 (1941).
66 KULAS, H. P.; MARGGRAF, W. D.; KOCH, M. A., and DIRINGER, H.: Comparative studies of lipid content and lipid metabolism of normal and transformed mouse cells. Z. Physiol. Chem. *353:* 1755–1760 (1972).
67 LANDSTEINER, K. and SCHEER, J. VAN DER: Antigens of red blood corpuscles: flocculation reactions with alcoholic extracts of erythrocytes. J. exp. Med. *42:* 123–142 (1925).
68 LIS, H.; SELA, B.-A.; SACHS, L., and SHARON, N.: Specific inhibition by N-acetyl-D-galactosamine of the interaction between soybean agglutinin and animal cell surfaces. Biochim. biophys. Acta *211:* 582–585 (1970).

69 MARCHESI, V. T. and ANDREWS, E. P.: Glycoproteins – isolation from cell membranes with lithium diidosalicylate. Science *174:* 1247–1248 (1971).
70 MARCUS, D. and CASS, L.: Glycosphingolipids with Lewis blood group activity: uptake by human erythrocytes. Science *164:* 553–555 (1969).
71 MASAMUNE, H. and HAKOMORI, S.: On the glycoprotein and mucopolysaccharides of cancerous tissue, in particular references on 'cancer blood group substances'. Symp. Cell Chem. *10:* 37–43 (1960).
72 MASAMUNE, H.; KAWASAKI, H.; ABE, S.; OYAMA, K., and YAMAGUCHI, Y.: Molisch positive mucopolysaccharides of gastric cancers as compared with the corresponding components of gastric mucosa. First report. Tohoku J. exp. Med. *68:* 81–91 (1958).
73 MASAMUNE, H.; YOSIZAWA, Z.; OH-UTI, K.; MATUDA, Y., and MASUKAWA, A.: Biochemical studies on carbohydrates: on sugar components of hexosamine containing carbohydrates from gastric cancers. Tohoku J. exp. Med. *56:* 37–42 (1952).
74 MCKIBBIN, J. M.: The composition of glycolipids in dog intestine. Biochemistry *8:* 679–685 (1969).
75 MORA, P. T.; BRADY, R. O.; BRADLEY, R. M., and MCFARLAND, V. W.: Gangliosides in DNA virus-transformed and spontaneously transformed tumorigenic mouse cell lines. Proc. nat. Acad. Sci., Wash. *63:* 1290–1296 (1969).
76 NAIRN, R. C.; FOTHERGILL, J., and MCENTEGART, H.: Loss of gastrointestinal specific antigen in neoplasia. Brit. med. J. *i:* 1791–1793 (1962).
77 NICOLSON, G. L.: Difference in topology of normal and tumour cell membranes shown by different surface distributions of ferritin-conjugated concanavalin A. Nature, Lond. *233:* 244–246 (1971).
78 OH-UTI, K.: Polysaccharides and a glycidamin in the tissue of gastric cancer. Tohoku J. exp. Med. *51:* 297–304 (1949).
79 PANGBORN, M. C.; ALMEIDA, J. O.; MALATANER, A. M.; SILVERSTEIN, A. M., and THOMPSON, W. R.: Cardiolipin antigen. Monogr. Ser. Wld Hlth Org. No. 6, 2nd ed., pp. 9–17 (1955).
80 PHILIPPART, M.: Etude biochimique des mucopolysaccharides et sphingolipidoses. 12e Congr. de l'Ass. des Pédiatres de Langue française, Strasbourg 1969, pp. 1–41.
81 PRENDERGAST, R. C.; TOT, P. D., and GARGIULO, A. W.: Reactivity of blood group substances of neoplastic oral epithelium. J. dent. Res. *47:* 306–310 (1968).
82 RADIN, N. S.: Discussion – glycolipid chromatography. Fed. Proc. *16:* 825–826 (1957).
83 SCHACHTER, H.; MICHAELS, M. A.; CROOKSTON, M. C.; TILLEY, C. A., and CROOKSTON, J. H.: A quantitative difference in the activity of blood group A-specific N-acetylgalactosaminyltransferase in serum from A_1 and A_2 human erythrocytes. Biochem. biophys. Res. Commun. *45:* 1011–1018 (1971).
84 SCHIFF, F. und ADELSBERGER, L.: Über blutgruppenspezifische Antikorper und Antigene. Z. ImmunForsch. *40:* 335–367 (1924).
85 SHEN, L. and GINSBURG, V.: Sugar analysis of cells in culture – determination of the sugars of HeLa cells by isotope dilution. Arch. Biochem. Biophys. *122:* 474–480 (1967).

86 SIDDIQUI, B. and HAKOMORI, S.: Change of glycolipids in minimal deviation hepatoma. Cancer Res. *30:* 2930–2936 (1970).
87 SIDDIQUI, B. and HAKOMORI, S.: A revised structure for the Forssman glycolipid hapten. J. biol. Chem. *246:* 5766–5769 (1971).
88 SIMMONS, D. A. R. and PERLMANN, P.: Carcinoembryonic antigen and blood group substances. Cancer Res. *33:* 313–322 (1973).
89 SLOMIANY, A. and HOROWITZ, M. I.: Blood group A active glycolipids of hog gastric mucosa. J. biol. Chem. *248:* 6232–6238 (1973).
90 SLOMIANY, B. L.; SLOMIANY, A., and HOROWITZ, M. I.: Blood group A active ceramide hexasaccharide lacking N-acetylglucosamine isolated from hog stomach mucosa. Biochim. biophys. Acta (in press).
91 SMITH, E. L. and MCKIBBIN, J. M.: Separation of dog intestine glycolipids into classes according to sugar content by thin-layer chromatography. Anal. Biochem. *45:* 608–616 (1972).
92 SNYDER, F. and WOOD, R.: Alkyl and alk-1-enyl ethers of glycerol in lipids from normal and neoplastic human tissues. Cancer Res. *29:* 251–257 (1969).
93 STEINER, S.; BRENNAN, P. J., and MELNIK, J. L.: Fucosylglycolipid metabolism in oncoranovirus-transformed cell lines. Nature, Lond. *245:* 19–21 (1973).
94 STELLNER, K.; HAKOMORI, S., and WARNER, G. A.: Enzymic conversion of H_1 glycolipid to A or B glycolipid and deficiency of these enzyme activities in adenocarcinoma. Biochem. biophys. Res. Commun. *55:* 439–445 (1973).
95 STELLNER, K.; WATANABE, K., and HAKOMORI, S.: Isolation and characterization of glycosphingolipids with blood group H specificity from membranes of human erythrocytes. Biochemistry *12:* 656–661 (1973).
96 STEPANOV, A. V.; KUSIN, Z.; MAKAJEVA, Z., and KOSSJAKOW, P. N.: Spetsificheskiya polsakharidi krovi. Biokhimiya *5:* 547–550 (1940).
97 SUZUKI, C.; MAKITA, A., and YOSIZAWA, Z.: Glycolipids isolated from porcine intestine. Arch. Biochem. Biophys. *127:* 140–149 (1968).
98 TELLEM, M.; PLOTKIN, H. R., and MERANCE, D. R.: Studies of blood group antigens in benign and malignant human breast tissue. Cancer Res. *23:* 1528–1531 (1963).
99 TOMITA, M.; KUROKAWA, T.; OSAWA, T.; SAKURAI, Y., and UKITA, T.: Effect of trypsin treatment of mouse fibroblasts and their SV40-transformed cells on the agglutin ability by several phytoagglutinins having different sugar-binding properties. Gann *63:* 269–271 (1972).
100 WARREN, L.; FUHRER, J. P., and BUCK, C. A.: Surface glycoproteins of normal and transformed cells: a difference determined by sialic acid and a growth-dependent sialyl transferase. Proc. nat. Acad. Sci., Wash. *69:* 1838–1842 (1972).
101 WARREN, L.; FUHRER, J. P., and BUCK, C. A.: Surface glycoproteins before and after transformation by oncogenic viruses. Fed. Proc. *32:* 80–85 (1973).
102 WHERRETT, J. R. and HAKOMORI, S.: Characterization of blood group B-glycolipid accumulating in the pancreas of a patient with Fabry's disease. J. biol. Chem. *248:* 3046–3051 (1973).
103 WITEBSKY, E.: Zur Methodik der Gruppenbestimmung in menschlichen Blutflecken. Münch. med. Wschr. *74:* 1581–1582 (1927).
104 WU, H.; MEEZAN, E.; BLACK, P. H., and ROBBINS, P. W.: Comparative studies

on the carbohydrate-containing membrane components of normal and virus-transformed mouse fibroblasts. II. Separation of glycoproteins and glycopeptides by Sephadex chromatography. Biochemistry 8: 2518–2524 (1969).
105 YAMAKAWA, T. and IIDA, T.: Immunochemical study on the red blood cells. I. Globoside, as the agglutinogen of the ABO system on erythrocytes. Jap. J. exp. Med. 23: 327–331 (1953).
106 YAMAKAWA, T. and IRIE, R.: On the mucolipid nature of ABO group substance of erythrocytes. J. Biochem. 48: 919–920 (1960).
107 YAMAKAWA, T.; IRIE, R., and IWANAGA, M.: The chemistry of lipid of posthemolytic residue or stroma of erythrocytes. IX. Silicic acid chromatography of mammalian stroma glycolipids. J. Biochem. 48: 490–507 (1960).
108 YAMAKAWA, T.; MATSUMOTO, M., and SUZUKI, S.: The chemistry of lipid of posthemolytic residue or stroma of erythrocytes. VIII. The nature of hexosamine and fatty acids of blood cell sphingolipids. J. Biochem. 43: 63–72 (1956).
109 YAMAKAWA, T.; MATSUMOTO, M.; SUZUKI, S., and IIDA, T.: The chemistry of lipid of posthemolytic residue or stroma of erythrocytes. VI. Sphingolipids of erythrocytes with respect to blood group activities. J. Biochem. 43: 41–52 (1956).
110 YAMAKAWA, T.; NISHIMURA, S., and KAMIMURA, M.: The chemistry of lipid of posthemolytic residue or stroma of erythrocytes. XIII. Further studies on human red cell glycolipids. Jap. J. exp. Med. 35: 201–207 (1965).
111 YAMAKAWA, T. and SUZUKI, S.: The chemistry of lipid of posthemolytic residue or stroma of erythrocytes. III. Globoside, the sugar-containing lipid of human blood stroma. J. Biochem. 39: 393–402 (1952).
112 YANG, H.-J. and HAKOMORI, S.: A sphingolipid having a novel type ceramide and 'lacto-N-fucopentaose III'. J. biol. Chem. 246: 1192–1200 (1971).
113 YAU, T. M. and WEBER, M. J.: Changes in acyl group composition of phospholipids from chicken embryonic fibroblasts after transformation by Rous sarcoma virus. Biochem. biophys. Res. Commun. 49: 114–120 (1972).

Author's address: SEN-ITIROH HAKOMORI, Department of Pathobiology, School of Public Health and Department of Microbiology, School of Medicine, University of Washington, *Seattle, WA 98195* (USA)

The Role of Cholesteryl 14-Methylhexadecanoate in Gene Expression and its Significance for Cancer

J. HRADEC

Department of Biochemistry, Oncological Institute, Prague

Contents

I. Introduction 197
II. Cholesteryl 14-Methylhexadecanoate and Protein Synthesis 199
   A. Formation of the Aminoacyl-tRNA Complex 200
   B. Transcription of the Genetic Message 208
   C. Translation of the Genetic Message 208
III. Cholesteryl 14-Methylhexadecanoate and Malignant Growth 217
IV. The Significance of Cholesteryl 14-Methylhexadecanoate for Malignant Growth 219
V. References 222

I. Introduction

Early experiments in our laboratory suggested that an unknown factor capable of affecting protein synthesis is present in animal tissues. Saline extracts of freeze-dried tissue derived from tumor-bearing rats, when added to rat liver homogenates *in vitro*, consistently stimulated incorporation of labeled methionine into protein more than corresponding preparations from tissues of normal rats. These results indicated that a higher quantity of this material is contained in tumor-bearing animals but that this factor is also present in normal tissues [18]. Preliminary isolation experiments provided evidence that the active substance is probably of lipid nature [22]. Further characterization and, in particular, the elucidation of the chemical structure of the active substance required a richer source than rat tissues for isolation of larger amounts of this factor.

A systematic search for a suitable starting material gave evidence that egg yolks may be useful for these purposes [20]. Comparative experiments on the isolation of the active factor from this material and tissues of different animals revealed that all these various materials contain a very similar, if not the same substance stimulating protein synthesis. Evidence was also obtained that the active substance is a sterol ester with some admixture of phospholipids [41]. This material showed slight carcinogenic activity when administered subcutaneously to rats. Because of its lipid nature and carcinogenic activity, this substance was given the provisional name of carcinolipin [37]. The carcinogenic activity of carcinolipin was recently confirmed by SHABAD et al. [59] who obtained a significantly increased percentage of tumors in offspring of A mice administered purified cholesteryl 14-methylhexadecanoate.

Studies on the chemical constitution of carcinolipin required the development of methods for its final purification. Liquid-solid chromatography on silicic acid provided further evidence that the active substance is a sterol ester. Reversed-phase liquid-liquid chromatography resulted in an extensive purification of the active material and it became obvious that this substance is an ester of cholesterol with some uncommon fatty acid [39]. Since no possibility seemed to exist for the separation of individual cholesteryl esters, another approach was chosen in further purification steps of carcinolipin. The pure cholesteryl ester fraction was hydrolyzed and the resulting fatty acid mixture was separated into individual components. Purified fatty acids were then reesterified with cholesterol and the activity of such semisynthetic preparations was tested on protein synthesis in cell-free systems. The clathrate chromatography of fatty acids, newly developed in this laboratory, yielded a fatty acid mixture highly enriched in branched-chain fatty acids. Only this single fraction showed a significant protein synthesis-stimulating effect when esterified with cholesterol. This mixture of branched-chain fatty acids was further separated into several fractions by gas-liquid chromatography. Only one of these components was active in protein synthesis after esterification with cholesterol [38].

The pure preparation of carcinolipin obtained in this way was then subjected to analyses, the results of which led ultimately to the final elucidation of its chemical constitution. The presence of cholesterol in its molecule was confirmed by a comparison of the sterol moiety of carcinolipin and its derivatives with authentic cholesterol. The fatty acid present in carcinolipin was subjected to oxidative splitting followed by the gas-liquid chromatographic identification of its degradation products, and to mass spectrometry. Results of these analyses provided unequivocal evidence that the fatty acid moiety of carcinolipin is 14-methylhexadecanoic acid. This opinion was confirmed by chemical synthesis. Synthetic cholesteryl 14-methylhexadecanoate showed the same activity in protein synthesis as the natural product [31].

The elucidation of the chemical structure of carcinolipin was followed by the development of a method for the quantitative determination of cholesteryl 14-methylhexadecanoate (CMH) in biological materials [26]. In this procedure, lipids are extracted with chloroform-methanol (2:1) and cholesteryl esters are separated by thin-layer chromatography on silicic acid. The cholesteryl ester fraction is eluted and esters are subjected to alkaline hydrolysis. Fatty acids are converted into methyl esters and separated by gas-liquid chromatography. Known quantities of methyl margarate and methyl 14-methylhexadecanoate are added to samples before chromatography as internal standards. This procedure is very sensitive, specific and proved valuable for the quantitative determination of CMH in materials where only minute amounts of this ester are present (highly purified enzymes).

II. Cholesteryl 14-Methylhexadecanoate and Protein Synthesis

Results of preliminary experiments suggested that carcinolipin is capable of affecting protein synthesis, as indicated by its stimulating effect on the incorporation of labeled amino acids into proteins of tissue homogenates *in vitro* [24]. This opinion was supported by the fact that addition of this substance to incubation mixtures containing rat liver slices stimulated the net production of serum albumin in these systems [19]. Experiments with better defined cell-free systems from rat liver demonstrated that carcinolipin stimulates the incorporation of labeled amino acids into proteins in incubation mixtures composed from microsomes, cell sap and all necessary cofactors. A stimulation of amino acid activation and transfer of amino acyl-tRNA to microsome was also obtained in the presence of this substance [23]. These results indicated that carcinolipin affects protein synthesis at more than one of its steps. After the chemical constitution of this substance had been elucidated and purified CMH became available, more detailed studies were performed in our laboratory on the mode of action of this compound in gene expression.

A. Formation of the Aminoacyl-tRNA Complex

Results of Zamecnik's group provided the first experimental evidence that an important intermediate, aminoacyl-tRNA, is involved in protein synthesis [16]. They demonstrated that the charging of tRNA with amino acids is catalyzed by specialized enzyme-amino acid-tRNA ligases. These enzymes and tRNA are present in the cellular fraction of pH 5 enzymes which may be obtained by acidification of the high speed supernatant of tissue homogenates [15, 17].

In experiments of HRADEC and DUŠEK [32], pH 5 enzymes were isolated from rat liver, freeze-dried and extracted with organic solvents. Such preparations showed a significantly decreased charging of tRNA with various amino acids when compared with the corresponding nonextracted preparations. Normal activity of extracted pH 5 enzymes was obtained if extracted lipids were added to the incubation mixtures in quantities corresponding to those removed by the extraction. If individual lipid groups were separated from extracts of pH 5 enzymes and tested for their effect on the charging of tRNA with amino acids, only cholesterol and the cholesteryl ester fraction were found to be active in this respect (table I). The same reactivation of extracted pH 5 enzymes could be also obtained after the addition of pure CMH. The reactivating effect of free cholesterol was explained by the assumption that CMH could be synthesized from free cholesterol and the 14-methylhexadecanoic acid present in the fatty

Table I. Charging of tRNA with amino acids in the presence of extracted pH 5 enzymes and individual lipid fractions [32]

Lipid fraction added	Radioactivity incorporated[1]	
	nonextracted enzymes	extracted enzymes
None	311	124
Phospholipids	239	91
Cholesterol	261	295
Fatty acids	259	85
Triglycerides	255	81
Cholesteryl esters	270	308

[1] cpm/mg of tRNA.

acid pool of pH 5 enzymes by cholesterol-esterifying enzymes in this cellular fraction. The presence of such enzymes in rat liver cell sap has been established by SWELL and LAW [64]. This opinion was supported by the fact that free cholesterol does not stimulate the incorporation of labeled amino acids into proteins of rat tissue homogenates *in vitro* [21, 25]. Moreover, if the free sterol were so active, then all cholesteryl esters might be expected to give the same result as CMH. Only this ester, however, of all cholesteryl esters present in the calf liver showed a stimulating effect on protein synthesis [38]. Different quantities of CMH were extracted from pH 5 enzyme preparations with different organic solvents and the decreased activity of such preparations in charging of tRNA with amino acids was roughly proportional to the amount of the ester extracted (table II).

Similarly, different quantities of cholesteryl 14-methylhexadecanoate were extracted with different organic solvents from freeze-dried post-microsomal supernatant of rat liver. In the presence of such extracted preparations, no incorporation of labeled amino acids into proteins could be demonstrated in systems containing ribosomes, cell sap and all necessary cofactors. A complete reactivation of these systems could be obtained by the addition of extracted lipids or CMH.

These results indicated that lipids are involved in the synthesis of aminoacyl-tRNA complex and that CMH is apparently the only lipid active in this respect. They also indicated that this compound may directly

Table II. Quantities of CMH extracted from different pH 5 enzyme preparations and the residual activity of such preparations [32]

Preparation No.	pH enzymes extracted with			
	light petroleum	benzene	ether	chloroform
1. CMH-extracted[1]	43.7	43.7	67.8	81.7
Activity[2]	62	70	41	20
2. CMH-extracted	42.8	42.0	59.3	74.4
Activity	66	69	44	23
3. CMH-extracted	48.0	49.3	75.1	88.2
Activity	55	59	36	14

1 μg CMH extracted/g of enzyme protein.
2 Percent of the value obtained with the control nonextracted preparation.

affect the function of enzymes required for these processes. This possibility was tested in further experiments.

Amino acid-tRNA ligases (EC6. 1. 1.-) catalyze the formation of an aminoacyl-tRNA complex which is a substrate required for translation of the genetic message. They operate in the following way [2]:

$$\text{Amino acid} + \text{ATP} + \text{enzyme} \rightarrow \text{aminoacyl-AMP-enzyme} + \text{PP}_i, \quad (1)$$
$$\text{Aminoacyl-AMP-enzyme} + \text{tRNA} \rightarrow \text{aminoacyl-tRNA} + \text{AMP} + \text{enzyme}. \quad (2)$$

Specific amino acid-tRNA ligases were prepared from various animal tissues and the CMH contents were determined in fractions obtained during the purification procedures [33]. The CMH contents increased significantly during the purification of all enzymes used. This was true not only for L-tyrosine-tRNA ligase (table III) and for the L-alanine-tRNA ligase (table IV) but also for enzymes synthesizing the L-tryptophanyl-tRNA and L-threonyl-tRNA complex. Increase in CMH content during the purification procedure was roughly proportional to the increased specific activity. Extraction of purified L-threonine-tRNA ligase with organic solvents significantly decreased the activity of this enzyme in

Table III. Specific activity and CMH contents of fractions obtained during the purification of L-tyrosine-tRNA ligase [33]. Enzyme was purified according to Schweet [58]

Step No.	Fraction	Specific activity[1]		CMH content[2]
		tRNA	hydroxamate	
1	Postmicrosomal supernatant	2,500	54	0.88
2	Calcium phosphate gel eluate	26,000	76	2.17
3	First $(NH_4)_2SO_4$ precipitate	33,200	85	3.29
4	Second $(NH_4)_2SO_4$ precipitate	225,000	2,580	9.13
5	DEAE-cellulose column eluate	465,000	5,920	40.70

1 cpm/mg of tRNA or cpm in hydroxamate incorporated per mg of protein.
2 nmol/mg of protein.

Table IV. Specific activity and CMH contents of fractions obtained during the purification of L-alanine-tRNA ligase [33]. Enzyme was purified by the method of WEBSTER [68]

Step No.	Fraction	Specific activity[1]	CMH content[2]
1	Crude supernatant	18	0.090
2	pH 5 enzymes	310	0.219
3	Second $(NH_4)_2SO_4$ precipitate	830	0.934
4	Calcium phosphate gel eluate	1,472	2.030
5	DEAE-cellulose eluate	35,000	19.250

1 cpm/mg of tRNA incorporated per mg of protein.
2 nmol CMH/mg of protein.

Table V. Effect of extraction with various solvents on the activity of L-threonine-tRNA ligase [33]. The enzyme was purified according to ALLENDE *et al.* [2]

Solvent	Radioactivity in tRNA[1]		CMH extracted[2]
	− CMH	+ CMH	
Ether	137	340	8.93
Chloroform	56	312	18.20
n-Hexane	298	397	6.80
Benzene	164	400	14.90
Cyclohexane	227	423	5.68
Isooctane	267	416	5.44
Ethylene glycol monoethyl ether	151	398	15.25
Control value (unextracted enzyme)	445	−	−

1 cpm/mg of tRNA.
2 nmol/mg of protein.

catalyzing L-threonyl-tRNA complex formation. Normal activity could be obtained after the addition of CMH (table V). Apparently aminoacyl adenylate formation is also affected by the extraction of ligases with organic solvents as indicated by results obtained with L-tryptophan-tRNA ligase (table VI). No exact agreement was found, however, between the results of the hydroxamate assay which measures the first step and of the aminoacyl-tRNA formation which assays the second step of the reaction

Table VI. Effect of the extraction with various solvents on the activity of L-tryptophan-tRNA ligase [33]. The enzyme was purified by the method of DAVIE [8]

Solvent	Radioactivity in hydroxamate[1]		
	− CMH	+ CMH	CMH extracted[2]
Benzene	0	143	6.20
Ether	0	173	8.92
Chloroform	0	147	10.80
Carbon tetrachloride	0	152	12.95
n-Hexane	54	169	5.64
Control value (unextracted enzyme)	199	—	—

1 cpm/incubation mixture.
2 nmol/mg of protein.

catalyzed by amino acid-tRNA ligases. It thus seemed that in multifunctional enzymes catalyzing more than one reaction, the function of CMH may be limited to one specific reaction only. Apparently both functions of amino acid-tRNA ligases are largely independent of each other, the activity for the aminoacyl-tRNA formation being more sensitive to different treatments. This opinion is supported by the results published by PAPAS and MEHLER [56]. The time course of formation of aminoacyl-tRNA complex with native and extracted enzymes indicated that a certain amount of time is required for the recombination of added CMH with enzyme protein before complete reactivation of the enzyme is obtained (fig. 1).

These results provided evidence that CMH forms part of the molecule of amino acid-tRNA-ligases required for normal enzymatic activity. Although some quantitative differences were found with different enzymes in this respect, this conclusion seems to be generally valid for most, if not all, amino acid-tRNA ligases.

This opinion seems to be supported by the results of BANDYOPADHYAY and DEUTSCHER. These authors reported that all amino acid-tRNA ligases and a large proportion of tRNA of rat liver can be isolated as a high molecular weight complex [3]. Furthermore, they have presented evidence that this complex contains a high proportion of lipids (up to 20–25 % of

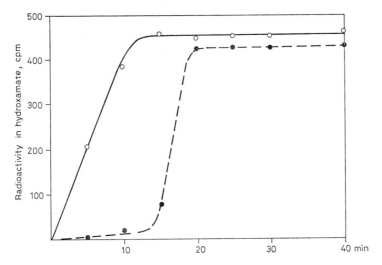

Fig. 1. Time course of formation of L-tyrosyl hydroxamate in the presence of nonextracted (○) and extracted L-tyrosine-tRNA ligase with CMH added (●) [33].

the protein present) and that these lipids are almost exclusively cholesteryl esters. One of the major fatty acid components that they have identified from the cholesteryl esters is 14-methylhexadecanoic acid. Since the removal of lipids from the ligase complex leads to its dissociation, BANDYOPADHYAY and DEUTSCHER [4] conclude that cholesteryl esters may be involved in maintaining its structure.

Formation of the aminoacyl-tRNA complex provides an important intermediate required absolutely for protein synthesis, not only in mammalian but also in more primitive prokaryotic cells [51]. It thus seemed of interest to establish whether lipids are involved in the formation of this complex in simpler synthetic pathways in bacteria. Experiments of HRADEC and DUŠEK [34] with subcellular systems isolated from *E. coli* provided evidence that a similar participation of lipids in the aminoacyl-tRNA synthesis operates in prokaryotic as in eukaryotic cells. Moreover, a cross-reactivity has been established between lipids contained in bacterial enzymes and CMH, indicating that the active lipid in *E. coli* may have a chemical structure similar to that of CMH. The chemical constitution of this bacterial lipid has not been determined as yet but it seems to be also an ester of 14-methylhexadecanoic or a similar branched-chain fatty acid.

The function of CMH in the formation of the aminoacyl-tRNA complex is apparently highly specific and a close relation obviously exists between the chemical structure of this compound and its activity. Esters of cholesterol with straight-chain fatty acids are quite inactive in this respect [38]. Of several esters of cholesterol with branched-chain fatty acids synthesized in this laboratory, only cholesteryl 13-methylpentadecanoate and 15-methylheptadecanoate, besides CMH, were capable of reactiving extracted preparations of L-leucyl-tRNA synthetase (table VII) [40]. Bearing in mind that these two active cholesteryl esters contain even-numbered anteiso-acids which do not occur in biological materials [1], it appears that a rather close relationship exists between the chemical structure and the biological activity of CMH. Results of these and other experiments indicate that the presence of a fatty acid with a definite chain-length and the anteiso-type of branching in the molecule is essential for the activity in protein synthesis. The free 14-methylhexadecanoic acid is inactive but the presence of cholesterol is apparently not essential for the activity. The methyl ester of 14-methylhexadecanoic acid showed an activity comparable with that of the cholesteryl ester [Hradec, unpublished results]. The presence of an ester bond in the molecule is thus obviously the other condition required for the activity of CMH in protein synthesis.

Involvement of CMH as a cofactor in the aminoacyl-tRNA complex formation may have important consequences for the regulation of gene

Table VII. Effect of cholesteryl esters with different branched-chain fatty acids on the charging of rat liver tRNA with [^{14}C]leucine [40]

Carbon atoms	Fatty acid in the cholesteryl ester branching	name	pH 5 enzymes[1] nonextracted	extracted
C_{14}	*anteiso*	11-methyltridecanoic	28.4	17.7
C_{15}	*anteiso-*	12-methyltetradecanoic	28.9	17.2
C_{16}	*iso-*	14-methylpentadecanoic	29.4	17.3
	anteiso-	13-methylpentadecanoic	39.5	28.6
C_{17}	*iso-*	15-methylhexadecanoic	27.3	18.1
	anteiso-	14-methylhexadecanoic (CMH)	40.2	29.3
C_{18}	*iso-*	16-methylheptadecanoic	28.7	17.5
	anteiso-	15-methylheptadecanoic	40.3	29.1
C_{19}	*anteiso-*	16-methyloctadecanoic	28.1	18.3
–	–	control value	29.1	17.6

[1] pmol L-leucine incorporated into the trichloroacetic acid-insoluble portion of the sample.

expression. Some authors presented evidence that the intracellular concentration and availability of aminoacyl-tRNA is a factor limiting gene expression at its translational level [10]. The charging of tRNA with amino acids is directly dependent on the function of amino acid-tRNA ligases and these enzymes are thus indirectly involved in the modulation of protein synthesis.

Experiments of KOMÁRKOVÁ and HRADEC [45] provided further evidence for the important role of CMH in aminoacyl-tRNA synthesis. If this ester was administered to rats, an accumulation of CMH in the liver cell sap appeared some time after the injection and was accompanied by a significant increase in amino acid-tRNA ligase activity. At different intervals after the administration, different concentrations of CMH were found in the high-speed supernatant of the liver tissue and differences in leucine-tRNA ligase activity proportional to these changes in CMH contents were demonstrated (fig. 2).

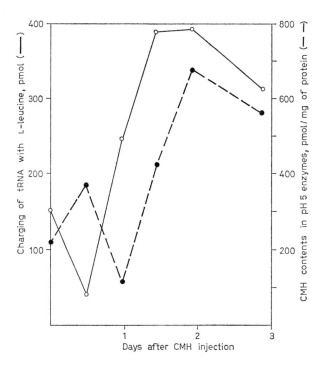

Fig. 2. Effect of CMH administration on the activity of L-leucine-tRNA ligase (○) and CMH contents in the pH 5 enzymes (●) of rat liver [45].

B. Transcription of the Genetic Message

The key enzyme involved in the transcription into mRNA of the genetic message stored in nuclear DNA is DNA-dependent RNA polymerase (nucleoside triphosphate: RNA nucleotidyl transferase, EC 2.7.7.6.) [61]. KOMÁRKOVÁ and HRADEC [46] demonstrated that the administration of CMH to rats enhances the activity of this enzyme in the nuclei of rat liver tissue. Addition of the ester to incubation mixtures containing solubilized liver nuclei significantly stimulated the activity of RNA polymerase *in vitro*. If the enzyme was purified from rat liver nuclei according the methods of MATTHAEI's group [44], the amount of CMH bound to enzymic protein gradually increased during the purification procedure. Extraction of the purified enzyme with organic solvents resulted in a highly decreased activity of such preparations and the extracted enzyme could be reactivated by addition of appropriate amounts of CMH [48]. These results, very similar to those obtained with amino acid-tRNA ligases, strongly indicate that CMH plays an important role in RNA polymerase activity comparable with that demonstrated for amino acid-tRNA ligases [33, 34, 45]. Also, MENON [54] presented evidence that lipids may be important for the function of RNA polymerase.

According to the theory of JACOB and MONOD [43], the quantity of proteins produced in the cell is dependent on the quantity of mRNA available for the translation of the genetic message. The activity of these translational processes is thus directly related to the activity of RNA polymerase which produces mRNA required for protein synthesis. Since CMH is involved in the activity of this latter enzyme, apparently as an essential cofactor, this compound may thus also mediate the regulation of gene expression at its transcriptional level.

C. Translation of the Genetic Message

Translation of the genetic message which takes place on the ribosome and is directed by mRNA, may be divided into three principal phases: initiation, elongation and termination of the peptide chain [52].

Insufficient data are available to determine the possible role of CMH in peptide initiation. Nevertheless, experiments described by KOMÁRKOVÁ and HRADEC [47] indicate that this compound affects the binding of informational macromolecules to rat liver ribosomes. Administration of CMH to rats was followed by a significant increase in the binding of tritiated mRNA to liver ribosomes in 24 h after the injection (table VIII). These results seem to suggest that CMH is involved in the combination of mRNA with ribosome which represents the starting reaction in peptide initiation.

Peptide elongation is initiated by the binding of aminoacyl-tRNA to ribosome. This step is catalyzed by the binding enzyme [53] or transferase I [42]. HRADEC et al. [35] presented evidence that during the purification of this enzyme from rat liver or human tonsils the quantity of CMH bound to enzymic protein increases and that this increase is roughly proportional

Table VIII. Binding of [^{3}H]nRNA to the liver ribosomes of rats injected with CMH [47]

Hours after injection	nRNA bound to ribosomes[1]	p
12	1,110 ± 176 (5)	0.1
24	1,295 ± 63 (5)	0.001
36	785 ± 152 (5)	0.001
48	568 ± 284 (5)	0.001
72	942 ± 187 (5)	0.8
Controls	968 ± 41 (5)	—

[1] cpm of [^{3}H]nRNA bound/mg of ribosomes. Values are means ± SD with the number of observations in parentheses.

Table IX. CMH contents of fractions obtained during the purification of peptide elongation factor from human tonsils [35]

No.	Fraction	Relative purification		CMH contents[1]	
		trans-ferase I	trans-ferase II	trans-ferase I	trans-ferase II
II	Postmicrosomal supernatant	1	1	0.492	0.492
III	(NH$_4$)$_2$SO$_4$ precipitate	4.8	4.8	0.472	0.472
IV	Sephadex G-200 eluate	40	25	2.98	1.40
V	Sepharose 4 B or DEAE-cellulose eluate	115	87	8.00	3.56

[1] nmol of CMH/mg of protein.

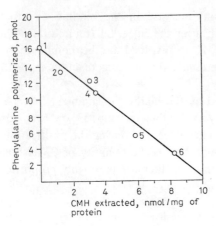

Fig. 3. Relation between the quantity of CMH extracted and the residual activity of extracted transferase I. Point 1 represent the activity of the control non-extracted binding enzymes of human tonsils. The other enzyme preparations were extracted with: (2) n-pentane; (3) ethyl ether; (4) toluene; (5) benzene, and (6) chloroform [35].

Table X. Effect of extraction with various organic solvents on the binding activity of rat liver transferase I [35]

Solvent	CMH extracted[2]	[³H]phenylalanine bound to ribosome[1]	
		ester extracted	ester replaced
Ether	2.20	5.7	11.8
Chloroform	3.58	3.9	13.8
Isooctane	4.26	1.7	11.0
Carbon tetrachloride	3.08	4.6	11.6
Benzene	2.06	5.8	11.3
Control value (unextracted enzyme)	–	11.6	

1 pmol of phenylalanine bound per reaction mixture.
2 nmol/mg of protein.

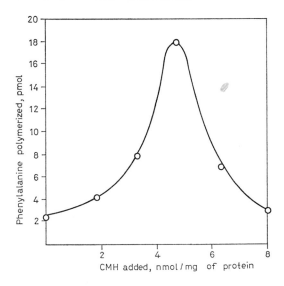

Fig. 4. Effect of different doses of CMH on the activity of extracted transferase I [35].

to the increase in specific activity of the enzyme (table IX). If purified preparations of transferase I were extracted with different organic solvents, different amounts of the ester were removed from a given enzymic preparation. The decrease in activity of such extracted enzymes was directly proportional to the quantity of CMH extracted (fig. 3). Enzymes from both rat liver and human tissues could be fully reactivated by the addition of CMH (table X). Only quantities of the ester replacing exactly those removed by the extraction were fully active, lower as well as higher doses of CMH having a lesser effect (fig. 4). These results seemed to support the opinion that CMH is present in the molecule of transferase I and that its presence is essential for the normal function of this enzyme, as with amino acid-tRNA ligases. No detailed data could be obtained in these experiments on the exact mode of action of this compound. However, the relatively low contents of CMH in both aminoacyl-tRNA synthetase and transferase I may indicate that this compound has a catalytic rather than stoichiometric function.

Transferase I is a typical multifunctional enzyme. As revealed by the experiments of HRADEC [27, 28], this enzyme forms intermediary complexes with aminoacyl-tRNA and with GTP before binding the aminoacyl-tRNA to ribosome. This enzyme was thus very suitable for a detailed study on the function of CMH in the binding reaction in respect to whether it is involved in all steps catalyzed by the binding enzyme or whether its function is limited to some particular specific step only. In experiments described by HRADEC [29], transferase I was extracted with isooctane and the activity of such extracted preparations was then tested in various intermediate reactions catalyzed by transferase I. These experiments revealed that the only step affected by the extraction of the enzyme was the formation of a complex between transferase I and the aminoacyl-tRNA. The extraction of the enzyme did not alter its binding capacity for GTP nor the formation of the ternary complex enzyme-aminoacyl-tRNA-GTP with ribosome. No binding of aminoacyl-tRNA to the ribosome is apparently possible without the prior combination of aminoacyl-tRNA with the binding enzyme [27]. This reaction thus represents the limiting step of the whole binding reaction. Since CMH affects this reaction step, its association with the binding enzyme may represent a regulatory mechanism in the supply of ribosomes with aminoacyl-tRNA.

Another enzyme required for peptide elongation is transferase II [62] or peptidyl-tRNA translocase [60]. CMH is apparently again a component essential for the normal function of this enzyme. As with transferase I, the relative concentration of this compound (per mg of protein) increases during enzyme purification. Preparations of purified translocase extracted with different organic solvents again showed a variable decrease of activity and this decrease in a given enzyme preparation was directly proportional to the quantity of CMH extracted. Extracted enzymes may be again fully reactivated by CMH (table XI). In systems containing extracted preparations of both elongation factors (transferase I and II), peptide synthesis was limited by the residual activity of the less active transfer factor, although these systems were fully reactivated by appropriate additions of CMH (table XII). Evidence was also obtained in these experiments that CMH is essential not only for the normal function of enzymes involved in peptide elongation but that it also plays a role in the function of ribosomal binding sites required for peptide elongation.

Extraction of ribosomes and ribosomal subunits adsorbed on a cellulosic ion-exchanger and freeze-dried, revealed that even highly purified particles contain small but constant amounts of CMH [36]. Based on the results of chloroform-methanol (2:1) extractions, the smaller ribosomal subunit contains two molecules of CMH while in the larger subunit one molecule of this ester is present. Unlike the elongation enzymes, CMH is apparently firmly bound to the ribosomal structure and very effective

Table XI. Effect of extraction with various organic solvents on the activity of rat liver translocase [35]

Solvent	[^{3}H]phenylalanine polymerized[1]		
	CMH extracted[2]	ester extracted	ester replaced
Benzene	1.39	7.50	15.25
Ether	1.12	8.60	15.05
Isopropyl ether	0.87	11.85	14.70
Isooctane	2.25	0.15	14.50
Control value (unextracted enzyme)	—		14.80

1 pmol/incubation mixture.
2 nmol/mg of protein.

Table XII. Polyphenylalanine synthesis with both extracted transfer factors from human tonsils [35]

Extracting solvent		Residual activity[1]		[^{14}C]phenylalanine polymerized[2]	
transferase I	transferase II	trans-ferase I	trans-ferase II	ester extracted	ester replaced
Ethyl ether	ethyl ether	73	65	4.34	8.52
n-Pentane	isooctane	80	12	1.25	8.35
Chloroform	ethyl ether	28	60	2.54	8.65
Toluene	ethyl ether	32	63	2.30	9.30
Isopropyl ether	isooctane	85	12	1.78	8.54
Ethyl ether	isopropyl ether	65	80	4.00	9.05
Benzene	isooctane	35	6	0.41	8.24
Control value (unextracted enzymes)		100	100	8.48	

1 Percent of the control value.
2 pmol/reaction mixture.

solvent mixtures must be used for its extraction. The extracted smaller ribosomal subunit binds significantly more aminoacyl-tRNA than the corresponding nonextracted particle. After the addition of CMH into systems containing extracted smaller subunits, the binding activity decreases and values corresponding to those obtained in control nonextracted systems are obtained (table XIII). With extracted ribosomes also,

Table XIII. Binding of [³H]phenylalanine-tRNA to extracted ribosomal smaller subunits [36]

Solvent	[³H]phenylalanine bound[1]	
	no CMH added	CMH replaced
Isooctane	0.38	0.29
Chloroform-methanol (2:1)	0.41	0.25
Control value (unextracted subunits)	0.27	

1 pmol/reaction mixture.

the activity of the transferase II – dependent binding of GTP is significantly increased and normal values may be obtained if CMH is added to these systems. The binding of the initiating aminoacyl-tRNA to ribosomes or smaller subunits is affected by the extraction of the particles in the same way as the binding of elongation aminoacyl-tRNA [HRADEC, unpublished results]. Extraction of ribosomes with organic solvents inhibits almost completely the peptide elongation reactions in the presence of native transferases. Unlike other ribosomal functions, no complete reactivation could the obtained after the addition of CMH, although some stimulation was noted.

Peptidyl transferase, which catalyzes the formation of the peptide bond and which is an integral component of the larger ribosomal subunit [66], was the last enzyme of the peptide elongation cycle whose activity was tested in our laboratory with respect to the possible involvement of CMH in its function [36]. Extracted ribosomes showed *in vitro* a significantly increased activity of the A site of peptidyl transferase (fig. 5). It cannot be excluded, however, that this increased activity was induced by an enhanced binding of initiating aminoacyl-tRNA to extracted particles as mentioned above. Anyway, normal activity was obtained after the addition of CMH to such extracted systems. On the other hand, a decreased activity of the P site of peptidyl transferase was demonstrated with extracted ribosomes or larger ribosomal subunits (fig. 6). Addition of CMH increased the peptidyl transferase activity in such systems up to normal levels if quantities of the ester were used replacing exactly those extracted. An important function of CMH in peptidyl transferase activity

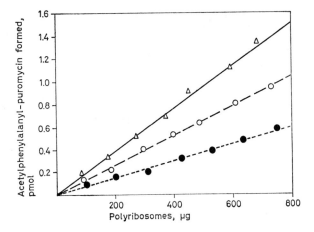

Fig. 5. The activity of the A site of peptidyl transferase in nonextracted polysomes (●) and particles extracted with ether (○) and isooctane (△) [36].

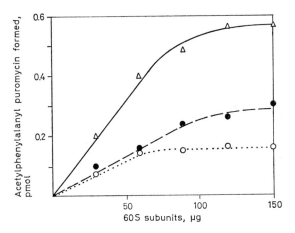

Fig. 6. Peptidyl transfer with nonextracted 60S ribosomal subunits (△) and particles extracted with ether (●) and isooctane (○) [36].

was also indicated by the results reported by KOMÁRKOVÁ and HRADEC [47]. The administration of CMH to rats induced significant changes in the activity of both binding sites of this enzyme at definite intervals after the injection.

Results obtained with ribosomes indicate that the ester affects not only the function of the A site localized on the smaller ribosomal subunit

[67] but also some functions of the larger ribosomal subunit. According to RICHMAN and BODLEY [57], there are at least four different functional regions on the larger subunit: a portion of the P site, a portion of the A site, peptidyl transferase site and a GTPase site. Our experiments provided good evidence that CMH affects the function of the A site to which aminoacyl-tRNA becomes bound [67] and the peptidyl transferase site as indicated by changes in peptidyl transferase activity of extracted ribosomes and larger subunits. Changes in the binding of GTP to extracted ribosomes indicate that the ester may also affect the GTPase site. Only indirect evidence, however, could be presented for the effect of CMH on the ribosomal P site. At the moment we may conclude that a partially irreversible 'uncoupling' of the ribosomal A and P sites may be induced by the extraction and that the peptidyl-tRNA translocation is thus affected and this leads to a decreased peptide elongation.

It is probable that the ester present in both ribosomal subunits is bound to some ribosomal proteins in much the some way as it is apparently bound to enzymic proteins [33, 35]. Evidence has been presented by KURLAND [49] that some ribosomal proteins play a structural role in ribosome assembly while others are required for ribosomal functions. It seems possible that CMH is bound to some of these specific functional proteins and may in this way affect ribosomal activity. Alternatively, the ester could be bound to some other protein and affect the configuration of the ribosome as a whole and thus affect function.

Unlike enzymes from which CMH has been removed by extraction and which always showed a decreased activity, the extraction of this ester from ribosomes is in some instances followed by an enhanced activity of such particles. This seems to indicate that the ester may have some regulatory role and that some ribosomal functions may be modulated by the quantity of the ester present.

The initial reaction in peptide elongation, i.e. the binding of aminoacyl-tRNA to ribosome, is apparently another step of gene expression modulating the intensity of protein synthesis and determining the quantity of newly synthesized protein. This is supported by the fact that the decreased protein synthesis in muscle cells of diabetic rats is due to a decreased activity of the binding enzyme [50]. Evidence has been also presented [11, 12] that the increased production of proteins in kidney cells of nephrotic rats is induced by an increased activity of transferase I. A similar mechanism is apparently involved in the increased protein synthesis in rat spleen during the immunity reaction [69]. Since the activity of the binding enzyme is obviously highly dependent on the presence of CMH, this ester may thus modulate the activity of transferase I and by this mechanism affect the quantity of newly synthesized protein.

III. Cholesteryl 14-Methylhexadecanoate and Malignant Growth

Preliminary experiments on the isolation of CMH from different animal tissues suggested that higher quantities of this compound are present in tumor tissue than in other tissues [22, 41]. Detailed studies on the distribution of CMH in the tumor and other tissues of tumor-bearing mice indicated that cancer tissue contains 5–10 times more CMH than other tissues. Thus, an average of about 20 mg CMH/100 g of wet tissue was found in 4 different mouse tumors (sarcoma 37, sarcoma 180, Ehrlich carcinoma and Krebs carcinoma) whereas only about, 4.5 mg CMH/100 g wet tissue is present in normal mouse liver and even less in other mouse tissues (spleen, kidney and lung) [30].

No detailed data are available so far on the distribution of cholesteryl esters in tumors but some evidence may nevertheless be found that tumor growth is accompanied by changes in cholesterol metabolism. Evidence has been presented by SNYDER et al. [63] that hepatomas contain more cholesteryl esters than normal liver tissue. Moreover, the ester concentration is apparently related to tumor malignancy since the ester content increases as hepatomas become more deviated. As reported by DAY et al. [9], tumor tissue contains large quantities of cholesterol. An accumulation of this sterol in tumors was also found by other authors [70]. Differences in total cholesterol content were also found in subcellular particles of tumors and normal tissues. Thus, microsomes of hepatomas cells contain more total cholesterol than corresponding particles from normal liver cells [65]. The total cholesterol content of tumors is apparently dependent on the growth rate of a particular tumor, as indicated by an increase of total cholesterol content in Walker tumor tissue [6]. Moreover, malignant growth in rats appears to affect the cholesterol content of tissues of host animals [7]. All these reports indicate that a close relationship exists between the metabolism of cholesterol and cholesteryl esters and tumor growth.

A relatively large variability was found in CMH content of tissues of tumor-bearing mice and rats at different stages of tumor-growth, indicating that changes in the distribution of this ester may be related intimately to the tumor growth. A series of experiments with rats bearing different transplantable tumors, in which CMH content was determined in the blood serum, liver tissue and, in some instances in the tumor tissue, revealed that very similar and highly characteristic changes of CMH metabolism occur during tumor growth in these animals [30]. With all transplantable tumors tested (Walker 256 carcinoma, Yoshida ascites tumor, and the Dutch ovarian tumor), a sudden decrease of CMH level in serum and liver tissue appeared a few hours after tumor transplantation. Depending on the growth rate of the tumor used, this period of decreased CMH content lasted 2 (Yokhida tumor) to 8 days (Dutch tumor). A gradual increase of CMH level in both serum and liver tissue appeared just at the period when the tumor started to grow progressively and significantly increased CMH content persisted for the whole period of the rapid tumor growth. At the terminal stages of the tumor growth, the content in serum and liver decreased to significantly subnormal levels. Very similar changes in CMH level were also found to occur during chemical carcinogenesis. After the administration of benzo(a)pyrene, a significant decrease of CMH level in liver tissue and serum occurred, lasting for several weeks. A few days before the appearance of the first tumors, a sudden increase of CMH content, first in the serum and later on in the tumor tissue, was found. Again a period of decreased CMH content followed the ultimate stages of tumor growth.

Since virtually identical changes in CMH content were demonstrated with different transplantable tumors and during chemical carcinogenesis and, moreover, since the duration of individual periods of decreased or increased levels of CMH were strictly dependent on the growth rate of the tumor used, the conclusion seems justified that the changes described are characteristic of malignant growth. This opinion was supported by further results. In rats transplanted with the Dutch tumor, in which a spontaneous regression of tumors occurred, followed by an inflammatory reaction, significant differences were found, compared with animals bearing progressively growing tumors. Instead of the characteristic drop in CMH level in the ultimate period of tumor growth, in this case the ester content remained significantly increased over the normal value during the whole duration of the inflammatory reaction. Experiments with rats in which a chronic inflammatory reaction was induced by administration of turpentine suggested that chronic inflammation is apparently accompanied by an increase of CMH level in the serum and liver. Unlike the effect of tumor growth, a sudden increase of CMH concentration in the liver and serum appears in the acute phase of the inflammation. During healing of the inflammatory process, the level of CMH gradually returns to its normal value. Thus, during chronic inflammation, no periods of decreased concentration of the ester are found as during tumor growth. The impairment of liver function is also accompanied by a large increase of CMH content in the liver and serum, as revealed by experiments with rats administered carbon tetrachloride or subjected to a partial hepatectomy. No decreased CMH levels, characteristic of tumor growth, were found in these experiments.

All results presented in this chapter seem to indicate that tumor growth in animals is accompanied by significant changes in CMH metabolism. In general, a higher demand for CMH seems to exist in the organism of tumor-bearing animals. This is indicated by (a) a very high content of this ester in the liver of tumor-bearing animals, which may be the result of its intensified synthesis (see below), and (b) increased levels in the blood serum of these animals, reflecting enhanced processes for its transport to other tissues, probably mainly into the tumor tissue. Since CMH, like other cholesteryl esters, is apparently transported in lipoproteins bound to the protein moiety, the finding of an enhanced production of lipoproteins in tumor-bearing animals seems to be in good agreement with this assumption [55]. The subject of lipoproteins in relation to cancer is reviewed in another chapter of this volume by BARCLAY and SKIPSKI [5].

IV. The Significance of Cholesteryl 14-Methylhexadecanoate for Malignant Growth

The tumor as a rapidly growing tissue must have a more rapid turnover of protein and hence more intensive protein synthesis than normal tissues. Moreover, protein synthesis is apparently also enhanced in other tissues of tumor-bearing animals, at least during early stages of tumor growth. Evidence has been obtained recently in this laboratory [DUŠEK, unpublished results) that the activity of several enzymes involved in the translation of the genetic message increases significantly in the liver of Walker tumor-bearing animals. The activity of ribosomes seems also to be increased.

No evidence has been presented as yet that there are qualitative differences in mechanisms involved in gene expression between normal and tumor cells. Some alterations of tRNA were found in neoplasia, but these changes seem to be related to cell differentiation rather than specifically to malignant growth [71]. The results obtained so far indicate that the same general mechanisms are operative in the biosynthesis of proteins in all cellular types. Although many more studies are obviously required to settle this question, the present evidence does not indicate that the mechanism of protein synthesis in cancer cells differs in any fundamental aspect from that of normal cells (see GRIFFIN and BLACK [13] for a recent review). It thus seems that the results obtained in this laboratory on the role of CMH in gene expression in normal liver cells are generally applicable to malignant cells also.

The well-documented fact [33, 35, 48] that CMH is present as an integral component of the molecules of several enzymes required for protein synthesis and that its presence is essential for the normal function of these systems seems, in particular, to be of utmost importance. It follows from this fact that a correlation should exist between CMH content and the intensity of protein synthesis in a particular tissue. Several results seem to support this conclusion. Thus, significantly more CMH is present in tissues actively engaged in protein synthesis (liver) than in other tissues less active in this respect (spleen, lung) [30]. The liver tissue of tumor-bearing animals at initial phases of tumor growth shows, as mentioned above, an enhanced protein synthesis and contains significantly more CMH than normal liver [30]. Administration of CMH to rats induces an increase of aminoacyl-tRNA synthetase activity which is accompanied by a proportional increase of CMH in pH 5 enzymes in the liver tissue [45]. Similarly, enhanced peptide elongation reactions induced by the administration of the ester to rats are accompanied at the same time after the injection by an increase of CMH content in the liver cell sap which is roughly proportional to the increased enzymic activity [47]. Thus, rather strong evidence is available that there is a close relation between the quantity of CMH present in a given enzyme fraction and the activity of this enzyme.

The regulating and modulating role of CMH in enzymes required for gene expression is indicated by the fact that enzymic activities may be increased or decreased simply by varying the amount of the ester added to a particular system. In some instances, the presence of the ester alone can regulate and limit the enzymic activity, as demonstrated by some enhanced ribosomal functions in particles from which CMH has been removed [36]. Besides this regulatory activity at the level of enzyme molecules, CMH is also involved in the regulation of protein synthesis as a whole, being present in enzymes which catalyze limiting and modulating steps of gene expression. Transferase I may be taken as an example of this.

No data can be presented as yet on the molecular mechanisms by which addition of CMH to a particular enzyme fraction induces an enhancement of enzymic activity. Two general mechanisms may be used for the explanation of this fact. First of all, the quantity of CMH present in the enzyme molecule may increase. Another probably more appropriate explanation would be the assumption that some quantity of preformed reserve protein molecules free of CMH may exist in the enzyme fraction which are readily able to combine with an excess of CMH to produce additional molecules of fully active enzyme. Further experiments are required to clear this point.

It follows from these assumptions that a demand must exist in all animal tissues for a constant supply of CMH. It seems that this requirement cannot be autonomously fully met by the cells of all tissues. Recent results obtained in this laboratory by KvíčALA [unpublished results] indicate that the cell sap of tumor cells is capable of esterifying cholesterol only to a very limited extent, although the liver high-speed supernatant is very active in this respect. The high demand for CMH that apparently exists during the initial stages in rapidly growing tumor tissue must hence be met by a supply of preformed CMH from outside. The main site of cholesteryl ester production, aside from blood serum, is apparently the liver tissue. A gradual increase in cholesterol esterification was found in this laboratory in liver homogenates during the initial phases of growth of the Walker tumor in rats [KvíčALA, unpublished results). Separation and quantitative determination of individual radioactive cholesteryl esters produced by biosynthesis *in vitro,* using methods recently developed in this laboratory [14], provided evidence that CMH is the only ester whose production is significantly enhanced in the liver of tumor-bearing animals at the initial phases of tumor growth. On the other hand, at the terminal stages of tumor growth the intensity of CMH biosynthesis is significantly below the normal level [unpublished results]. These results are in good agreement with the finding of changes in CMH levels in blood serum and the liver tissue during the growth of transplantable tumors [30].

It may be concluded that the presence of tumor requiring relatively high amounts of CMH for its intensive protein synthesis activates an enzymic system required for CMH synthesis in the liver. This activation is apparently only gradual and it may well happen that the liver tissue immediately after tumor transplantation is not able to meet fully these requirements of the tumor tissue for CMH. This can lead to a depletion of the CMH pool in the liver tissue and result in a decreased quantity of the ester in liver tissue at initial phases of the tumor growth. During the next period, the intensified production of CMH in the liver is apparently sufficient to supply the tumor tissue fully with the quantities of CMH required and an excess of the ester may even be produced, resulting in an increase of CMH content in the liver and serum over the normal level. Changes found in CMH levels which are characteristic for tumor-bearing animals may thus reflect the disturbed balance between the production and utilization of CMH.

Evidence presented here seems to indicate that CMH may play an important role in the regulation of protein synthesis and hence in the growth of malignant tumors. Several aspects of this problem, however, await a more detailed elucidation before a final decision is possible.

V. References

1 ABRAHAMSON, S.; STÄLLBERG-STENHAGEN, S., and STENHAGEN, E.: The higher saturated branched-chain fatty acids, p. 100 (Pergamon Press, Oxford 1963).
2 ALLENDE, C. C.; ALLENDE, J. E.; GATICA, M.; CELIS, J.; MORA, G., and MATAMALA, M.: The aminoacyl ribonucleic acid synthetases. I. Properties of the threonyladenylate-enzyme complex. J. biol. Chem. 241: 2245–2251 (1966).
3 BANDYOPADHYAY, A. K. and DEUTSCHER, M. P.: Complex of aminoacyl-transfer RNA synthetases. J. molec. Biol. 60: 113–122 (1971).
4 BANDYOPADHYAY, A. K. and DEUTSCHER, M. P.: Lipids associated with the aminoacyl-transfer RNA synthetase complex. J. molec. Biol. 74: 257–261 (1973).
5 BARCLAY, M. and SKIPSKI, V. P.: Lipoproteins in relation to cancer: in CARROLL Lipids and tumors. Progr. biochem. Pharmacol., vol. 10, pp. 76–111 (Karger, Basel 1975).
6 BOYD, E. M. and MCEWEN, H. D.: The concentration and accumulation of lipids in the tumor component of a tumor-host organism, Walker carcinoma 256 in albino rats. Canad. J. med. Sci. 30: 163–172 (1952).
7 CARRUTHERS, C.: The influence of transplantable rat mammary carcinomas on the chemical composition of the host. Oncology 23: 241–256 (1969).
8 DAVIE, E. W.: Tryptophan-activating enzyme; in COLOWICK and KAPLAN Methods in enzymology, vol. 5, pp. 718–722 (1962).
9 DAY, E. A.; MALCOM, G. T., and BEELER, M. F.: Tumor sterols. Metabolism 18: 646–651 (1969).
10 EARL, D. C. N. and HINDLEY, S. T.: The rate-limiting step of protein synthesis *in vivo* and *in vitro* and the distribution of growing peptides between the puro-

mycin-labile and puromycin-non-labile sites on polyribosomes. Biochem. J. 122: 267–276 (1971).
11 GIRGIS, G. R. and NICHOLLS, D. M.: Control of the increased protein synthesis in kidney of nephrotic rats by supernatant factors. Biochem. biophys. Acta 247: 335–347 (1971).
12 GIRGIS, G. R. and NICHOLLS, D. M.: Protein synthesis limited by transferase I. Biochem. biophys. Acta 269: 465–476 (1972).
13 GRIFFIN, A. C. and BLACK, D. D.: Protein biosynthesis; in BUSCH Methods in cancer research, vol. 6, pp. 189–251 (Academic Press, New York 1971).
14 HELMICH, O. and HRADEC, J.: Radio gas chromatography of cholesteryl 14-methylhexadecanoate. J. Chromat. 91: 505–512 (1974).
15 HOAGLAND, M. B.; KELLER, E. B., and ZAMECNIK, P. C.: Enzymatic carboxyl activation of amino acids. J. biol. Chem. 218: 345–358 (1956).
16 HOAGLAND, M. B.; STEPHENSON, M. L.; SCOTT, J. F.; HECHT, L. I., and ZAMECNIK, P. C.: A soluble ribonucleic acid intermediate in protein synthesis. J. biol. Chem. 231: 241–257 (1958).
17 HOAGLAND, M. B.; ZAMECNIK, P. C., and STEPHENSON, M. L.: Intermediate reactions in protein biosynthesis. Biochem. biophys. Acta 24: 215–216 (1957).
18 HRADEC, J.: Effect of added tissue homogenates on the incorporation of labeled methionine into the proteins of Ehrlich ascites carcinoma in vitro. Z. Krebsforsch. 63: 4–11 (1959).
19 HRADEC, J.: Metabolism of serum albumin in tumour-bearing rats. Brit. J. Cancer 12: 290–304 (1958).
20 HRADEC, J.: Nature of the carcinogenic substance in egg yolks. Nature, Lond. 182: 52–53 (1958).
21 HRADEC, J.: Effect of carcinogens and related compounds on the growth of Ehrlich ascites carcinoma and its possible mechanism. Brit. J. Cancer 13: 336–347 (1959).
22 HRADEC, J.: A protein synthesis-stimulating factor in tissues of tumour-bearing animals. Acta Un. int. Cancr. 16: 1146–1149 (1960).
23 HRADEC, J.: Effect of carcinolipin on protein synthesis in cell-free systems. Biochem. biophys. Acta 47: 149–157 (1961).
24 HRADEC, J.: Carcinolipin – an essential factor in protein synthesis. Acta Un. int. Cancr. 20: 926–928 (1964).
25 HRADEC, J.: Effect of some polycyclic aromatic hydrocarbons on protein synthesis in vitro. Biochem. J. 105: 251–259 (1967).
26 HRADEC, J.: A chromatographic method for the quantitative determination of cholesterol-14-methylhexadecanoate (carcinolipin) in biological materials. J. Chromat. 32: 511–518 (1968).
27 HRADEC, J.: Intermediate reactions in the binding of aminoacyl-transfer ribonucleic acid to rat liver ribosomes. Formation and properties of an aminoacyl-transfer ribonucleic acid-transferase I complex. Biochem. J. 126: 923–931 (1972).
28 HRADEC, J.: Intermediate reactions in the binding of aminoacyl-transfer ribonucleic acid to rat liver ribosomes. The role of guanosine triphosphate. Biochem. J. 126: 933–943 (1972).
29 HRADEC, J.: Intermediate reactions in the binding of aminoacyl-transfer ribonu-

cleic acid to rat liver ribosomes. The interaction of cholesteryl 14-methylhexadecanoate. Biochem. J. *126:* 1225–1229 (1972).

30 Hradec, J.: The possible role of cholesteryl 14-methylhexadecanoate in tumor growth; in Wood Tumor lipids: biochemistry and metabolism, pp. 54–65 (Amer. Oil Chem. Soc. Press, Champaign, Ill. 1973).

31 Hradec, J. and Dolejš, L.: The chemical constitution of carcinolipin. Biochem. J. *107:* 129–134 (1968).

32 Hradec, J. and Dušek, Z.: Effect of lipids, in particular cholesteryl-14-methylhexadecanoate, on the incorporation of labeled aminoacids into transfer ribonucleic acid in vitro. Biochem. J. *110:* 1–8 (1968).

33 Hradec, J. and Dušek, Z.: Effect of cholesteryl 14-methylhexadecanoate on the activity of some amino acid-transfer ribonucleic acid ligases from mammalian tissues. Biochem. J. *115:* 873–880 (1969).

34 Hradec, J. and Dušek, Z.: Effect of lipids on aminoacyl-tRNA synthesis in *Escherichia coli*. FEBS Letters *6:* 86–88 (1970).

35 Hradec, J.; Dušek, Z.; Bermek, E., and Matthaei, H.: The role of cholesteryl 14-methylhexadecanoate in peptide elongation reactions. Biochem. J. *123:* 959–966 (1971).

36 Hradec, J.; Dušek, Z., and Mach, O.: Influence of cholesteryl 14-methylhexadecanoate on some ribosomal functions required for peptide elongation. Biochem. J. *138:* 147–154 (1974).

37 Hradec, J. and Kruml, J.: Carcinolipin – an endogenous carcinogenic substance. Nature, Lond. *185:* 55 (1960).

38 Hradec, J. and Menšík, P.: Purification of the fatty acid present in carcinolipin. J. Chromat. *32:* 502–510 (1968).

39 Hradec, J. and Sommerau, J.: Isolation of carcinolipin by combined liquid-solid and liquid-liquid chromatography. J. Chromat. *32:* 230–242 (1968).

40 Hradec, J. and Sommerau, J.: Effect of cholesteryl esters with different saturated fatty acids on aminoacyl-tRNA synthesis in rat liver. FEBS Letters *9:* 161–163 (1970).

41 Hradec, J. and Štroufová, A.: Studies on a protein synthesis-affecting substance from biological materials. Biochem. biophys. Acta *40:* 32–43 (1960).

42 Ibuki, F. and Moldave, K.: Evidence for the enzymatic binding of aminoacyl transfer ribonucleic acid to rat liver ribosomes. J. biol. Chem. *243:* 791–798 (1968).

43 Jacob, F. and Monod, J.: Genetic regulatory mechanisms in the synthesis of proteins. J. molec. Biol. *3:* 318–356 (1961).

44 Kaufmann, R.: DNA-abhängige RNA-Polymeraseaktivität aus Zellkernen von humaner Plazenta; naturwiss. Diss. Göttingen (1970).

45 Komárková, E. and Hradec, J.: Changes of aminoacyl-tRNA synthesis in the liver of rats administered cholesteryl 14-methylhexadecanoate. FEBS Letters *14:* 130–132 (1971).

46 Komárková, E. and Hradec, J.: Effect of cholesteryl 14-methylhexadecanoate on the RNA polymerase activity of rat liver nuclei *in vivo* and *in vitro*. FEBS Letters *18:* 109–111 (1971).

47 KOMÁRKOVÁ, E. and HRADEC, J.: Protein synthesis in the liver of rats injected with cholesteryl 14-methylhexadecanoate. Biochem. J. *129:* 367–372 (1972).
48 KOMÁRKOVÁ, E. and HRADEC, J.: Evidence for the role of cholesteryl 14-methylhexadecanoate in the activity of DNA-dependent RNA polymerase from rat liver. FEBS Letters (in press).
49 KURLAND, C. G.: The proteins of bacterial ribosome; in MCCONKEY Protein synthesis, vol. 1, pp. 179–228 (Dekker, New York 1971).
50 LEADER, D. P.; WOOL, I. G., and CASTLES, J. J.: Aminoacyl-transferase I-catalyzed binding of phenylalanyl-transfer ribonucleic acid to muscle ribosomes from normal and diabetic rats. Biochem. J. *124:* 537–541 (1971).
51 LOFTFIELD, R. B.: The aminoacylation of transfer ribonucleic acid; in MCCONKEY Protein synthesis, vol. 1, pp. 2–88 (Dekker, New York 1971).
52 MATTHAEI, H.; SANDER, G.; SWAN, D.; KREUZER, T.; CAFFIER, H. und PARMEGGIANI, A.: Reaktionsschritte der Polypeptidsynthese an Ribosomen. Mechanismen der Proteinsynthese X. Naturwissenschaften *55:* 281–284 (1968).
53 MCKEEHAN, W. L. and HARDESTY, B.: Purification and partial characterization of the aminoacyl transfer ribonucleic binding enzyme from rabbit reticulocytes. J. biol. Chem. *244:* 4330–4339 (1969).
54 MENON, I. A.: A possible role of lipids in RNA polymerase from mammalian cells. Canad. J. Biochem. *7:* 807–817 (1972).
55 NARAYAN, K. A. and MORRIS, H. P.: In vitro synthesis of rat serum lipoproteins and proteins by Morris hepatoma 7777. FEBS Letters *27:* 311–315 (1972).
56 PAPAS, T. S. and MEHLER, A. H.: Modification of the transfer function of proline transfer ribonucleic acid synthetase by temperature. J. biol. Chem. *243:* 3767–3769 (1968).
57 RICHMAN, N. and BODLEY, J. W.: Ribosomes cannot interact simultaneously with elongation factors EF Tu and EF G. Proc. nat. Acad. Sci., Wash. *69:* 686–689 (1972).
58 SCHWEET, R. S.: Tyrosine-activating enzyme from hog pancreas; in COLOWICK and KAPLAN Methods in enzymology, vol. 5, pp. 722–726 (Academic Press, New York 1962).
59 SHABAD, L. M.; KOLESNICHENKO, T. S., and SAVLUCHINSKAYA, L. A.: Transplacental effect of carcinolipin in mice. Neoplasma *20:* 347–348 (1973).
60 SILER, J. and MOLDAVE, K.: Studies on the kinetics of peptidyl transfer RNA translocase from rat liver. Biochem. biophys. Acta *195:* 138–144 (1969).
61 SILVESTRI, L.: RNA-polymerase and transcription (North Holland, Amsterdam 1970).
62 SKOGERSON, L. and MOLDAVE, K.: Evidence for the role of aminoacyltransferase II in peptidyl transfer ribonucleic acid translocation. J. biol. Chem. *243:* 5361–5367 (1968).
63 SNYDER, F.; BLANK, M. L., and MORRIS, H. P.: Occurrence and nature of O-alkyl and O-alk-l-enyl moieties of glycerol in lipids of Morris transplanted hepatomas and normal rat liver. Biochem. biophys. Acta *176:* 502–510 (1969).
64 SWELL, L. and LAW, M. D.: Synthesis of serum and sub-cellular liver cholesterol esters in fasted and fed rats. J. Nutr. *95:* 141–147 (1968).

65 THEISE, H. und BIELKA, H.: Die Lipidzusammensetzung der Mikrosomenfraktion von Leber und Hepatom. Arch. Geschwulstforsch. *32:* 11–19 (1968).
66 VASQUEZ, D.; BATTANER, E.; NETH, R.; HELLER, G., and MONRO, E. R.: The function of 80S ribosomal subunits and effects of some antibiotics. Cold Spr. Harb. Symp. quant. Biol. *34:* 369–375 (1969).
67 WATSON, J. D.: The synthesis of proteins upon ribosomes. Bull. Soc. Chem. biol. *46:* 1399–1425 (1964).
68 WEBSTER, G. C.: Isolation of an alanine-activating enzyme from pig liver. Biochem. biophys. Acta *49:* 141–152 (1961).
69 WILLIS, D. B. and STARR, J. L.: Protein biosynthesis in the spleen. III. Aminoacyltransferase I as a translational regulatory factor during the immune response. J. biol. Chem. *246:* 2828–2834 (1971).
70 WINDISCH, F.; DITTMANN, J.; NORDHEIM, W. und GERHARDT, U.: Cholesterol und Cholesteride, ihre elektronenübertragende Energiewirksamkeit und Akkumulation im Krebsgewebe mit zunehmender Malignität. Z. ges. inn. Med. *15:* 235–239 (1960).
71 ZAMECNIK, P. C.: Summary of symposium on transfer RNA and transfer RNA modification in differentiation and neoplasia. Cancer Res. *31:* 716–721 (1971).

Author's address: JAN HRADEC, M. D., D. Sc., Department of Biochemistry, Oncological Institute, *18000 Prague 8* (Czechoslovakia)

Sterols and Other Lipids in Tumors of the Nervous System

JOSEPH F. WEISS

Department of Neurosurgery, New York University Medical Center, New York, N.Y.

Contents

I. Introduction	228
II. Developmental Changes in Neural Lipids	230
A. Characteristic Neural Lipids	230
B. Sterol Synthesis and Composition	231
C. Desmosterol	234
D. Sterol Esters	236
E. Effect of Hypocholesteremic Agents on the Developing Nervous System	237
III. Human Tumors of the Nervous System	238
A. Classification of Tumors	238
B. Changes in Lipid Classes, Neutral Lipids, and Fatty Acid Composition	239
C. Sterol Composition and Synthesis and the Effect of Drugs	240
D. Phospholipids	241
E. Glycolipids	242
F. Cerebrospinal Fluid	243
1. Comparison of Blood and Cerebrospinal Fluid Lipids	245
2. Cerebrospinal Fluid Lipids in Neoplasia	245
3. Use of Triparanol to Augment CSF Desmosterol	246
IV. Models for Brain Tumor Research	248
A. Primary Tumors	248
1. Induction of Tumors in Animals	248
2. Sterol Metabolism in Nitrosourea-Induced Tumors	250
B. Transplanted Tumors	252
1. Lipid Composition and Synthesis	252
2. Sterol Metabolism and the Effect of Drugs	253
C. Tissue Culture	254
1. Lipid Metabolism	254
2. Sterols and the Effect of Drugs	255
V. Summary	260
VI. Acknowledgments	260
VII. References	261

I. Introduction

Mature brain and nerve have a high and characteristic lipid content. The major neural lipids (cholesterol, phospholipids, glycolipids) occur mainly in membrane structures, and especially in myelin, which is characterized by a lower metabolic activity than other membranes.

The changes that occur in brain lipids during neoplasia have been variously described as chemical dedifferentiation, a loss of constancy, or as simulating developing brain. Comparison of tumor lipids with those in normal tissue is difficult because of the heterogeneous cell types in brain and the relative quantitative importance of the lipids localized in myelin. It might be preferable to compare tumors of various cellular origin with their corresponding normal cell types. Only recently, techniques have become available for the isolation of neuronal and glial cell fractions [42, 68, 77] and subcellular membranes, such as plasma membranes [44], in relatively pure form and in sufficient quantity for biochemical analysis. Tumor samples, themselves, may be contaminated with normal nervous tissue, which in general has a higher lipid content, or with necrotic, cystic, or vascular areas. Changes may occur in tissue surrounding the tumor, although the tissue appears histologically normal or only slightly edematous. Significant changes, which may be related to demyelination, have been found in the lipid composition of tissue surrounding tumor [103, 135]. Tumor models, such as subcutaneously grown animal tumors and cells grown in tissue culture, may eliminate difficulties in obtaining pure tumor samples. Although very useful for studying biochemical alterations in neoplasia, these models may not reflect adequately the unique problems presented by intracerebral site of tumor growth ('blood-brain barrier' phenomenon, differences in enzymatic and energy requirements, immunologic privileges). In comparing intracerebrally and subcutaneously grown tumors, environmental factors are an important determinant of the mode of metabolism of the tumor [59].

It is not known whether changes in lipid composition or metabolism of brain tumors are directly involved in cancerous growth or are secondary reflections of neoplasia. Changes in lipids at the molecular level affecting the spatial requirements of the molecule could theoretically affect the properties of the membranes of the cell. There is ample evidence of changes in the surface properties of tumor cells [120]. Alterations in plasma membrane components, including lipids, may be reflected in the cell's interactions with other cells, its response to its environment, and the acquisition or presentation of antigens. There is some evidence that a human glial antigen of the carcinoembryonic variety is in part a lipoprotein [118], and characteristic brain lipids (cerebrosides and gangliosides) are known to have immunological activity [84]. Few studies have been carried out on the effects of lipids on promoting tumor growth in the nervous system. For example, the finding that there is a higher incidence of intracranial tumors in mice fed high fat and cholesterol diets [115, 116] should be investigated further.

Differences in lipid composition of neoplastic nervous tissue compared to normal tissue may be exploitable for diagnostic or chemotherapeutic purposes. Brain tumor chemotherapy has been based mainly on affecting nucleic acid or protein synthesis, and altering lipid metabolism as a possible form of chemotherapy has hardly been investigated. Analysis of tumor samples or cerebrospinal fluid (CSF) of patients with tumors may be useful adjunctive methods for tumor diagnosis, classification, or grading of malignancy. Serial analyses of CSF may be useful in the follow-up of patients undergoing therapy for brain tumors. Although changes in CSF lipids are more probable in patients with brain tumors, it is not inconceivable that changes in blood lipids may also occur [69].

Recent research on sterol metabolism in tumors of the nervous system, effects of drugs on tumor sterols, and the use of newer experimental tumor models for these studies will be included in this chapter. Lipid metabolism in the developing nervous system will also be emphasized because of some similarities found in lipid synthesis and composition of neoplastic and developing nervous tissue. Published work on lipids in normal and neoplastic growth of the nervous system will be selectively reviewed. Reviews by GROSSI PAOLETTI, FUMAGALLI, P. PAOLETTI and R. PAOLETTI have been published recently on brain sterols [74], drugs acting on brain tumor sterols [72], and lipids in brain tumors [38]. These include discussions of methodology and the differences between lipid metabolism in brain tumors and nonneural tumors. The topic of phospholipids in brain tumors has been covered by WHITE [130]. General reviews on the biochemistry of brain tumors [61, 133], including discussions of enzymatic changes related to lipid metabolism, have also appeared recently. In the present paper, an attempt has been made to duplicate the material and references included in previous reviews as little as possible.

II. Developmental Changes in Neural Lipids

Studies of lipid metabolism in developing tissues may indicate whether there are distinct or common changes related to growth and neoplasia [134]. Some similarities exist between the changes that occur in the major lipid classes in brain (sterols, phospholipids, and perhaps in glycosphingolipids) during neoplasia and development. If similarities in lipid synthesis or uptake of lipids from blood exist in these two conditions of rapid growth, it is possible that the developing animal brain can also be used as a model for testing agents that could interfere with lipid metabolism in brain tumors.

A. Characteristic Neural Lipids

Cholesterol and some of the complex lipids are characteristic of brain in the sense that they occur there in greater concentrations than in other tissues. Lipid changes during brain development are associated primarily with the progressive deposition of myelin, characterized by an increase in its most representative lipids [91]. These include the glycolipids, cerebroside and sulfatide; the major phosphosphingolipid, sphingomyelin; and ethanolamine plasmalogen. About 70 % of the cholesterol in adult brain is localized in myelin.

The major phospholipids of mature brain in quantity are phosphatidyl ethanolamine, including phosphatidyl ethanolamine plasmalogen, followed by phosphatidyl choline, phosphatidyl serine, and phosphatidyl inositols. Minor phospholipid components include phosphatidic acid, phosphatidyl glycerols, other plasmalogens (vinyl ether phosphoglycerides) and alkyl ether phosphoglycerides. A major change in the pattern of phospholipid composition during development is an increase in the ratio of ethanolamine phospholipids to phosphatidyl choline as the brain matures [14, 91]. Gangliosides are generally considered to be more characteristic of neurons and grey matter [113], but the pattern of individual gangliosides in glial and neuronal fractions appears to be similar [42]. Myelin has a low concentration of gangliosides, characterized by a major GM1 fraction [42]. There are specific changes in the ganglioside pattern of whole brain during development, most notably in the GD1[a] fraction [113]. There are also characteristic changes in the fatty acid moieties of brain phospholipids and glycolipids during maturation [83]. The deposition of fatty acids in developing brain can be influenced by dietary fatty acids and essential fatty acid deficiency [32].

The turnover of structural lipid, especially in myelin, is low once maturity is reached. Differences exist in lipid turnover between the peripheral nervous system (PNS) and central nervous system (CNS), which are probably related to the different characteristics and origin of peripheral nerve and brain myelin [86]. Normal glial cell and Schwann cell membranes must be related in some way to myelin since myelin is considered to be an extension of the cell surface membrane: CNS myelin from oligodendroglial cells and PNS myelin from Schwann cells. Neoplastic glial and Schwann cells lose their ability to form myelin and this must be considered when comparing lipids in gliomas and Schwannomas to those in myelin or whole brain.

Neutral lipids, other than cholesterol, are very minor components of nervous tissue after maturation is reached [91]. They include cholesterol esters, glycerides, hydrocarbons, glyceryl ethers, and free fatty acids.

Excellent and thorough reviews of recent research on lipids of the nervous system, including studies on developmental changes, have been presented by RAMSEY and NICHOLAS [83] and ROUSER et al. [91].

B. Sterol Synthesis and Composition

The rate of sterol synthesis is much higher during brain development than in normal, mature brain [53, 74]. The pattern of sterol synthesis and deposition is also different in developing brain compared to other immature tissues. In figure 1, the incorporation *in vivo* of labeled mevalonate into the sterol fraction of chick embryo brain, liver and yolk sac membranes is compared at different days of incubation and after hatching. The relative independence of brain sterol synthesis from that in nonneural tissues at certain stages of development is evident. Sterol synthesis is high at an earlier stage of development in liver and yolk sac membranes and then decreases, possibly related to the accumulation of esterified cholesterol derived from yolk cholesterol observed in these tissues. As incubation progresses, a higher proportion of radioactivity from mevalonate is found in nonsterol isoprenoid compounds in liver and yolk sac membranes, while in brain a higher proportion is found in the desmosterol and cholesterol fractions [125]. It is difficult to determine the exact amount of brain sterol derived from peripheral sources during development. For chick

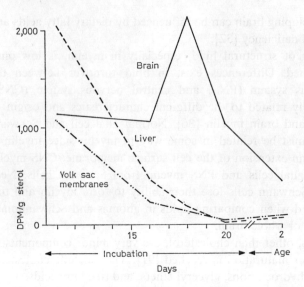

Fig. 1. Incorporation of 2-^{14}C-mevalonate into sterols of developing chick embryo tissues. 2-^{14}C-mevalonate (2 μCi) injected via the yolk sac of embryos at 11, 16, 18, and 20 days of incubation and in 2-day-old chick. Values determined on pools of 3 tissues. GROSSI PAOLETTI and WEISS, unpublished.

brain, it was calculated that *de novo* synthesis in the brain accounts for most of the cholesterol content (up to 90 %) throughout the incubation period [15]. However, FISH *et al.* [26], using labeled acetate, have calculated that more cholesterol arises from extraneural sources than is synthesized in the brain during the early part of incubation.

A large number of sterols metabolically related to cholesterol have been detected in developing and adult brain (table I). Most of the sterols identified thus far have been isolated by a combination of chromatographic techniques, and identified by gas-liquid chromatography-mass spectrometry [34, 81, 82, 126]. Using labeled mevalonate, it has been shown that many of these sterols are metabolically active precursors of cholesterol in developing and adult animal brain [126], adult human brain [34], and glioblastoma [33]. Apparently, the same preferred pathway of cholesterol synthesis after squalene cyclization functions in developing, mature, and neoplastic brain. After lanosterol (8,24-cholestadien-4,4,14α-trimethyl-3β-ol) is formed from squalene, two major pathways leading to cholesterol formation are evident. Sterols with one double bond in the nucleus and one in the side chain have a larger degree of labeling, and the

Table I. Schematic diagram of sterols identified in nervous tissue and possible pathways of brain cholesterol synthesis. Proposed major route of cholesterol synthesis in brain is underlined

		Squalene		
$C_{30}\Delta^8$		$C_{30}\Delta^{8,\ 24}$		$C_{30}\Delta^7$
$C_{29}\Delta^8$	$C_{29}\Delta^{14,\ 24\ a}$	$C_{29}\Delta^{8,\ 24}$		$C_{29}\Delta^7$
$C_{28}\Delta^8$	$C_{28}\Delta^{14,\ 24\ a}$	$C_{28}\Delta^{8,\ 24}$	$C_{28}\Delta^{7,\ 24}$	$C_{28}\Delta^7$
$C_{27}\Delta^8$	$C_{21}\Delta^{14\ a}$	$C_{27}\Delta^{8,\ 24}$	$C_{27}\Delta^{7,\ 24}$	$C_{27}\Delta^{7\ b}$
			$C_{27}\Delta^{5,\ 7,\ 24\ c}$	$C_{27}\Delta^{5,\ 7\ c}$
			$C_{27}\Delta^{5,\ 24}$	
			$C_{27}\Delta^5$ Cholesterol	
			$C_{27}\Delta^0$ Hydroxy sterols and oxygenated products	

[a] Detected after *in vitro* incubation.
[b] Not yet positively identified in fresh tissue.
[c] Detected after administration of Δ^7-reductase inhibitor.
Further details on sterol precursors of cholesterol in brain are given in the literature [25, 26, 28, 34, 56, 81, 82, 126].

pathway in which the double bond in the lateral chain is retained until desmosterol (5,24-cholestadien-3β-ol) is formed, appears to be the preferred pathway [35]. The last step in brain cholesterol synthesis is the reduction of the double bond in the 24 position of desmosterol to cholesterol. The possible pathways for cholesterol biosynthesis in different tissues appear to be qualitatively similar but quantitatively different [70]. For example, the pathway of cholesterol biosynthesis passing through desmosterol is probably of lesser importance in liver than in brain [28]. Other discussions of possible pathways of brain cholesterol synthesis have been presented by PAOLETTI *et al.* [76] and RAMSEY [80].

In normal mature human brain, cholesterol represents 99 % of the total sterols and cholestanol is second in concentration, occurring in quantities of less than 0.5 % [34]. Cholestanol also occurs in brain during development, although it is considered to be a catabolite of cholesterol [126]. A fraction of steroidal acid(s) possibly derived from cholesterol, has been detected in developing brain [65], indicating that in conditions of rapid growth of nervous tissue, end products, as well as precursors, may be present.

C. Desmosterol

The major quantitative difference in sterol synthesis in developing brain compared to mature brain is the accumulation of desmosterol, rather than any of the other sterol precursors occurring in brain [126]. The desmosterol fraction can be measured easily by gas chromatographic analysis [75, 129]. Cholesterol precursors with the same degree of unsaturation as desmosterol may also occur in very small quantities in this fraction. During the premyelination and early myelination stages of brain development, an increase in desmosterol has been observed in all vertebrate species studied [75]. Demosterol has also been detected in early myelin or premyelin [3, 30]. The peak amount of desmosterol varies in different species: the maximum amount reported in human brain is approximately 11 % of the total sterols [19, 75], while in rodents, the highest levels are three times the amount found in humans [56, 75]. The desmosterol level in developing rat spinal cord was found to be lower than that in brain [56]. The curve of desmosterol accumulation is very constant when plotted against rat age and can be considered a biochemical constant of developing rat brain [31, 56, 75].

A definite difference exists between desmosterol accumulation in developing rat CNS and PNS. Cholesterol accumulation with time in rat cerebrum (CNS) and trigeminal nerve (PNS) is shown in figure 2. The cholesterol concentration per wet weight is a biochemical index of the progression of myelination. In agreement with the biochemical evidence, histological staining showed that myelination in rat cerebrum begins after 10 days of age, while myelination in the trigeminal nerve is already advanced at 10 days of age, beginning at about 6 days after birth [124]. The desmosterol level in rat cerebrum reached a peak of approximately 34 % of the total sterols at 6 days of age during the premyelination stage of brain development (fig. 3). Desmosterol did not accumulate to more than 3 % of the total sterols in developing trigeminal nerve. Another difference was noted between the sterol composition of PNS and CNS. As desmosterol diminished in maturing trigeminal nerve, another sterol, occurring in quantities approximately equal to that of desmosterol,

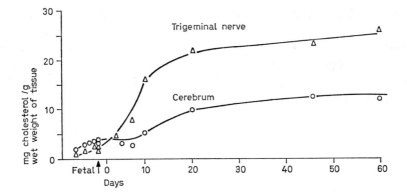

Fig. 2. Comparison of cholesterol concentration in developing CNS and PNS of Sprague-Dawley (CFE) rats. From WEISS *et al.* [123] and unpublished data.

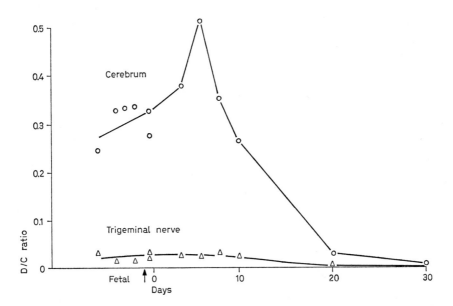

Fig. 3. Comparison of desmosterol/cholesterol ratios (D/C) in developing CNS and PNS of Sprague-Dawley (CFE) rats. From WEISS *et al.* [123] and unpublished data.

became evident by gas chromatographic analysis. The retention time of this sterol or group of sterols is similar to that of sterols with a double bond in the 7-position of the steroid nucleus. Analysis of other rat tissues at 21 days of gestation indicated a low desmosterol fraction in nonneural tissues, except for skin, which had a level approaching that in the PNS. Desmosterol has also been detected in adult rat skin, but other cholesterol precursors occur in larger quantities [48].

An increase in the desmosterol/cholesterol ratio in brains of undernourished rats has been reported [99], although the ratios found in the brains of control rats at different days after birth were, in general, significantly lower than previously reported values [31, 56, 75]. No difference was found in the curve of desmosterol accumulation in moderate undernutrition, when brains of Wistar rats from small (6 rats) and large litters (15 rats) were compared [WEISS et al., unpublished]. In the latter group, there was a significant decrease in body weight from the 15th to 35th day after birth, but not in brain weight. The curve of desmosterol accumulation in developing chick embryo brain was shifted by manipulating the thyroid activity and hatching time of the chick embryo [75]. The desmosterol/cholesterol ratio was found to be unaltered in mice with genetic disorders affecting myelination compared to developing mice undergoing normal myelination [54].

It has been suggested that desmosterol induces its own reduction in brain by high substrate levels, thus increasing cholesterol synthesis and myelin formation [45]. The low levels of desmosterol found in the developing PNS at a time when there is an increased rate of cholesterol deposition (fig. 2 and 3) indicate that a high desmosterol level is not inherently related to the myelination sequence or rapid growth of nervous tissue. The increased levels of desmosterol in CNS and PNS may be in part related to the predominant growing cell type seen at the time of study. However, more specific factors that might influence desmosterol accumulation in nervous tissue, such as levels of enzymes or cofactors, or influence of peripheral circulation, are far from clear.

D. Sterol Esters

Cholesterol esters are found in brain in increased quantities during the metabolically active states of development, malignancy, and demyelination [74]. In developing rat brain, there is a peak period of cholesterol ester increase, which corresponds to the period of active myelination [23]. Another separate event appears to affect the level of cholesterol esters in developing rat brain. An underlying developmental change is characterized on a unit weight basis by the highest level of cholesterol esters immediately after birth and a steady decline to the adult level by 30 days of age (approximately 10 μg/g brain). On the basis of whole brain, cholesterol esters steadily increase throughout development. A cholesterol-esterifying enzyme and cholesterol hydrolases that undergo developmental changes are also present in developing rat brain [24]. Analysis of the fatty acids of the cholesterol esters indicated that the esters probably were synthesized primarily in the brain, since the fatty acids were not similar to serum cholesterol ester fatty acids [23]. Although the desmosterol level is also high during the period when the cholesterol ester level is high, most of the esterified sterol was found to be cholesterol in developing rat [56] and chick brain [27]. In chick embryo brain, there is a definite peak period of cholesterol ester accumulation before adult levels are reached [37], analogous to that found in developing rat brain [23]. The major change in sterol metabolism during chick embryo incubation is the progressive esterification of yolk cholesterol, which seems to occur in yolk sac membranes as the yolk cholesterol is transported to the embryonic tissues [67]. A surge of esterified cholesterol into the embryonic tissues, especially liver, occurs toward the end of the incubation period [66]. In contrast to other developing tissues, the proportion of brain sterol that is esterified, is small. Further studies on cholesterol esters in developing nervous tissue may indicate whether these compounds have a role in regulating cholesterol synthesis or deposition into membranes, or in the metabolism of fatty acids or phosphoglycerides.

E. Effect of Hypocholesteremic Agents on the Developing Nervous System

Some investigators have detected changes in cholesterol concentration and synthesis in mature animals treated with various agents [53]. Hypocholesteremic agents that block the conversion of cholesterol precursors to cholesterol have only slight effects on mature brain, while causing further changes in the characteristic sterol composition of developing brain. For example, triparanol (1-[p-(β-diethylaminoethoxy)-phenyl]-1-(p-tolyl)-2(p-chlorophenyl) ethanol or 20,25-diazacholesterol, when administered during development, cause further accumulation of desmosterol in developing brain, including myelin [3, 30, 93]. AY-9944 (*trans*-1,4-bis[2-chlorobenzylaminomethyl]cyclohexane dihydrochloride) causes the accumulation of 7-dehydrocholesterol in developing brain, and to a lesser extent of 7-dehydrodesmosterol [28]. Either desmosterol or 7-dehydrocholesterol are slowly replaced by cholesterol in the myelin fraction [4, 30]. These hypocholesteremic agents may also act at an earlier stage of cholesterol synthesis and affect the total amount of sterol available for formation of structural membranes [30]. Effects of various hypocholesteremic drugs on the morphology of normal nervous tissue have been studied [95]. Morphological changes are greatest when the drugs are administered to developing animals and include retardation of myelination in peripheral nerve [87], and reversible degeneration of oligodendroglia in rat CNS [112] after AY-9944 administration.

III. Human Tumors of the Nervous System

A. Classification of Tumors

The classification and incidence of human tumors of the nervous system will be discussed briefly. For the biochemist, the preferred classification is probably a cytogenetic one, that is a correlation of tumors with the embryonic cell of origin. This type of classification is difficult, since

Table II. Major types of primary intracranial tumors and approximate incidence. Data from WECHSLER [121] and ZIMMERMAN [137]

Neuroectodermal tumors
Of central neural tissues
   Neuronal series (5%): neuroblastoma, medulloblastoma, ganglioneuroma
   Glial series: astrocytoma – glioblastoma multiforme (40–50%), oligodendroglioma (5%), ependymoma (5%)
Of peripheral nerves: Schwannoma or neurinoma (5–10%)
Of pituitary gland (5–10%)
Of pineal gland (<5%)

Mesenchymal tumors of CNS
Of meninges: meningioma (10–20%), sarcoma (5%)
Of blood vessels (5%): hemangioblastoma
Of embryonic remnants (5%): craniopharyngioma, teratoma

it is conceivable that precursor cells have been changed and/or transformed, or that subsequent alterations of neoplastic cells occur [121]. Table II is a simplified classification showing the most common primary intracranial tumors occurring in man. The most important tumor from a clinical point of view is the malignant glioma, which accounts for about one half of the cases of primary intracranial tumors. Tumors of the spinal cord and extracranial peripheral nerves are less common than intracranial tumors in humans. In addition to primary brain tumors, secondary malignant tumors originating mainly from metastases from lung or breast tumors are found intracranially. There is disagreement over the classification of some tumor types and analysis for biochemical characteristics of neural tissue may assist in classification, e.g. in determining whether a tumor is of neuroectodermal or mesenchymal origin [121].

B. Changes in Lipid Classes, Neutral Lipids, and Fatty Acid Composition

The total lipid content of human tumors of neuroectodermal origin is lower than that of normal brain, so that in general the major lipid classes (cholesterol, phospholipids, glycolipids) are also decreased [9, 36, 71]. Meningiomas appear to have a lower total lipid content than tumors of glial origin [57], but a higher lipid content than normal meninges [10]. There is an increase in lipids that occur only in trace amounts in mature human brain (triglycerides, cholesterol esters, free fatty acids). These changes appear to some degree in all major types of brain tumors: meningiomas, metastatic carcinomas, neurinomas, and gliomas [9, 36, 104]. The lipid pattern of lower grade astrocytomas is often similar to normal brain. In one study, free sterols represented about 19 % of the total lipid in gliomas and meningiomas, while esterified sterols represented an additional 7–9 % [36]. In neurinomas, 16–17 % of the total lipid was esterified sterols and only 10 % was free sterol [36]. Meningiomas are characterized by an increased squalene content and a saturated hydrocarbon fraction that differs in component distribution from values found in meninges [10]. Compared to the changes in the major neutral lipid classes, changes in minor neutral lipids are usually not of notable quantity in intracranial tumors [104]. Not enough brain tumor samples have been analyzed for glyceryl alkyl and alk-1-enyl ethers to determine if there is a significant increase in these neutral lipids [105].

The percentage of linoleic acid in the total fatty acids of brain tumors is higher than in mature brain and much higher than in developing brain [38, 71, 108]. Meningiomas have a lower linoleic acid content than gliomas but a higher level of arachidonic acid, and the fatty acid composition appears to be more related to that of normal leptomeninges than to dura mater [111]. In the cholesterol ester fraction, an increase in monounsaturated and polyunsaturated fatty acids was found in various types of tumors compared to that in normal brain [36]. Compared to the cholesterol esters in plasma, there are more saturated fatty acids and monoenoic fatty acids and less polyunsaturated fatty acids in intracranial brain tumors [38, 131]. The difference between the fatty acid composition of cholesterol esters in glioblastoma and plasma suggested to some workers that the sterol ester component of glioblastoma does not originate in the blood [131].

C. Sterol Composition and Synthesis and the Effect of Drugs

Desmosterol was detected in levels of 1–2 % of the total sterols in human glioblastomas and oligodendrogliomas, but not in astrocytomas or in nonglial tumors by gas chromatographic analysis [38]. A detailed analysis of the sterols in a human glioblastoma was made using a combination of chromatographic techniques and gas-liquid chromatography-mass spectrometry [33]. Essentially the same sterols were found as were present in developing and normal brain (table I). There was a definite increase in desmosterol and oxygenated products of cholesterol catabolism compared to a normal sample of human brain [33].

Increased levels of desmosterol in intracranial tumors can be obtained by administering triparanol to patients [72]. The normal brain tissues analyzed, both cerebral and cerebellar, did not contain desmosterol, except for the only sample of normal dura available, which had a very high level of desmosterol compared to that in the plasma of the patient. The desmosterol content of two neurinomas was also relatively high compared to other tumor types and in relation to the plasma content after triparanol administration. The desmosterol concentration in glial tumors never reaches that of plasma which, on the other hand, is in equilibrium with liver sterols.

In a patient given labeled mevalonate, the distribution of radioactivity in the cholesterol and desmosterol fractions of glioblastoma was different from that in plasma, both before and after triparanol administration [72].

Studies *in vitro* showed that the incorporation of mevalonate is even more pronounced in the sterol precursor, being higher in the desmosterol fraction than in the cholesterol fraction [72]. The block of desmosterol transformation into cholesterol is particularly high for the most malignant gliomas. The pattern of mevalonate incorporation into isoprenoid compounds other than sterols appeared to be quantitatively similar in normal brain and brain tumors [52].

AZARNOFF et al. [2] also showed that human gliomas, meningiomas, and neurinomas synthesize cholesterol from labeled acetate much more readily than normal mature nervous tissue. The possibility of interfering with brain tumor growth through a block of lipid synthesis was first tested by this group using vanadium which inhibits cholesterol synthesis from acetate. The administration of this drug before surgery to two patients with gliomas resulted in the inability of the tumors to synthesize cholesterol from acetate *in vitro*. The authors noted that preliminary trials of diammonium oxytartrate vanadate as a chemotherapeutic agent were unsuccessful. Other than the high toxicity of this compound, perhaps it acts too early in the biosynthetic pathway of cholesterol synthesis and interferes with other very basic biochemical mechanisms.

D. Phospholipids

Phospholipid synthesis measured by incorporation of ^{32}P, is higher in brain tumors than in normal brain [98], although the total phospholipid is decreased considerably in intracranial tumors compared to normal tissue [71, 98, 130]. The major differences in phospholipid composition are an increase in the phosphatidyl choline fraction (PC) and a decrease in ethanolamine-containing phospholipids (PE) in most types of intracranial tumors, including gliomas, meningiomas, neurinomas, and metastatic carcinomas [12, 71, 130]. An increase in the PC/PE ratio was also found in oligodendroglioma samples in one study [130], but another study indicated that the distribution of phospholipid fractions in oligodendrogliomas is similar to that in average white matter or myelin, where the ethanolamine fraction is in highest concentration [11]. There were also relatively more inositol phosphoglycerides in metastatic carcinomas than in normal brain phospholipids [130]. It is not yet clear whether there are significant changes in minor brain phospholipids, such as phosphoglyceryl ether components, during neoplasia [103, 105].

The same increase in choline-containing phospholipids was also found in fractionated microsomes from glial tumors compared to microsomes of normal cortex of white matter [12]. When different regions of a glioblastoma were analyzed, no phosphatidyl ethanolamine was found in the center [12]. The higher PC/PE ratio in tumors is analogous to the situation in fetal brain [14], and it has been suggested that this indicates a low degree of cellular differentiation and organization. CHRISTENSEN LOU et al. [12] also suggested that the changes in the PC/PE ratio might explain some of the changes in surface properties of neoplastic cells. Utilizing Sephadex filtration, they found the protein affinity of PE is stronger than that of PC and the protein affinity of the lipids produced in brain tumors is decreased in relation to normal tissue [12]. There still is little information on the composition of normal neural cells compared to tumor cells. Since PE is more characteristic of myelin, it is possible that a high PC/PE ratio is a normal characteristic of glial cells. There is some indication from analyses of isolated rat glial cells that this may be the case [42, 68, 77].

WHITE [130] has discussed the changes in phospholipid fatty acids in detail. Changes in the fatty acids of ethanolamine and choline phosphoglycerides included an increase in linoleic and arachidonic acids and a decrease in C_{22} fatty acids [130].

E. Glycolipids

Cerebrosides and sulfatides have been detected in varying amounts in intracranial tumors [12, 71, 103]. Although gangliosides were not detected in brain tumors by some workers, levels up to 3 % of the total lipids were detected in glioblastomas by SLAGEL et al. [103]. The central portion of one glioblastoma did not contain cerebroside [12], and this may account for the failure to detect polar lipids in some tumor samples. Oligodendrogliomas were found to have a higher concentration of ceramide monohexosides than other glial tumors, but more malignant oligodendrogliomas had lower levels than those classified as benign [11]. According to CHRISTENSEN LOU and CLAUSEN [11], the polar lipid distribution of oligodendrogliomas is not far from that of normal white matter during the myelination process, in accordance with the theory that the myelin sheath in the CNS is derived from oligodendroglial plasma membranes. In a study of free and protein-bound cerebrosides (calculated as galactose per weight of protein) the ratio of free to combined cerebrosides was found to be lower in brain tumors than in normal brain [78]. In this study, the total cerebroside content was lower than normal, and similar in astrocytomas of varying degrees of malignancy and in oligodendrogliomas. In a later study, these authors detected cerebrosidase activity in astrocytomas and oligodendrogliomas but not in other types of brain tumors or cerebral tissue [79].

There have been reports of increases in specific glycosphingolipids in different types of brain tumors compared to normal brain. Despite the discrepancies or lack of verification of some of the findings, the results are summarized in table III. A dihexose ceramide constituted 16 % of the sialic acid-free glycolipids in glioblastoma samples [12]. This compound had thin-layer chromatographic properties of cytolipin H, a lipid hapten found in extraneural tumors [85]. Monosialosyl-lactosyl-ceramide was determined to be characteristic of meningiomas [97], but it was not found in high levels in meningiomas in another study [55]. It was also found to increase in gliomas in one study [46], but not in another [55]. Disialosyl-lactosylceramide was determined to be the major glioma ganglioside [96], but was not found in increased quantities in other studies [46, 55]. There is agreement on the increase of the simple ganglioside, sialosyl-galactosylceramide in meningiomas [46, 55], although normal meningeal tissue has not been analyzed. This ganglioside apparently has no direct metabolic relationship with either the major brain gangliosides or adult brain cerebroside [100]. A decrease in the more polar gangliosides and an increase in the less polar gangliosides was found in malignant gliomas [55]. This may be analogous to the difference in ganglioside composition in developing brain compared to mature brain [113]. It is possible that in brain tumors, as in some other neoplastic or transformed cells, glycolipids with 'incomplete carbohydrate' chains are increased [41]. In any case, increases in minor gangliosides or less polar gangliosides appear to occur in brain tumors. UHLENBRUCK and GIELEN [119] have discussed the immunological implications of alterations in brain tumor glycolipids.

F. Cerebrospinal Fluid

The possibility exists that analysis of CSF lipids may be of practical value for the diagnosis or classification of brain tumors and in the follow-

Table III. Glycosphingolipids reported to be increased in brain tumors compared to normal brain

Structure	Generic term	Notations [113]	Notations [132]	Occurrence
Gal(β,1→4)Glu(1→1)Cer	lactosylceramide (cytolipin H)			gliomas [12]
NAN(2→3)Gal(β,1→4)Glu(1→1)Cer	monosialosyl-lactosylceramide	GM3	G Lact[1]	meningiomas [97] gliomas [46]
NAN(2→8)NAN(2→3)Gal(β,1→4)Glu(1→1)Cer	disialosyl-lactosylceramide	GD3	G Lact[2]	gliomas [96] unconfirmed [46, 55]
Gal(β,1→3)GalNAc(β,1→4)Gal(β,1→4)Glu(1→1)Cer 3 (↑) 2 NAN	monosialosyl-N-tetraglycosylceramide	GM1	G NT[1]	gliomas [55]
Gal(β,1→3)GalNAc(β,1→4)Gal(β,1→4)Glu(1→1)Cer 3 (↑) 2 NAN NAN	disialosyl-N-tetraglycosylceramide	GDIa	G NT[2a]	gliomas [55]
NAN(2→3)Gal(1→1)Cer	sialosyl-galactosylceramide		G Gal	meningiomas [46,55]

up of patients undergoing treatment for tumors. In this section, the possibilities for lipid analysis, current procedures, and drawbacks of CSF analysis will be discussed. It would be worthwhile to find tumor-specific lipids or characteristic brain lipids that are increased in quantity during neoplasia as biochemical indicators of tumor activity. There are known to be non specific changes that occur in CSF components, including lipids, during various neurological disorders [58]. Changes in the blood-brain or blood-CSF barriers during neoplasia may also result in changes related to blood lipid composition.

1. Comparison of Blood and Cerebrospinal Fluid Lipids

In general, the lipid pattern of the CSF reflects that of the blood. There are some noteworthy differences between normal CSF lipids and blood lipids, including higher proportions of phosphatidyl ethanolamine, and lower proportions of esterified cholesterol and linoleate [51]. An exchange of lipids between lipoproteins of CSF and the surrounding brain tissue, especially of the phospholipid fraction, might account for some differences [51]. The concentration of phospholipid in normal CSF is approximately 3% of that in serum and the concentration of CSF cholesterol is approximately 2% of the serum concentration. SCHRAPPE [94] has suggested that the esterified cholesterol/phospholipid ratio in CSF is a relevant and sensitive indicator of CSF lipid changes, since this ratio is independent of serum patterns. Cerebrosides have been detected in microgram quantities in normal CSF, and cerebrosides from normal CSF and brain were practically identical in composition but differed from those of plasma [63, 92]. Major brain gangliosides at an approximate total concentration of 1 μg/ml have been detected in normal CSF [8]. The lipids in plasma are almost entirely associated with protein. No immunological difference was found between α-lipoprotein and β-lipoprotein of CSF and serum, although it is possible that the lipid moiety of the CSF lipoprotein is different from that of serum [114].

2. Cerebrospinal Fluid Lipids in Neoplasia

While α-lipoproteins were always found in CSF, β-lipoproteins were only found in cases of neurological disorders. An increase in the β-lipoprotein fraction was found most consistently in patients with brain tumors; there was an increase in about one half of the patients with brain tumors studied [18]. It was suspected that the β-lipoproteins originated in the blood.

An increase in CSF cholesterol is often an indication of an organic neurological disorder, including neoplasia [58]. The cholesterol ester fraction may also increase, especially in cases of spinal tumors [117]. Desmosterol, a characteristic lipid of gliomas, has been detected in the CSF of approximately one third of the patients with gliomas studied [73, 129]. It was found in small but measurable quantities using 5 ml of CSF for gas chromatographic analysis.

Increases in the cerebroside fraction, a characteristic lipid fraction of brain, occur in the CSF in varying degrees in a number of neurological disorders, including neoplasia [13]. Because of the possibility of increases in specific glycolipids in brain tumors of different types, it is suggested that increases of specific glycolipids might be detectable by immunological techniques.

3. Use of Triparanol to Augment CSF Desmosterol

A test for the diagnosis of brain tumors based on the augmentation of CSF desmosterol after a short oral treatment (5 days) with triparanol has been proposed [73]. CSF desmosterol concentrations higher than 0.1 μg/ml or a desmosterol/cholesterol ratio (100 D/C) greater than 3 indicated the presence of a brain tumor. Using these criteria in an analysis of 74 patients without tumors and 91 patients with intracranial tumors, PAOLETTI and co-workers [29, 73] were able to determine the presence of a tumor with 75 % accuracy or the absence of a tumor with 80 % accuracy. A false-positive diagnosis was very rare. A correlation was seen between degree of malignancy of astrocytomas and level of desmosterol, comparing astrocytoma grades I and II to glioblastomas. On the other hand, high levels of demosterol were found in the CSF of patients with acoustic neurinomas and meningiomas and in spinal tumors. The main disavantage of this test is that it gives no indication of the site of the neoplastic process. It is not known whether plasma desmosterol also contributes to the increased desmosterol levels in the CSF of patients with brain tumors after triparanol treatment, but there is no correlation between high CSF desmosterol and blood levels [129]. Since desmosterol can be found in the CSF of some patients with tumors, even without triparanol treatment [73, 129], it seems to originate from synthesis within the tumor in these cases. MARTON et al. [60] failed to detect desmosterol in the free sterol fraction of 1 ml of CSF in a group of 23 patients with CNS tumors and 5 given triparanol. In a preliminary study, we have observed a higher level of desmosterol in the sterol ester fraction of pooled CSF from patients given triparanol.

In our studies [128, 129], triparanol has been administered to patients with malignant gliomas at various stages of treatment (before and after surgery and chemotherapy). Results are summarized in table IV. Looking at the clinical data up to this point, it is apparent that all patients do not respond to triparanol administration with an elevation of CSF desmosterol. In those patients that do respond initially, many more must be studied in order to determine whether serial determination of CSF desmosterol can be used as a biochemical indicator of tumor activity. Figure 4 shows the use of serial measurements of CSF desmosterol after triparanol administration in the follow-up of a patient with a malignant glioma. In this patient, an increase in CSF desmosterol was observed on tumor recurrence.

Table IV. CSF sterols in patients before and after triparanol administration, 250 mg b.i.d. × 5 days [128, 129, and unpublished data]

Group	Number of patients		Cholesterol average, $\mu g/ml$	Desmosterol average, $\mu g/ml$
Non neoplastic diseases	16	no triparanol	4.75 (2.81–6.84)	n.d.
Non neoplastic diseases	7	triparanol	4.09 (2.57–4.89)	0.02 (0.006–0.03)
Glioma patients various stages of treatment	36	no triparanol	11.96 (3.50–52.12)	0.01 (n.d.–0.10)
Glioma patients inoperable and recurrence with no recent surgery	13	triparanol	8.68 (2.84–40.23[1])	0.09 (0.005–5.52[1])
Glioma patients after surgery	19	triparanol	18.00 (4.99–36.91)	0.25 (0.005–1.02)
Glioma patients after surgery and chemotherapy	16	triparanol	11.97 (2.79–55.32)	0.08 (0.005–0.29)

n.d. = Not detectable.
Figures given in parentheses indicate range.
[1] Not included in average.

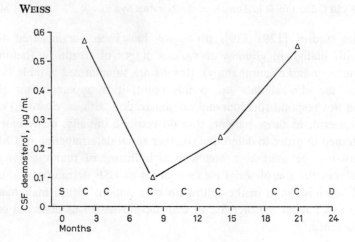

Fig. 4. Use of triparanol during the follow-up of a patient (M47) receiving treatment for a glioma. S = Surgery; C = chemotherapy, 1,3-bis(2-chloroethyl)-1-nitrosourea; D = died. From WEISS *et al.* [129].

Other diagnostic drugs may also be useful in causing an increase in cholesterol precursors in tumors and in CSF. Studies of the sterol composition of tumor cells grown in tissue culture, indicate that 20, 25-diazacholesterol is very effective in replacing tumor cell cholesterol with desmosterol [122].

IV. Models for Brain Tumor Research

A. Primary Tumors

1. Induction of Tumors in Animals

Progress in brain tumor research is dependent in large measure on development of adequate animal models. Recently, most types of intracranial tumors have been induced in experimental animals with oncogenic viruses [50]. The most common animal models that have been used for biochemical studies are mouse brain tumors originally obtained by direct implantation of carcinogens (mainly higher aromatic hydrocarbons) in the brain [137]. Naturally occuring conditions for the production of brain tumors are simulated more closely by resorptive carcinogens (extracranial route administration). The most successful resorptive carcinogens for selective induction of malignant tumors of the nervous system with a reproducible high rate of incidence were found to be ethylnitrosourea (ENU) and methylnitrosourea (MNU) by DRUCKREY et al. [20]. Tumors can be obtained in rats by administering ENU or MNU by various routes [17, 39, 127] and one of the easiest and most effective methods is administration of ENU during the last quarter of pregnancy (table V). The pattern of tumor induction in the nervous system varies in different strains and may be used to advantage for obtaining large numbers of tumors at approximately the same site, and also for studying tissue during 'preneoplastic' stages. In the CFE Sprague-Dawley strain, approximately one half of the male rats treated transplacentally with ENU developed malignant neurinomas of the PNS – Schwann cell tumors of the trigeminal nerve [17]. In human pathology, the occurence of malignant neurinomas is rare, especially at this site. A pattern that more closely resembles that in humans is obtained in CDF rats which have a high incidence of malignant gliomas of the CNS (table V). There is a basic correspondence in the morphology of the tumors of the CNS induced in the rat with nitrosourea derivatives and some of the naturally occuring

Table V. Transplacental carcinogenesis by ethylnitrosourea in rats given 20 mg/kg body weight during last five days of pregnancy. Data from CRAVIOTO et al. [17] and unpublished work

Rats (offspring) with	CFE Sprague-Dawley (n=100)	BD-IX (n=40)	CD Fischer (n=28)	Long-Evans (n=10)	DPPD[1] treated (n=10)	vitamin E[2] treated (n=9)
PNS tumors, %	81	75	11	60	10	22
CNS tumors, %	11	28	72	40	70	44
Nonneural tumors, %	8	0	0	20	30	22
No tumors, %	14	10	25	10	10	11
Median latency period for tumor formation, days	198	224	346	306	408	300

1 N,N'-diphenyl-p-phenylenediamine, 0.05% of diet.
2 dl-α-tocopheryl acetate, 0.2% of diet.
DPPD and vitamin E added to pellets and fed throughout life beginning at weaning. All other rats were fed standard laboratory chow.

human tumors, although the tissue architectures are somewhat different [20].

Table V also shows a study of the effect of feeding lipid antioxidants on the tumor incidence in Long-Evans rats. Because of the small number of animals studied, this subject will not be discussed in detail. The results suggest that rats treated transplacentally with ENU and fed DPPD throughout life have a lower incidence of tumors of the PNS which develop before intracerebral gliomas.

2. Sterol Metabolism in Nitrosourea-Induced Tumors

The desmosterol content of nitrosourea-induced tumors, both of the PNS and CNS, was found to be greater than in any other human or animal tumor of the nervous system [127]. The sterol composition of a larger series of tumors has been studied, taking advantage of the predilection for certain sites of tumor development. The cholesterol content per wet weight of tissue is decreased in the tumors and a high level of cholesterol in a tumor sample may indicate the presence of normal tissue. The cholesterol concentration in neoplastic trigeminal nerve is 4–5 times lower than in normal tissue, and in cerebrum the concentration is 2–3 times lower in tumors. In table VI, the ratio of desmosterol to cholesterol is compared in normal and neoplastic CNS or PNS. A definite difference was found in this study in the average amount of desmosterol in malignant intracerebral gliomas and in malignant neurinomas of the PNS (trigeminal nerve).

The amount of desmosterol in malignant neurinomas of the trigeminal nerve increased to a consistent level, approximately 2 % of the total sterols. Higher amounts of desmosterol were found in some neurinoma samples in two previously reported studies [39, 127]. A possible reason for this could be the proximity of tumors of both the CNS and PNS in some animals. The desmosterol content in gliomas was more variable than in trigeminal neurinomas, but reached high levels in some tumors (table IV). One oligodendroglioma sample contained as much desmosterol as cholesterol, the highest amount of desmosterol ever detected in tissue. The amount of desmosterol in tumors of the PNS and brain in this study parallels the amounts in corresponding developing nervous tissue (fig. 3), and may be related to the predominant cell type seen at the time of study (Schwann cell or glial cell).

Trigeminal nerves were also analyzed before the expected occurrence of tumors (preoplastic state), but no increase in desmosterol was seen

Table VI. Ratio of desmosterol/cholesterol in rat nervous tissue. From WEISS *et al.* [123] and unpublished data

	Desmosterol/cholesterol	
	average	range
Peripheral nervous system		
Grossly normal mature		
trigeminal nerve	0.005 (15)	0.001–0.009
Malignant trigeminal		
neurinomas (Schwannomas)	0.018 (32)	0.003–0.054
Central nervous system		
Grossly normal		
mature cerebrum	0.004 (9)	0.001–0.013
Intracerebral tumors		
Total	0.113 (39)	
Oligodendrogliomas	0.204 (10)	0.008–1.023
Astrocytomas	0.063 (6)	0.008–0.151
Mixed gliomas	0.090 (17)	0.008–0.515
Undetermined	0.075 (6)	0.011–0.190

Number of samples in parentheses.

until the nerves were clearly neoplastic [124]. GROSSI PAOLETTI *et al.* [39] also observed that desmosterol was only present in high concentrations in well-developed tumors and was detected in only one of the glial foci examined in spite of the low total sterol content in the brain areas that included the foci. The earliest sign of tissue deviation was the incorporation of thymidine, which could be detected even in the absence of other histological changes in the oligodendroglial foci [39].

In order to determine whether the sterol metabolism of intracranial tumors is influenced by circulating sterols, animals with signs of intracranial tumors and control animals were injected with ^{14}C-cholesterol and sacrificed after two days. The uptake of the isotope was determined in neoplastic and normal tissue after the tissues were perfused with saline. The results indicate that gliomas and intracranial neurinomas take up circulating cholesterol more readily than normal nervous tissue (table VII). Recently it has been suggested that there is a pool of cholesterol in the brain from which all metabolizing structures draw their cholesterol supplies and there is a continuous exchange of cholesterol between the brain pool and the blood [107]. The rate of this exchange may be related to the rate of blood flow through the tissue.

Table VII. Uptake of ^{14}C-cholesterol from peripheral circulation by intracranial nervous tissue after two days. From WEISS et al. [124] and unpublished data

	Specific activity of tissue cholesterol/ specific activity of serum cholesterol	
	average	range
Grossly normal trigeminal nerves 8 samples; 6 animals	0.015	0.002–0.044
Trigeminal nerve tumors 2 samples; 2 animals	0.065	0.053–0.076
Grossly normal cerebrum 10 samples; 8 animals	0.012	0.001–0.023
Intracerebral gliomas 6 samples; 3 animals	0.171	0.052–0.503

B. Transplanted Tumors

1. Lipid Composition and Synthesis

Chemically and virally induced tumors of the nervous system have been maintained by subcutaneous passage and used for biochemical studies, including studies on lipid metabolism. Nitrosourea-induced tumors can be readily cultured *in vitro* and the cells transplanted to rats, either subcutaneously or intracerebrally [5, 7, 17].

Clonal cell lines produced from tumors induced in rats with methylnitrosourea, have been grown in newborn hamsters and the biochemical characteristics studied and used as a model for astrocyte composition [22]. The total lipid content of the tumors was accounted for by the structural lipids: water-insoluble glycolipids, cholesterol, and phospholipids. The tumors contained as much phospholipid as cholesterol on a molar basis, and low concentrations of water-insoluble glycolipids, proteolipid protein, and gangliosides. It may be of interest in view of the discussion of gangliosides in section III. E, that the principal ganglioside found in the subcutaneously grown astrocytomas was hematoside, GM3, which is a minor brain ganglioside.

The lipid and fatty acid composition has been studied in methylcholanthrene-induced glial tumors (ependymomas) grown subcutaneously and intracerebrally. Subcutaneously grown Perese tumor contained triglyceride as its major lipid [109]. In this respect, the lipid pattern of this tumor appears to be substantially different from primary human tumors [36] and probably from astrocytomas grown subcutaneously in newborn hamsters [22].

Furthermore, although the transplanted mouse gliomas remain localized, they influence hepatic triglyceride metabolism [110]. A quantitative decrease in hepatic triglycerides and a decrease in the rate of incorporation of the nonesterified fatty acids into triglycerides was found. The pattern of neoplastic influence on hepatic lipids suggests host modification to support the growth of the tumor.

The fatty acid pattern of transplanted glial tumors was similar whether they were grown intracerebrally or subcutaneously [109]. The major change was an increase in linoleic acid; the triglyceride fraction being especially high in this fatty acid. Fatty acids are synthesized at a high rate from simple precursors in experimental ependymomas, both *in vivo* and *in vitro* [38].

2. Sterol Metabolism and the Effect of Drugs

Sterol synthesis is very active in subcutaneously transplanted glial tumors. Incorporation of labeled precursors *in vivo* or *in vitro* into sterols was much higher than in normal brain [38].

When mice bearing subcutaneously grown glial tumors were injected with tritiated cholesterol, equilibration between plasma and tumor specific activities occurred after seven days [38]. Normal brain had a much slower uptake, while liver and intestine showed a faster and higher uptake during the time intervals studied. Uptake of labeled cholesterol by the subcutaneously grown tumors appears to be more efficient than uptake by primary induced intracranial tumors (table VII). Studies on hepatomas demonstrated that care must be taken in interpreting results of studies on sterol metabolism in transplantable tumors, which may be unrelated to the situation in primary tumors, simply due to such factors as alterations in the blood supply [49].

The effects of drugs interfering at various stages with the synthesis of cholesterol have been tested on a subcutaneously transplanted Zimmerman ependymoblastoma [40]. SKF-525A caused a large reduction in total plasma sterol, while tumor and brain sterol levels and composition were not affected and no interference with tumor growth could be demonstrated. This tumor contains desmosterol at a maximum amount of 5% of the total sterols. The sterol composition can be changed further with inhibitors of later stages in the pathway of sterol synthesis, such as inhibitors of desmosterol reductase and 7-dehydrocholesterol reductase. Triparanol and diazacholesterol treatment caused an increase in the desmosterol content of both plasma and tumor up to more than 40 % of the total sterols. Although subcutaneous treatment with triparanol resulted in some inhibition of tumor growth, treatment with 20,25-diazacholesterol did not. Since both of these compounds block cholesterol synthesis at the same level, desmosterol accumulation is probably not related to growth inhibition of the tumor. The greatest inhibition was caused by dietary AY-9944, which was the only drug that caused histological changes in the tumor structure. AY-9944 treatment reduced total plasma sterols and resulted in the accumulation of 7-dehydrocholesterol (30–50 % of the total sterols). Corresponding concentrations of 7-dehydrocholesterol were found in tumors. Other sterols also accumulated, including 7-dehydrodesmosterol. Although AY-9944 was quite toxic, as seen by its effect on mouse body weight, the relative reduction in tumor weight was greater. It is not certain whether inhibition of tumor growth by AY-9944 in these experiments is related to blockage of cholesterol biosynthesis, to accumulation in the tumor of certain sterols, or to nonspecific toxic effects. When a mouse glial tumor was transplanted intracerebrally, the survival time of animals treated with AY-9944 increased compared to untreated mice with tumors [102]. The above experiments point out the possibility of new chemotherapeutic approaches to the treatment of brain tumors.

C. Tissue Culture

1. Lipid Metabolism

There is an increasing number of studies on sterols [89] and other lipids [106] in normal and malignant cells in tissue culture, although there are few published studies on lipid metabolism in cultured neural cells. Human and experimental brain tumors are readily adaptable to growth in tissue culture [5, 7, 16] and established cell lines are available, e. g. C_6, a glial cell line derived from a nitrosourea-induced rat astrocytoma, and cloned cell lines of mouse neuroblastoma, C1300. Extensive studies are being carried out on defining differentiated biochemical functions of neuronal and glial cells in tissue culture [7, 88].

Lipid biosynthesis from ^{14}C-labeled glucose or acetate and $^{32}P_i$ was studied in cell cultures of normal immature astrocytes and astrocytoma cells, C_6 [43]. Glucose was the best precursor for total lipids. Triglyceride was the highest labeled neutral lipid and cholesterol and cholesterol esters contained only small amounts of isotope. Cerebrosides and sulfatides, present at most in minute quantities, were virtually free of ^{14}C. Phospholipid contained the bulk of the radioactivity, with lecithin being the most highly labeled. Astrocytoma cells and neuroblastoma cells grown in monolayers incorporated comparable amounts of $^{32}P_i$, but the patterns of incorporation into phospholipids were substantially different [21]. Polyphosphoinositides of high metabolic activity were found in cell cultures of normal newborn hamster astrocytes and astrocytoma [43].

The ganglioside and neutral glycosphingolipid composition of clonal lines of mouse neuroblastoma, C1300, were studied [136]. Marked differences in the glycosphingolipid patterns, particularly in the case of gangliosides, were observed in various lines. The variations in the glycosphingolipid profile were not lethal as far as cell growth in culture was concerned. No appreciable change in the profile of glycosphingolipids was observed in a clonal line grown in suspension as 'undifferentiated' neuroblasts or in monolayers as 'differentiated' neurons. In comparison to the solid tumor grown subcutaneously, the total amounts of gangliosides or neutral glycosphingolipids were higher in the lines in tissue culture.

The fatty acid composition of C_6 rat glial cells was determined after growth in tissue culture media with and without vitamin B_{12} [6]. In a vitamin B_{12}-deficient state, increasing amounts of unbranched fatty acids with 15 and 17 carbon atoms arose. The authors noted that the fatty acid pattern of cells grown in a serum-free medium was identical to that in a medium supplemented with normal serum, although the pattern of polyunsaturated fatty acids in the glial cells was not shown [6]. The fatty acid composition of other mammalian cells (and especially the polyunsaturated fatty acid content) is affected by fatty acids in the serum or media [106]. The ability of cells in culture to convert linoleate to arachidonate was studied [62]. Three types of cells possessing different capacities for desaturating polyunsaturated fatty acids were found: the HeLa and L cells did not have this ability, heart and CHL-1 cells converted a small amount of linoleate to arachidonate, and glial cells (derived from a rat glioma) did so to a much greater extent. The factors controlling this capacity are not known.

2. Sterols and the Effect of Drugs

Answers to what determines the sterol composition in tumors of the nervous system, and more specifically desmosterol accumulation, might be

Table VIII. Desmosterol in cells grown in monolayers in 90% synthetic medium, 10% delipidized fetal calf serum. From WEISS *et al.* [122] and unpublished data

Cell line or primary tumor	Desmosterol/cholesterol ratio
L cell mouse fibroblast line (L-929)	8.271 X/C = 0.792
C_6 glial cell line (from rat astrocytoma)	average of 6: 0.044 range: 0.027–0.071
E-970, rat oligodendroglioma	average of 6: 0.117 range: 0.002–0.520
EA-285, rat mixed glioma	0.012, 0.0250, 0.0907
E-757, rat trigeminal neurinoma	average of 4: 0.003 range: 0.001–0.005
E-837, rat trigeminal neurinoma	0.002
EA-296, human astrocytoma	0.017, 0.046
EA-361, human astrocytoma	average of 6: 0.085 range: 0.009–0.435
EA-351, human astrocytoma	0.071, 1.350 X/C = 0.021, 1.287

X = Sterol more polar than desmosterol.
Grown in Neuman-Tytell medium [64] with delipidized serum [1].
Values were determined after various passages and growth periods.

provided by studying cells grown in tissue culture under various conditions. Further questions about the significance of desmosterol arose with the finding of ROTHBLAT *et al.* [90] that L cell mouse fibroblasts, when grown in a medium containing serum from which the lipids had been extracted, contained desmosterol as the major cell sterol. They did not detect desmosterol in eight other normal and malignant cell lines, including a human fibroblast line and that line transformed by simian virus [89]. Desmosterol was not detected in harvested cells of rat or human glial tumors grown in serum-containing medium, even after the addition of triparanol [60]. Desmosterol was detected at significant levels in harvested cells only when cell lines or tumors were grown in monolayers in a 90-percent synthetic medium and 10-percent fetal calf serum, delipidized by treatment with cold solvents (table VIII). The ratio of desmosterol to cholesterol was determined in harvested cells of various nitrosourea-induced tumors of the rat nervous system, human astrocytomas, the C_6 glial cell line and L cell mouse fibroblasts. Desmosterol was present in highest amounts in L cells. In agreement with previous studies on L cell mouse fibroblasts [89], another sterol more polar than desmosterol (probably 7-dehydrodesmosterol) was also found in the cells. Not enough samples have been analyzed to indicate whether there is a definite difference between the desmosterol content of cultured neurinomas and gliomas. Desmosterol was detected in some cultures of human astrocytomas in high levels. Another sterol more polar than desmosterol was also found in one cultured astrocytoma. It has not yet been determined why higher levels of desmosterol are seen in some experiments or passages and not in others. Some possible factors that might influence desmosterol content in cells of nervous system tumors are type of cell, growth phase, cell density, or adaptation to new growth conditions. The desmosterol content of L cell mouse fibroblasts is consistently high in this medium, and it has been suggested that the enzyme, desmosterol-reductase is missing or inoperative in L cells [89].

Nitrosourea-induced tumors were cultured in the presence of hypocholesteremic agents. Triparanol was added to the cultures in alcohol, since it is water-insoluble. The control flasks contained regular or delipidized serum plus alcohol. There was a further increase in the desmosterol content in the harvested cells of a glioma cultured in the delipidized serum medium after various periods of growth (table IX). The drug was inactive in this respect when the cells were grown in medium containing regular serum. Diazacholesterol, a water-soluble inhibitor of desmosterol reductase, was more effective in causing replacement of cell cholesterol by desmosterol both in a cultured rat glioma (table IX) and neurinoma. Essentially complete replacement by desmosterol was seen after 2 weeks growth in medium containing delipidized serum and a nontoxic level of diazacholesterol (0.001 mM).

Some cultured cells [89], including normal and malignant glial cells [43], synthesize sterols at a low rate in the presence of serum or exogenous sterol. Experiments utilizing desmosterol-reductase inhibitors indicate an operative feedback control of cholesterol synthesis in neural cells *in vitro* which is sensitive to exogenous sources of cholesterol (table IX). The presence of desmosterol in the cells should reflect *de novo* sterol synthesis, which appears to be very active in the neural cells grown in the

Table IX. Effect of desmosterol-reductase inhibitors (0.001 mM) and media on desmosterol content of cells of a rat glioma (E-970). From WEISS *et al.* [122] and unpublished data

Medium	Growth period, weeks	Ratio of desmosterol/cholesterol
10% serum + 0.5% alcohol	1	trace
10% serum + 0.5% alcohol + triparanol	1	0.005
10% delipidized serum + 0.5% alcohol	1	0.022
	2	0.118
	3	0.060
10% delipidized serum + 0.5% alcohol + triparanol	1	0.685
	2	0.639
	3	0.917
10% serum + 20,25 diazacholesterol	1	0.012
10% delipidized serum + 20,25 diazacholesterol	1	13.00
	2	35.23

Values are averages of 2 determinations; media described in table VIII.

Table X. Effect of AY-9944 and media on the sterol composition of glial cells (C$_6$ cell line) at one week of growth. From WEISS *et al.* [122] and unpublished data

Drug concentration, mM	Ratio of sterol precursor/cholesterol	
	10% serum	10% delipidized serum
0	0.001	0.040
0.0001	0.049[1]	3.560[1]
0.001	0.052[1]	4.284[1]
0.01	no growth	no growth

Values are averages of 2 determinations; media described in table VIII.
[1] Assumed to be mainly 7-dehydrocholesterol.

delipidized serum medium. The results also indicate that the toxicity of higher doses of desmosterol-reductase inhibitors added to serumless or serum-containing media is probably not related to desmosterol accumulation. Diazacholesterol can be used as a tool for estimating *de novo* sterol synthesis in cells grown in a chemically defined medium, similar to the use of other cholesterol inhibitors in measuring cholesterol synthesis *in vivo* [101]. Diazacholesterol may also be a useful diagnostic tool when given to patients with brain tumors for augmenting the desmosterol content of tumor cells and CSF.

The effect of AY-9944 and media on the sterol composition of harvested cells of a rat glioma and the C_6 astrocytoma cell line is shown in table X. With this drug, there was a small accumulation of cholesterol precursor, assumed to be 7-dehydrocholesterol, even in the presence of regular serum. With delipidized serum medium, a greater amount of inhibition was obtained. AY-9944 affected cell growth at lower concentrations than diazacholesterol, and also appeared to cause a decrease in the total sterol content per cell at a low concentration [122].

The results *in vitro* correlated with previous experiments on the effect of hypocholesteremic agents on the growth of a transplanted mouse glioma. Of the drugs tested *in vivo*, AY-9944 was the most effective in retarding tumor growth [40]. ROTHBLAT [89] has suggested that all cholesterol precursors may not support cellular growth even though desmosterol seems to be an effective substitute for cholesterol in cells. Primary diploid human skin fibroblasts were unable to grow in the absence of exogenous sterol, but would grow when cholesterol or desmosterol, but not 7-dehydrocholesterol, was added to the growth medium [47].

It is not clear whether control mechanisms similar to those in cultured neural tumors are also present in primary or transplanted tumors. Glial tumor cells (EA-285) and C_6 cells were transplanted either subcutaneously or intracerebrally in young CDF rats (irradiated in the case of C_6 cells), and the desmosterol content of the transplanted tumors was determined. Preliminary studies indicate that the level of desmosterol is much lower than in primary nitrosourea-induced rat glial tumors or in the cells grown in tissue culture in a cholesterol-free medium. This may indicate that environmental factors, such as availability of exogenous cholesterol, can affect the sterol composition of tumor cells *in vivo*. On the other hand, it has been suggested that sterol biosynthesis in subcutaneously grown mouse glial tumors is affected very little by cholesterol feeding [38]. Much remains to be learned about the possible control mechanisms affecting the various stages of cholesterol biosynthesis in both normal and neoplastic nervous tissue.

V. Summary

Considerable differences are found in the lipid composition of tumors of the nervous system compared to normal tissue; these include differences related to the metabolism of structural lipids (cholesterol, phospholipids, glycosphingolipids). Some of these differences in composition may be due to a decreased myelin content of tumor tissue and therefore reflect a relative increase of nonmyelin cellular lipids. Specific differences in the lipid composition of neoplastic and normal neural cells (e.g. glial cells) remain to be established. Some similarities exist between developing and neoplastic nervous tissue with respect to lipid composition and the increased synthesis of characteristic neural lipids.

Data from studies using models for brain tumor research (primary animal tumors, transplanted tumors and cells in tissue culture) have been compared with data on lipids in human tumors and in the cerebrospinal fluid of patients with brain tumors. The most comprehensive studies have been carried out on cholesterol metabolism in tumor models, the developing nervous system, human tumors and cerebrospinal fluid. This includes studies on the effects of drugs on sterol metabolism. Studies on the occurrence of desmosterol in tumors of the nervous system have been described in detail. Although the relationship between an increase in desmosterol in tumors of the nervous system and the neoplastic process is not clear, the change in the sterol composition and the ability to modify tumor sterol synthesis and composition by the administration of drugs may have diagnostic or chemotherapeutic value.

VI. Acknowledgments

The work presented here from our laboratory was done with HUMBERTO CRAVIOTO, HERBERT J. KAYDEN, JOSEPH RANSOHOFF, and ELVIRA DE C. WEISS, with the technical assistance of FRANK DRNOVSKY. This chapter is dedicated to the memory of ENRICA GROSSI PAOLETTI, who contributed greatly to the advancement of knowledge in this area, during her short research career.

This work was supported by US Public Health Service Research Grant CA-10887 from the National Cancer Institute.

VII. References

1 ALBUTT, E. C.: A study of serum lipoproteins. J. med. Lab. Technol. *23:* 61–82 (1966).
2 AZARNOFF, D. L.; CURRAN, G. L., and WILLIAMSON, W. P.: Incorporation of acetate-1-C^{14} into cholesterol by human intracranial tumors *in vitro.* J. nat. Cancer Inst. *21:* 1109–1115 (1958).
3 BANIK, N. L. and DAVISON, A. N.: Desmosterol in rat brain lipids. J. Neurochem. *14:* 594–595 (1967).
4 BANIK, N. L. and DAVISON, A. N.: Exchange of sterols between myelin and other membranes of developing rat brain. Biochem. J. *122:* 751–758 (1971).
5 BARKER, M.; HOSHINO, T.; GURCAY, O.; WILSON, C. B.; NIELSEN, S. L.; DOWNIE, R., and ELIASON, J.: Development of an animal brain tumor model and its response to therapy with 1,3-bis(2-chloroethyl)-1-nitrosourea. Cancer Res. *33:* 976–986 (1973).
6 BARLEY, F. W.; SATO, G. H., and ABELES, R. H.: An effect of vitamin B_{12} deficiency in tissue culture. J. biol. Chem. *247:* 1270–1276 (1972).
7 BENDA, P.; SOMEDA, K.; MESSER, J., and SWEET, W. H.: Morphological and immunochemical studies of rat glial tumors and clonal strains propagated in culture. J. Neurosurg. *34:* 310–323 (1971).
8 BERNHEIMER, H.: Zur Kenntnis der Ganglioside im Liquor cerebrospinalis des Menschen. Klin. Wschr. *47:* 227–228 (1969).
9 BRANTE, G.: Studies on lipids in the nervous system with special reference to quantitative chemical determination and topical distribution. Acta physiol. scand. *18:* suppl. 63 (1949).
10 CAIN, C. E.; BELL, O. E., jr.; WHITE, H. B., jr.; SULYA, L. L., and SMITH, R. R.: Hydrocarbons from human meninges and meningiomas. Biochim. biophys. Acta *144:* 493–500 (1967).
11 CHRISTENSEN LOU, H. O. and CLAUSEN, J.: Polar lipids of oligodendrogliomas. J. Neurochem. *15:* 263–264 (1968).
12 CHRISTENSEN LOU, H. O.; CLAUSEN, J., and BIERRING, F.: Phospholipids and glycolipids of tumors in the central nervous system. J. Neurochem. *12:* 619–627 (1965).
13 CHRISTENSEN LOU, H. O. and MATZKE, J.: Cerebroside and other polar lipids of the cerebrospinal fluid in neurological diseases. Acta neurol. scand. *41:* 445–447 (1965).
14 CLAUSEN, J.; CHRISTENSEN LOU, H. O., and ANDERSEN, H.: Phospholipid and glycolipid patterns of infant and foetal brain. Thin-layer chromatographic studies. J. Neurochem. *12:* 599–606 (1965).
15 CONNOR, W. E.; JOHNSTON, R., and LIN, D. S.: Metabolism of cholesterol in the tissues and blood of the chick embryo. J. Lipid Res. *10:* 388–394 (1969).
16 CRAVIOTO, H.; PALEKAR, L.; WEISS, E. DE C., and BENNETT, K.: Experimental neurinoma in tissue culture. Acta neuropath. *21:* 154–164 (1972).
17 CRAVIOTO, H.; WEISS, J. F.; WEISS, E. DE C.; GOEBEL, H. H., and RANSOHOFF, J.: Biological characteristics of peripheral nerve tumors induced with ethylnitrosourea. Acta neuropath. *23:* 265–280 (1973).

18 DENCKER, S. J. and SWAHN, B.: The diagnostic value of lipoprotein determinations in cerebrospinal fluid. Acta psychiat. neurol. scand. *36:* 325–335 (1961).
19 DENNICK, R. G.; DEAN, P. D. G., and ABRAMOVICH, D. A.: Desmosterol levels in human foetal brain- a reassessment. J. Neurochem. *20:* 1293–1294 (1973).
20 DRUCKREY, H.; IVANKOVIC, S.; PREUSSMANN, R.; ZÜLCH, K. J., and MENNEL, H. D.: Selective induction of malignant tumors of the nervous system by resorptive carcinogens; in KIRSCH, GROSSI PAOLETTI and PAOLETTI The experimental biology of brain tumors, pp. 85–147 (Thomas, Springfield 1972).
21 EICHBERG, J.; SHEIN, H. M., and HAUSER, G.: Effect of neurotransmitters and other pharmacological agents on the metabolism of phospholipids in pineal-gland cultures and cloned neuronal and glial cells. Biochem. Soc. Trans. *1:* 352–359 (1973).
22 EMBREE, L. J.; HESS, H. H., and SHEIN, H. M.: Biochemical structural components of cloned N-nitrosomethylurea- induced astrocytoma cells grown subcutaneously. Neurology, Minneap. *22:* 194–201 (1972).
23 ETO, Y. and SUZUKI, K.: Cholesterol esters in developing rat brain. Concentration and fatty acid composition. J. Neurochem. *19:* 109–115 (1972).
24 ETO, Y. and SUZUKI, K.: Developmental changes of cholesterol ester hydrolases localized in myelin and microsomes of rat brain. J. Neurochem. *20:* 1475–1477 (1973).
25 FIESER, L. F.: Cholesterol and companions. III. Cholestanol, lathosterol and ketone 104. J. amer. chem. Soc. *75:* 4395–4403 (1953).
26 FISH, W. A.; BOYD, J. E., and STOKES, W. M.: Metabolism of cholesterol in the chick embryo. III. Localization and turnover of desmosterol (24-dehydrocholesterol). J. biol. Chem. *237:* 334–337 (1962).
27 FUMAGALLI, R.; GRAFNETTER, D.; GROSSI, E. e MORGANTI, P.: Steroli liberi ed esterificati nel tessuto nervoso dell'embrione di pollo durante lo sviluppo, con particolare riguardo al desmosterolo. Atti Accad. Med. Lomb. *18:* 535–540 (1963).
28 FUMAGALLI, R.; NIEMIRO, R., and PAOLETTI, R.: Investigation of the biogenetic reaction sequence of cholesterol in rat tissues, through inhibition with AY-9944. J. amer. Oil Chem. Soc. *42:* 1018–1023 (1965).
29 FUMAGALLI, R. and PAOLETTI, P.: Sterol test for human brain tumors. Relationship with different oncotypes. Neurology, Minneap. *21:* 1149–1156 (1971).
30 FUMAGALLI, R.; SMITH, M. E.; URNA, G., and PAOLETTI, R.: The effect of hypocholesteremic agents on myelinogenesis. J. Neurochem. *16:* 1329–1339 (1969).
31 GALLI, C. and RE CECCONI, D.: Lipid changes in rat brain during maturation. Lipids *2:* 76–82 (1967).
32 GALLI, C.; WHITE, H. B., jr., and PAOLETTI, R.: Brain lipid modifications induced by essential fatty acid deficiency in growing male and female rats. J. Neurochem. *17:* 347–355 (1970).
33 GALLI, G.; GALLI-KIENLE, M.; CATTABENI, F.; FIECCHI, A.; GROSSI PAOLETTI, E., and PAOLETTI, R.: The sterol precursors of cholesterol in normal and tumor tissues. Adv. Enzyme Reg. *8:* 311–321 (1970).
34 GALLI, G.; GROSSI PAOLETTI, E., and WEISS, J. F.: Sterol precursors of cholesterol in adult human brain. Science *162:* 1495–1496 (1968).

35 GALLI, G.; GROSSI PAOLETTI, E., and WEISS, J. F.: Studies on the pathways of brain cholesterol biosynthesis after squalene cyclization. Abh. Dtsch. Akad. Wiss. Berl., kl. Med., No. 2, pp. 31–36 (1968).
36 GOPAL, K.; GROSSI, E.; PAOLETTI, P., and USARDI, M.: Lipid composition of human intracranial tumors; a biochemical study. Acta neurochir. *11:* 333–347 (1963).
37 GRAFNETTER, D.; GROSSI, E., and MORGANTI, P.: Occurrence of sterol esters in the chicken brain during prenatal and postnatal development. J. Neurochem. *12:* 145–149 (1965).
38 GROSSI PAOLETTI, E. and PAOLETTI, P.: Lipids of brain tumors; in KIRSCH, GROSSI PAOLETTI and PAOLETTI The experimental biology of brain tumors, pp. 299–329 (Thomas, Springfield 1972).
39 GROSSI PAOLETTI, E.; PAOLETTI, P.; PEZZOTTA, S.; SCHIFFER, D., and FABIANI, A.: Tumors of the nervous system induced by ethylnitrosourea administered either intracerebrally or subcutaneously to newborn rats. Morphological and biochemical characteristics. J. Neurosurg. *37:* 580–590 (1972).
40 GROSSI PAOLETTI, E.; SIRTORI, C. R.; WEISS, J. F., and PAOLETTI, R.: Effect of hypocholesteremic agents on an experimental brain tumor in mice. Adv. exp. Med. Biol., vol. 4, pp. 457–471 (Plenum Publishing, New York 1969).
41 HAKOMORI, S. and MURAKAMI, W. T.: Glycolipids of hamster fibroblasts and derived malignant-transformed cell lines. Proc. nat. Acad. Sci., Wash. *59:* 254–261 (1968).
42 HAMBERGER, A. and SVENNERHOLM, L.: Composition of gangliosides and phospholipids of neuronal and glial cell enriched fractions. J. Neurochem. *18:* 1821–1829 (1971).
43 HAUSER, G.; EICHBERG, J., and SHEIN, H. M.: Lipid biosynthesis in normal and neoplastic neuroglial cultures. Trans. amer. Soc. Neurochem. *1:* 48 (1970).
44 HEMMINKI, K. and SUOVANIEMI, O.: Preparation of plasma membranes from isolated cells of newborn rat brain. Biochim. biophys. Acta *298:* 75–83 (1973).
45 HINSE, C. H. and SHAH, S. N.: The desmosterol reductase activity of rat brain during development. J. Neurochem. *18:* 1989–1998 (1971).
46 HOGAN, E. L.; NEWELL, L. R., and WOOTEN, W. B.: Ganglioside composition of human brain tumors. Neurology, Minneap. *23:* 437 (1973).
47 HOLMES, R.; HELMS, J., and MERCER, G.: Cholesterol requirement of primary diploid human fibroblasts. J. Cell Biol. *42:* 262–271 (1969).
48 HORLICK, L. and AVIGAN, J.: Sterols of skin in the normal and triparanol-treated rat. J. Lipid Res. *4:* 160–165 (1963).
49 HORTON, B. J.; MOTT, G. E.; PITOT, H. C., and GOLDFARB, S.: Rapid uptake of dietary cholesterol by hyperplastic liver nodules and primary hepatomas. Cancer Res. *33:* 460–464 (1973).
50 IKUTA, F. and KUMANISHI, T.: Experimental virus-induced brain tumors; in ZIMMERMAN Progress in neuropathology, vol. 2, pp. 253–334 (Grune & Stratton, New York 1973).
51 ILLINGWORTH, D. R. and GLOVER, J.: The composition of lipids in cerebrospinal fluid of children and adults. J. Neurochem. *18:* 769–776 (1971).
52 JAYARAMAN, J.; WEISS, J. F., and GROSSI PAOLETTI, E.: Metabolism of sterols

and related compounds in human brain tumors. Proc. Fed. Europ. Biochem. Soc., Prague 1968, p. 221.
53 KABARA, J. J.: Brain cholesterol. The effect of chemical and physical agents. Adv. Lipid Res., vol. 5, pp. 279–327 (Academic Press, New York 1967).
54 KANDUTSCH, A. A. and SAUCIER, S. E.: Sterol and fatty acid synthesis in developing brains of three myelin-deficient mouse mutants. Biochim. biophys. Acta *260:* 26–34 (1972).
55 KOSTIĆ, D. and BUCHHEIT, F.: Gangliosides in human brain tumors. Life Sci. *9:* 589–596 (1970).
56 KRITCHEVSKY, D.; TEPPER, S. A.; DITULLIO, N. W., and HOLMES, W. L.: Desmosterol in developing rat brain. J. amer. Oil Chem. Soc. *42:* 1024–1028 (1965).
57 LINDLAR, F. und BINGAS, B.: Die Lipide von Hirntumoren. Dtsch. Z. Nervenheilk. *187:* 737–748 (1965).
58 LOWENTHAL, A.: Chemical physiopathology of the cerebrospinal fluid; in LAJTHA Handbook of neurochemistry, vol. 7, pp. 429–464 (Plenum Publishing, New York 1972).
59 MAKER, H. S.; LEHRER, G. M.; SILIDES, D. J., and WEISS, C.: Circulatory factors in the carbohydrate metabolism of an experimental glial neoplasm. Ann. N. Y. Acad. Sci. *159:* 461–471 (1969).
60 MARTON, L. J.; GORDON, G. S.; BARKER, M.; WILSON, C. B., and LUBICH, W.: Failure to demonstrate desmosterol in spinal fluid of brain tumor patients. Arch. Neurol., Chicago *28:* 137–138 (1973).
61 MASPES, P. E. and PAOLETTI, P.: Recent advances in chemical composition and metabolism of brain tumors. Progr. neurol. Surg., vol. 2, pp. 203–266 (Karger, Basel 1968).
62 MEAD, J. F. and HAGGERTY, D. F., jr.: Recent advances in the metabolism and function of the essential fatty acids. Proc. 8th Int. Congr. Nutrition, Prague 1969, pp. 76–80 (Excerpta Medica, Amsterdam 1970).
63 NAGAI, Y. and KANFER, J. N.: Composition of human cerebrospinal fluid cerebroside. J. Lipid Res. *12:* 143–148 (1971).
64 NEUMAN, R. E. and TYTELL, A. E.: Serumless medium for cultivation of cells of normal and malignant origin. Proc. Soc. exp. Biol. Med. *104:* 252–256 (1960).
65 NICHOLAS, H. J.: Cholesterol turnover in the central nervous system. J. amer. Oil Chem. Soc. *42:* 1008–1012 (1965).
66 NOBLE, R. C. and MOORE, J. H.: Studies on the lipid metabolism of the chick embryo. Canad. J. Biochem. *42:* 1729–1741 (1964).
67 NOBLE, R. C. and MOORE, J. H.: The partition of lipids between the yolk and yolk-sac membrane during the development of the chick embryo. Canad. J. Biochem. *45:* 949–958 (1967).
68 NORTON, W. T. and PODUSLO, S. E.: Neuronal perikarya and astroglia of rat brain. Chemical composition during myelination. J. Lipid Res. *12:* 84–90 (1971).
69 NYDEGGER, U. E. and BUTLER, R. E.: Serum lipoprotein levels in patients with cancer. Cancer Res. *32:* 1756–1760 (1972).
70 OLSON, J. A.: The biosynthesis of cholesterol. Ergebn. Physiol. *56:* 173–215 (1965).
71 ONODERA, Y.; MASAMICHI, H.; YUKIO, Y., and TETSURO, M.: Lipid and fatty

acid compositions of brain tumor. Brain Nerve, Tokyo 23: 785–792 (1971); cited. in Biol. Abstr. 53: 2083 (1972).

72 PAOLETTI, P.; FUMAGALLI, R., and GROSSI PAOLETTI, E.: Drugs acting on brain tumor sterols; in KIRSCH, GROSSI PAOLETTI and PAOLETTI The experimental biology of brain tumors, pp. 457–479 (Thomas, Springfield 1972).

73 PAOLETTI, P.; VANDENHEUVEL, F. A.; FUMAGALLI, R., and PAOLETTI, R.: The sterol test for the diagnosis of human brain tumors. Neurology, Minneap. 19: 190–197 (1969).

74 PAOLETTI, R.; GROSSI PAOLETTI, E., and FUMAGALLI, R.: Sterols; in LAJTHA Handbook of neurochemistry, vol. 1, pp. 195–222 (Plenum Publishing, New York 1969).

75 PAOLETTI, R.; FUMAGALLI, R.; GROSSI, E., and PAOLETTI, P.: Studies on brain sterols in normal and pathological conditions. J. amer Oil Chem. Soc. 42: 400–404 (1965).

76 PAOLETTI, R.; GALLI, G.; GROSSI PAOLETTI, E.; FIECCHI, A., and SCALA, A.: Some pathways and mechanisms in lanosterol-cholesterol conversion in mammalian tissues. Lipids 6: 1134–1138 (1971).

77 PODUSLO, S. E. and NORTON, W. T.: Isolation and some chemical properties of oligodendroglia from calf brain. J. Neurochem. 19: 727–736 (1972).

78 POPOVA, G. M. and PROMYSLOV, M. S.: A study of the cerebrosides in some neuroectodermal tumors of the human brain (Rus.). Bull. exp. Biol. Med. 53: 673–675 (1962).

79 POPOVA, G. M. and PROMYSLOV, M. S.: The cerebrosidase activity of the brain and cerebral tumors in man (Rus.). Vop. Neirokhir. 35: 28–31 (1971); cited in Excerpta med. 25: 101 (1972).

80 RAMSEY, R. B.: New concepts in brain cholesterol metabolism. Biochem. Soc. Trans. 1: 341–348 (1973).

81 RAMSEY, R. B.; AEXEL, R. T., and NICHOLAS, H. J.: Formation of methyl sterols in brain cholesterol biosynthesis. Sterol formation *in vitro* and *in vivo* in adult rat brain. J. biol. Chem. 246: 6393–6400 (1971).

82 RAMSEY, R. B.; AEXEL, R. T.; JONES, J. P., and NICHOLAS, H. J.: Formation of methyl sterols in brain cholesterol biosynthesis. Sterol formation *in vitro* in actively myelinating rat brain. J. biol. Chem. 247: 3471–3475 (1972).

83 RAMSEY, R. B. and NICHOLAS, H. J.: Brain lipids. Adv. Lipid Res., vol. 10, pp. 143–232 (Academic Press, New York 1972).

84 RAPPORT, M. M.: Immunological properties of lipids and their relation to the tumor cell. Ann. N. Y. Acad. Sci. 159: 446–450 (1969).

85 RAPPORT, M. M.; GRAF, L.; SKIPSKI, V. P., and ALONZO, N. F.: Cytolipin H, a pure lipid hapten isolated from human carcinoma. Nature, Lond. 181: 1803–1804 (1958).

86 RAWLINS, F. A. and SMITH, M. E.: Myelin synthesis *in vitro*. A comparative study of central and peripheral nervous tissue. J. Neurochem. 18: 1861–1870 (1971).

87 RAWLINS, F. A. and UZMAN, B. G.: Effect of AY-9944, a cholesterol biosynthesis inhibitor on peripheral nerve myelination. Lab. Invest. 23: 184–189 (1970).

88 ROSENBERG, R. N.: Neuronal and glial enzyme studies in cell culture. In Vitro *8:* 194–204 (1972).
89 ROTHBLAT, G. H.: Cellular sterol metabolism; in ROTHBLAT and CRISTOFALO Growth, nutrition, and metabolism of cells in culture, vol. 1, pp. 297–325 (Academic Press, New York 1972).
90 ROTHBLAT, G. H.; BURNS, C. H.; CONNER, R. L., and LANDREY, J. R.: Desmosterol as the major sterol in L-cell mouse fibroblasts grown in sterol-free culture medium. Science *169:* 880–882 (1970).
91 ROUSER, G.; KRITCHEVSKY, G.; YAMAMOTO, A., and BAXTER, C. F.: Lipids in the nervous system of different species as a function of age: brain, spinal cord, peripheral nerve, purified whole cell preparations, and subcellular particulates: regulatory mechanisms and membrane structure. Adv. Lipid Res., vol. 10, pp. 261–360 (Academic Press, New York 1972).
92 SAMUELSSON, K.: Separation and identification of cerebrosides in cerebrospinal fluid by gas chromatography-mass spectrometry. Scan. J. clin. Lab. Invest. *27:* 381–391 (1971).
93 SCALLEN, T. J.; CONDIE, R. M., and SCHROEPFER, G. J., jr.: Inhibition by triparanol of cholesterol formation in the brain of newborn mouse. J. Neurochem. *9:* 99–103 (1962).
94 SCHRAPPE, O.: Ergebnisse intraindividueller Vergleichsuntersuchungen der Lipide im Liquor cerebrospinalis mit den Serum-Lipiden. Klin. Wschr. *50:* 158–164 (1972).
95 SCHUTTA, H. S. and NEVILLE, H. E.: Effects of cholesterol synthesis inhibitors on the nervous system. A light and electron microscopic study. Lab. Invest. *19:* 487–493 (1968).
96 SEIFERT, H.: Über ein weiteres Hirntumorcharakteristisches Gangliosid. Klin. Wschr. *44:* 469–470 (1966).
97 SEIFERT, H. und UHLENBRUCK, G.: Über Ganglioside in Hirntumoren. Naturwissenschaften *52:* 190–191 (1965).
98 SELVERSTONE, B. and MOULTON, M. J.: The phosphorus metabolism of gliomas: a study with radioactive isotopes. Brain *80:* 362–375 (1957).
99 SHAH, S. N.: Effect of neonatal food restriction and hyperphenylalaninemia on desmosterol to cholesterol (d/c) ratio in developing rat brain. Lipids *7:* 628–630 (1972).
100 SIDDIQUI, B. and MCCLUER, R. H.: Lipid components of sialosyl galactosyl ceramide of human brain. J. Lipid Res. *9:* 366–370 (1968).
101 SIPERSTEIN, M. D.: Regulation of cholesterol biosynthesis in normal and malignant tissues. Curr. Top. cell. Reg., vol. 2, pp. 65–100 (Academic Press, New York 1970).
102 SIRTORI, C. R.; RODRIGUEZ, G. A.; GUARINO, M. J.; AZARNOFF, R. S., and BOULOS, B. M.: Effect of trans-1,4-bis-2-dichlorbenzyl-aminoethyl-cyclohexane (AY-9944) on the experimental brain tumor G26A. Arzneimittelforsch. *22:* 914–916 (1972).
103 SLAGEL, D. E.; DITTMER, J. C., and WILSON, C. B.: Lipid composition of human glial tumour and adjacent brain. J. Neurochem. *14:* 789–798 (1967).

104 SMITH, R. R. and WHITE, H. B., jr.: Neutral lipid patterns of normal and pathologic nervous tissue. Studies by thin-layer chromatography. Arch. Neurol., Chicago 19: 54–59 (1968).
105 SNYDER, F. and WOOD, R.: Alkyl and alk-l-enyl ethers of glycerol in lipids from normal and neoplastic human tissues. Cancer Res. 29: 251–257 (1969).
106 SPECTOR, A. A.: Fatty acid, glyceride and phospholipid metabolism; in ROTHBLAT and CRISTOFALO Growth, nutrition, and metabolism of cells in culture, vol. 1, pp. 257–296 (Academic Press, New York 1972).
107 SPOHN, M. and DAVISON, A. N.: Cholesterol metabolism in myelin and other subcellular fractions of rat brain. J. Lipid Res. 13: 563–570 (1972).
108 STEIN, A. A.; OPALKA, E., and PECK, F.: Fatty acid analysis of brain tumors by gas phase chromatography. Arch. Neurol., Chicago 8: 50–55 (1963).
109 STEIN, A. A.; OPALKA, E., and ROSENBLUM, I.: Fatty acid analysis of two experimental transmissible glial tumors by gas-liquid chromatography. Cancer Res. 25: 201–205 (1965).
110 STEIN, A. A.; OPALKA, E., and ROSENBLUM, I.: Hepatic lipids in tumor-bearing (glioma) mice. Cancer Res. 25: 957–961 (1965).
111 STEIN, A. A.; OPALKA, E., and SCHILP, A. O.: Fatty acid analysis of meningiomas by gas-phase chromatography. J. Neurosurg. 20: 435–438 (1963).
112 SUZUKI, K. and DE PAUL, L.: Cellular degeneration in developing central nervous system of rats produced by hypocholesteremic drug AY-9944. Lab. Invest. 25: 546–555 (1971).
113 SVENNERHOLM, L.: Gangliosides; in LAJTHA Handbook of neurochemistry, vol. 3, pp. 425–452 (Plenum Publishing, New York 1970).
114 SWAHN, B.; BRONNESTAM, R., and DENCKER, S. J.: On the origin of the lipoproteins in the cerebrospinal fluid. Neurology, Minneap. 11: 437–440 (1961).
115 SZEPSENWOL, J.: Brain nerve cell tumors in mice on diets supplemented with various lipids. Path. Microbiol. 34: 1–9 (1969).
116 SZEPSENWOL, J.: Intracranial tumors in mice of two different strains maintained on fat-enriched diets. Europ. J. Cancer 7: 529–532 (1971).
117 TICHY, J.: Cholesterol in the cerebrospinal fluid. An analysis of 447 neurological patients. Rev. czech. Med. 12: 265–271 (1966).
118 TROUILLAS, P.: Antigène d'un virus oncogène à A. R. N. dans les gliomes humains. Sa relation avec les antigènes carcinoembryonnaires. Nouv. Presse méd. 1: 1979–1982 (1972).
119 UHLENBRUCK, G. und GIELEN, W.: Hirntumorcharakteristische Glykolipoide. Med. Welt, Stg. 7: 332–335 (1969).
120 WALLACH, D. F. H.: Cellular membranes and tumor behavior. A new hypothesis. Proc. nat. Acad. Sci., Wash. 61: 868–874 (1968).
121 WECHSLER, W.: Old and new concepts of oncogenesis in the nervous system of man and animals. Progr. exp. Tum. Res., vol. 17, pp. 219–278 (Karger, Basel 1972).
122 WEISS, J. F.; CRAVIOTO, H.; BENNETT, K.; WEISS, E. DE C., and RANSOHOFF, J.: Cholesterol precursors in tumors of the nervous system grown in tissue culture with and without hypocholesteremic agents. J. Neuropath. exp. Neurol. 33: 179 (1974).

123 WEISS, J. F.; CRAVIOTO, H., and RANSOHOFF, J.: Desmosterol accumulation during normal and malignant growth of rat central and peripheral nervous system and in tumor cells in culture. 164th Nat. Meet. amer. Chem. Soc., New York 1972.
124 WEISS, J. F.; CRAVIOTO, H.; WEISS, E. DE C; PALEKAR, L., and RANSOHOFF, J.: Sterol metabolism in tumors of the trigeminal nerve induced by nitrosourea derivatives. J. Neuropath. exp. Neurol. *31:* 203–204 (1972).
125 WEISS, J. F.; GALLI, G., and GROSSI PAOLETTI, E.: Sterol metabolism in the brain of the developing chick embryo; in RICHTER Biochemical factors concerned in the functional activity of the nervous system, Addendum (Pergamon Press, Oxford 1969).
126 WEISS, J. F.; GROSSI PAOLETTI, E., and GALLI, G.: Sterols with 29, 28 and 27 carbon atoms metabolically related to cholesterol, occurring in developing and mature brain. J. Neurochem. *15:* 563–575 (1968).
127 WEISS, J. F.; GROSSI PAOLETTI, E.; PAOLETTI, P.; SCHIFFER, D., and FABIANI, A.: Occurrence of desmosterol in tumors of the nervous system induced in the rat by nitrosourea derivatives. Cancer Res. *30:* 2107–2109 (1970).
128 WEISS, J. F.; KAYDEN, H. J., and RANSOHOFF, J.: Evaluation of patients undergoing therapy for gliomas by examination of CSF sterols after triparanol treatment. Trans. amer. neurol. Ass. *96:* 59–61 (1971).
129 WEISS, J. F.; RANSOHOFF, J., and KAYDEN, H. J.: Cerebrospinal fluid sterols in patients undergoing treatment for gliomas. Neurology, Minneap. *22:* 187-193 (1972).
130 WHITE, H. B., jr.: Normal and neoplastic human brain tissues: phospholipid, fatty acid and unsaturation number modifications; in WOOD Tumor lipids: biochemistry and metabolism, pp. 75–88 (Amer. Oil Chem. Soc., Champaign 1973).
131 WHITE, H. B., jr. and SMITH, R. R.: Cholesteryl esters of the glioblastoma. J. Neurochem. *15:* 293–299 (1968).
132 WIEGANDT, H.: Ganglioside. Ergebn. Physiol. *57:* 190–222 (1966).
133 WOLLEMANN, M.: Biochemistry of brain tumors; in LAJTHA Handbook of neurochemistry, vol. 7, pp. 503–542 (Plenum Publishing, New York 1972).
134 WOOD, R.: Relationship between embryonic and tumor lipids. I. Changes in the neutral lipids of the developing chick brain, heart and liver. Lipids *7:* 596–603 (1972).
135 YANAGIHARA, T. and CUMINGS, J. N.: Lipid metabolism in cerebral edema associated with human brain tumor. Arch. Neurol., Chicago *19:* 241–247 (1968).
136 YOGEESWARAN, G.; MURRAY, R. K.; PEARSON, M. L.; SANWAL, B. D.; MCMORRIS, F. A., and RUDDLE, F. H.: Glycosphingolipids of clonal lines of mouse neuroblastoma and neuroblastoma XL cell hybrids. J. biol. Chem. *248:* 1231–1239 (1973).
137 ZIMMERMAN, H. M.: Brain tumors: their incidence and classification in man and their experimental production. Ann. N. Y. Acad. Sci. *159:* 337–359 (1969).

Author's address: JOSEPH F. WEISS, Department of Neurosurgery, New York University Medical Center, 550 First Ave., *New York, NY 10016* (USA)

Defective Control of Lipid Biosynthesis in Cancerous and Precancerous Liver

JOHN R. SABINE

Department of Animal Physiology, Waite Agricultural Research Institute, University of Adelaide, Adelaide, S. A.

Contents

I. Introduction and Significance	269
II. Defective Controls	270
A. Hepatoma and Cholesterol Biosynthesis	270
B. Hepatoma and Fatty Acid Biosynthesis	272
C. Lipid Biosynthesis in Other Tumors	275
D. Lipid Biosynthesis in Pretumorous Tissue	275
III. Mechanisms of Control	277
A. Hepatic Cholesterol Synthesis	278
B. Hepatic Fatty Acid Synthesis	283
C. Defective Controls in Hepatomas	286
IV. Defective Control and Carcinogenesis	290
A. Possible Correlative (Causative?) Mechanisms	290
B. Correlation with Other Precancerous Changes	293
C. Some Doubts	294
V. Summary	295
VI. References	296

I. Introduction and Significance

Attempts so far to define *'CANCER'* in biochemical terms, while lacking a great deal in precision, all contain one common element... 'lack of control' [113]. In this sense cancer can be regarded as the ultimate result of the loss of extracellular control over intracellular processes – one or more cells of a tissue have escaped from the intricate set of controls normally exercised by the organism as a whole over the metabolic activity of that particular tissue. Thus, if one's aim is to provide some of that biochemical precision which is so obviously lacking from current descriptions of the cancerous state, I believe it is pertinent to begin by asking three questions: (1) What are the extracellular controls over the intracellular activity of any tissue? (2) Do tumors derived from this tissue retain these particular controls? (3) If not, why not? This paper is a review of the efforts that have been made to answer these questions with regard to the regulation of lipid synthesis in the liver and the lack of this regulation in the liver tumor.

As will be seen, there is now very strong evidence in favor of ascribing an essential relationship, even a causative one, to defective control of lipid metabolism in the initiation of cancer in the liver. This area of research has received considerable attention over the past ten years and different aspects, particularly those relating to cholesterol metabolism, have been subjects for several recent reviews [124, 144, 147]. My aim in this present exercise is to summarize what is now known about the defective physiological control of both fatty acid and cholesterol metabolism in cancerous and precancerous liver, to suggest likely functional reasons for a correlation between faulty control mechanisms and hepatic carcinogenesis, and to indicate those lines of future research that seem likely to be the most productive.

II. Defective Controls

A. Hepatoma and Cholesterol Biosynthesis

To date, workers in at least 10 laboratories have studied the physiological regulation of cholesterol synthesis in a total of 24 hepatomas, both primary and transplanted. The results of their published efforts are summarized in table I.

Clearly, in every hepatoma yet examined (with one possible exception) in four different species, the rate of cholesterol synthesis is quite insensitive both to the amount of cholesterol absorbed from the digestive tract and to the effects of fasting. As far as I am aware this, and the comparable situation for fatty acid synthesis to be described below, are the only examples of specific physiological control mechanisms present in a particular normal tissue but missing from every one of a large number of tumors derived from that tissue. From this one can reasonably argue

Table I. Physiological control of cholesterol synthesis in hepatomas

Host	Hepatoma	Response to control[1]						References
		f'back	fast	re-fed	c'mine	rhythm	adrenx	
Normal liver[2]		−	−	+	+	±	+	
Mouse	BW7756	0	0	0	0			41, 85, 123 125, 126, 150
	H4	0	0					85
	primary	0						85
	primary	?						85
Rat	3683	0						152
	3924A	0						25, 152
	5121 t.c.	0						145
	5123	0						149
	5123C	0			0	±	?	57, 123, 129
	5123 t.c.	0						150
	7288C	0						152
	7316A	0						152
	7787	0						25, 68, 152
	7793	0						68, 152
	7794A	0				0	0	68, 129, 152
	7795	0						152
	7800	0			0	±		57, 152
	9121	0		0				25, 50, 145
	9618A	0			?	0		57, 129, 147
	9633	0						147
	H-35	0						152
Trout	primary	0						145
Man	primary	0						150
	primary	0						152

1 F'back: feedback, i.e. high-cholesterol diet; re-fed: re-fed after fasting; c'mine: cholestyramine in the diet; rhythm: diurnal rhythm; adrenx: bilateral adrenalectomy.
2 − = Decreased synthesis; + = increased synthesis; 0 = no change; no symbol = the appropriate experiment has not been reported.

[124, 147, 148, 152] that defective dietary control of cholesterol synthesis might be important or even essential in the carcinogenic process – at least in the liver. The one possible exception so far reported, spontaneous hepatomas in C3H-A^{vy} mice [85], demands further investigation, since even one definitive exception would destroy this hypothesis.

Table I also shows that not all aspects of the physiological control of cholesterol synthesis in the hepatoma necessarily differ significantly from those in normal liver. Depending upon the particular tumor examined the diurnal rhythm of synthesis and the effect of adrenalectomy may or may not be abolished. It thus seems most unlikely that a study of these parameters in tumor-bearing animals would contribute significantly to our understanding of the carcinogenic process. As we [71, 74] and others [57, 118, 121] have pointed out previously, the cellular mechanisms giving rise to diurnal rhythm and to the effects of adrenalectomy are apparently independent of dietary factors, although of course the ultimate expression of these mechanisms is modified by such factors.

Despite this large volume of work relating to the defective *physiological* control of cholesterol synthesis in the hepatoma and its apparent importance in carcinogenesis, little attention has been given to the *pharmacological* control of sterol synthesis in tumors. From studies *in vivo*, Triparanol (MER 29) inhibits cholesterol synthesis in the hepatoma just as it does in the liver, namely by blocking the enzymatic pathway immediately beyond desmosterol [25], but conversely Triton WR 1339 does not stimulate synthesis, although it does so in the liver [85]. Various steroids besides cholesterol can either stimulate or inhibit hepatic cholesterol synthesis when fed in the diet, but these fail to influence hepatomal synthesis [85]. In studies *in vitro* we have shown that both bile salts [123] and a number of plant-growth retardants [112] can influence cholesterol synthesis in the hepatoma in a manner analogous to their effect in the liver, particularly when added to tissue homogenates. Cyclic 3'5'-adenosine monophosphate (cAMP), however, inhibits cholesterogenesis by liver slices but not by tumor slices [23].

B. Hepatoma and Fatty Acid Biosynthesis

The physiological control of fatty acid synthesis has been examined in 10 hepatomas, all of them transplanted and only in rats and mice (table II). Just as was the case with cholesterol synthesis, the rate of formation of fatty acids in every one of these tumors is insensitive to dietary factors known to alter grossly synthesis in the normal liver – high fat, fasting and refeeding after fasting. By analogy with the argument used above for the defective control of cholesterol synthesis, one would have to postulate that a defective control of fatty acid synthesis might also be of primary impor-

Table II. Physiological control of fatty acid synthesis in hepatomas

Host	Hepatoma	Response to Control[1]			References
		f'back	fast	re-fed	
Normal liver[2]		−	−	+	
Mouse	BW7756	0	0	0	41, 126
Rat	5123C	0		0	127
	7777		0	0	101
	7793		0		127
	7794A	0			185
	7795	0	0	0	127
	7800	0		0	127
	9098	0			185
	9121			0	50
	9618A	0		0	101, 185

1 F'back: feedback, i.e. high-fat diet; re-fed: re-fed after fasting.
2 Symbols as in table I.

tance in carcinogenesis, at least in the liver. This may be true, but our results with precancerous liver [80], as discussed below, would seem to render this unlikely. Nevertheless, defective control over fatty acid synthesis may be highly significant in the subsequent development and maintenance of the tumor once the neoplastic process has started.

There appears to have been little work done on the effect of drugs and other chemicals on fatty acid synthesis in the tumor. BRICKER and LEVEY [23] have reported briefly that fatty acid synthesis *in vitro* by two hepatomas is not depressed by exogenous cAMP nor by its dibutyryl derivative, contrary to the case with synthesis by both normal and host liver. Some plant-growth retardants [112] and long-chain acyl coenzyme A (CoA) [101, 128] can lower synthesis in both liver and hepatoma, again *in vitro,* but the effects of other compounds have not been reported.

Some examination has been made of the particular fatty acids contained in and synthesized by the hepatoma. Work from ELWOOD's laboratory [185] has shown that the fatty acid composition of three hepatomas, in animals on high-fat and low-fat diets, is not grossly different from that of their respective host livers. Indeed, for the least 'deviated' hepatoma yet available, 9618A [115], the compositions are almost identical. On the other hand, an earlier paper from this same group [50] had reported that for another hepatoma the changes in fatty acid composition in total lipids in response to fasting and refeeding were markedly different between liver and tumor. Our results [127] indicate that during incubation *in vitro* slices of hepatoma incorporate acetate-1-^{14}C into fatty acids of chain length comparable to those formed by the liver, although in general the contribution of synthesis *de novo,* relative to chain *elongation* (as measured by total-^{14}C:carboxyl-^{14}C ratios), is less for the tumor than for the liver. This ratio is also less affected by diet in the tumor than in the liver, but it is doubtful whether these differences are physiologically significant. As ZUCKERMAN *et al.* [185] have pointed out, an examination of the lipid composition of individual cell organelles would be more fruitful than an analysis of total cellular lipid. This point has particular relevance to my later discussion about possible mechanisms of defective control.

As with the work on fatty acid composition, so also studies dealing with fatty acid oxidation and its dietary manipulation have not shown any pattern consistent enough to shed light on the basic defects of control. It appears that, depending on the tumor chosen, the oxidation of palmitate [127] and acetate [50, 185] by the tumor may or may not alter with diet.

The effects of various diets on the enzymes involved, directly or indirectly, in fatty acid synthesis do, however, reveal some interesting control features. In their elegant study of acetyl CoA carboxylase and fatty acid synthetase, MAJERUS *et al.* [101] have shown that the activity of neither of these enzymes in the hepatoma was altered by diet, i.e. just as would be expected from the known defects in control of fatty acid synthesis. On the other hand, we have examined the effects of various diets on a variety of glycolytic and Krebs cycle enzymes in several hepatomas [90, 127, 132] and could find no consistent differences in response, with perhaps two interesting exceptions, between liver and hepatoma, although of course the actual levels of enzyme activity may have been quite different in the two tissues. The possible exceptions were the glucose-ATP phosphotransferases. The predominant enzyme in the liver, and the adaptable one, is the 'high K_m' glucokinase (EC 2.7.1.2.) whereas in the hepatoma the 'low K_m' hexokinase (EC 2.7.1.1.) predominates, and is the adaptable one [90, 127]. This may be an important feature and is worth further investigation [127], particularly as EMMELOT and BOS [51] have reported that in the hepatoma but not the liver, considerable hexokinase activity may be bound to the plasma membrane.

It is important to note that the actual rate of lipid synthesis by hepatomas, at least *in vitro,* may vary considerably between tumors and may be more or less than or the same as that by normal or host liver. There seems little reason to associate this with the cancer process. With four rat hepatomas we could observe no relationship between rate of tumor growth and rate of fatty acid synthesis *in vitro* [127]. On the other hand, SIPERSTEIN's data [147] suggests some correlation between cholesterol synthesis and degree of differentiation, and both MCGARRY and FOSTER [104] and KANDUTSCH and HANCOCK [85] have also suggested that this could be so.

To me, comparative rates of synthesis between liver and liver tumor seem unimportant – as, for that matter, do comparative activities *in vitro* of any particular enzyme. What is important, as I see the problem, is that these rates of lipogenesis in the tumor, whatever they may be, are not altered by the dietary regimes that grossly alter liver synthesis rates.

C. Lipid Biosynthesis in Other Tumors

Other chapters in this volume cover many aspects of the biosynthesis of a variety of lipids by a variety of neoplastic tissues. My thesis, however, deals with the physiological control of such synthesis, and in this area very little study has been done, apart from that already discussed with regard to hepatomas. Of course, if a tumor is found to have a lipid composition different from that of its normal tissue of origin, then obviously some controls of lipid synthesis have altered. But so far, again apart from controls in hepatomas, little work seems to have been done to attack this problem directly.

SIPERSTEIN [147] has reported briefly that, in a highly malignant leukemia (L2C) of the guinea pig, the leukemic white cells lack feedback control of cholesterol synthesis, although such control is present in the lymphocytes of normal lymph nodes. Many normal extrahepatic tissues of the guinea pig, unlike those of the rat, do display dietary regulation of cholesterogenesis [165], and thus it will be particularly important to determine whether this control is missing in tumors derived from these tissues.

D. Lipid Biosynthesis in Pretumorous Tissue

By 'pretumorous' in this context, I mean either tissue from an animal treated with a compound known to be highly carcinogenic for that tissue,

Table III. Physiological control of lipid synthesis in precancerous liver

Species	Carcinogen[1] or hepatoma-prone strain	Response to control over synthesis of[2]							References
		cholesterol				fatty acids			
		f'back	fast	c'mine	rhythm	f'back	fast	re-fed	
Normal liver[2]		−	−	+	±	−	−	+	
Rat	Aflatoxin	0							76, 146
	Ethionine	0							81
	FAA	0	−	0	?	−	−	+	76, 79, 80
	3'-meDAB	0							77
Trout	Aflatoxin	0							145, 146
Mouse	Strain A^{y}a	0							145
	Strain aa	0							145
	Strain CExDBA/2WyDi-F$_1$	−							85
	Strain C3H-A^{vy}	−							85

1 FAA: N-2-fluorenylacetamide; 3'-meDAB: 3'-methyl-4-dimethylaminoazobenzene.
2 Abbreviations for controls and symbols as in tables I and II.

or tissue from an animal known to be particularly prone to 'spontaneous' tumors of that tissue, but in both cases prior to the appearance of any histologically discernable tumor. Probably the only difference between these two situations is that in the latter, that is the tumor-prone animals, we just do not know the actual carcinogen responsible.

Following the lead given by SIPERSTEIN [144] we have looked at the control of lipogenesis by livers of animals confronted with a variety of liver carcinogens. This work and SIPERSTEIN's, as well as that with hepatoma-prone animals, is summarized in table III.

The results with chemical hepatocarcinogens are clear-cut. In every case, with four chemically diverse compounds, there is a breakdown in dietary control of cholesterol synthesis following carcinogen treatment. Moreover, this breakdown occurs within a few days or weeks of the beginning of treatment, i.e. many weeks or even months before tumors would be expected to appear. Also, we have consistently observed that with all carcinogens tested control of liver cholesterol synthesis is eventually regained following initial loss, even sometimes in the face of continued carcinogen feeding. I have interpreted our results as indicating that whereas perhaps all cells of the liver can lose feedback control of cholesterogenesis, most might eventually recover, with those few that do not giving rise eventually to hepatomas [124]. At this point it is important to recall that regenerating liver displays normal control [150].

The position with regard to hepatoma-prone animals is less clear. In contrast to earlier findings reported briefly by SIPERSTEIN [144–146], KANDUTSCH and HANCOCK [85] were unable to detect any lack of dietary feedback control of cholesterol synthesis in hepatoma-prone mice. Our own work [SABINE and HORTON, unpublished observations] suggests that such control may be lost and then regained, in a manner analogous to that happening during chemical carcinogenesis. Our studies, however, have been complicated by the fact that in our laboratory the tumor-prone strains we have been using, C3H-A^{vy} and C3H-A^{vy}fB, display only approximately a 20 % incidence of hepatomas [131], and not the 100 % reported in the United States [70]. Thus, obviously, further work needs to be done in this critical area.

One further report bears directly on this point concerning control in tumor-prone liver. BISSELL and ALPERT [11] have reported recently that amongst a group of male patients in Mulago Hospital, Uganda (a country noted for its high incidence of primary hepatoma) a considerable number, perhaps 50 %, failed to display a normal feedback control of cholesterol synthesis.

Several other important points arise from a consideration of the results summarized in table III. First, although the rate of cholesterol synthesis fails to respond to fasting in both transplantable hepatomas so far tested (table II), this control is not necessarily lacking in the precancerous liver, even at a time when dietary feedback control is missing [80]. Similarly, dietary control of fatty acid synthesis, although lost from all 10 fully developed hepatomas tested, may nevertheless be present in the early precancerous state [80].

III. Mechanisms of Control

Work reviewed in the preceding section has shown that all liver tumors, and possibly other tumors, are clearly defective in the physiological control of their rate of synthesis of cholesterol and fatty acids. In this section and the next, I propose to examine reports concerning both the mechanisms of these defective controls and also their possible relationship to the carcinogenic process. But before we can determine the basis for the defective mechanisms found in tumors, we need to understand the mechanisms of control in normal tissue. Unfortunately, the gaps in our knowledge here are considerable.

A. Hepatic Cholesterol Synthesis

Basically, the rate at which the liver is synthesizing cholesterol at any one time depends upon the balance struck between at least three continuously operating but more-or-less independent controls, all exercised primarily but not necessarily entirely over the enzyme considered to be rate-controlling for that biosynthetic pathway, namely β-hydroxy-β-methylglutaryl CoA (HMG CoA) reductase: (1) cholesterol of both endogenous and exogenous origin, being continually absorbed from the digestive tract, exerts a 'negative-feedback' control, i.e. it tends to inhibit synthesis [181]; (2) there is a feeding/fasting variation, i.e. the rate of synthesis varies with time since the last meal [27, 28, 100, 119, 138, 155, 182], and (3) there is a diurnal rhythm of synthesis, in that the rate varies with the time of day, and under standard lighting conditions may be some 3–6 times higher at midnight than at midday [6, 40, 49, 58, 66, 72, 74, 87, 141, 155, 162].

Almost certainly some of this control is exerted via hormone action, but the literature relating endocrine activity to hepatic cholesterol synthesis is particularly confusing [71]. Various reports have indicated that cholesterol synthesis or HMG CoA reductase activity can be increased by adrenalin [48], noradrenalin [18], insulin [96] and thyroxine [65]. As yet, however, it is difficult to assess what relationship pharmacological doses of hormones bear to physiological control mechanisms. Perhaps the most promising work is that dealing with induction(?) of the reductase by insulin and the inhibition of this by glucagon [96], particularly since the increase and subsequent decrease in enzyme activity is rapid and thus consistent with both the short half-life of the enzyme [45, 49, 71, 72, 121] and the rapid rise and fall of the diurnal rhythm.

As indicated in table I, liver tumors have consistently lost both feedback control and fasting control, but not necessarily diurnal rhythm. Hence, in this section I shall deal only with the mechanisms of 'normal' feedback and fasting controls. First, feedback.

Although it has been believed for some time that this control operates primarily over the reductase enzyme [27, 60, 100, 151, 153], we are still unable to say whether this is a phenomenon of regulation of enzyme activity or of enzyme amount. Since the half-life of this enzyme is short, some 2–4 h [45, 49, 71, 72, 121], both fine and rapid control over the activity of the total pathway could be exercised by regulation of the rate of enzyme synthesis and/or degradation. Since there are no apparent differences in K_m values for the enzyme from normal and cholesterol-fed animals [141, 142], and since the pattern of the fall in activity immediately following cholesterol feeding is very similar to that seen in the falling phase of diurnal rhythm, when enzyme synthesis is negligible [72], RODWELL [121] has argued that cholesterol feeding probably inhibits reductase synthesis. There have been several reports of the solubilization and partial purification of the enzyme [69, 88, 99], but this is not an easy exercise. More work is obviously needed here, particularly because of the value a specific antibody to the enzyme would have in resolving some of the important questions regarding control mechanisms.

Concerning the nature and function of an *extracellular* regulator, SIPERSTEIN and his colleagues have presented evidence [137] that feedback control is exercised necessarily by cholesterol absorbed from the digestive tract, and not by cholesterol released from body stores. They have suggested that a cholesterol-lipoprotein complex is responsible [149], but the nature and mode of action of this have not been resolved. The reductase is predominantly membrane-bound [28, 88, 99, 151] and only this form of the enzyme is inhibited, while the cytoplasmic form is not [149, 151]. Inhibition of synthesis is not dependent upon an increase in absolute levels of cholesterol in either plasma or liver [118, 147] although the hepatic level of cholesterol, and particularly of cholesterol ester, may rise following cholesterol feeding [55, 68, 116, 140, 144], and indeed under certain defined conditions the rate of cholesterol synthesis may be significantly inversely correlated with cholesterol levels [59, 75], particularly in short-term experiments [142].

SIPERSTEIN's earlier work [144] and that of others more recently [55, 116, 168] has shown a significant increase in cholesterol-ester content of the endoplasmic reticulum, and a concomitant shift of the ribosomal profile from polysomes to monosomes, following cholesterol feeding. This feature could be important with regard to defective control in tumors, as will be discussed later. Obviously, further attempts should be made both to specifically localize 'dietary' cholesterol in the liver [160] and to determine its effects on intracellular membrane structures, following both short-term and long-term feeding.

Another important point, and one that is likely to receive increasing attention, is that various other sterols, of both plant and animal origin, can mimic the inhibitory effect of cholesterol, either given *in vivo* or added to cells in culture. KANDUTSCH and PACKIE [86] compared the effects of various C_{27}, C_{21} and C_{19} steroids upon both sterol synthesis and HMG CoA reductase activity in mouse liver. At low or moderate levels in the diet all steroids tested, including of course cholesterol, were inhibitory. On the other hand, only some of them, this time not including cholesterol, inhibited cholesterogenesis and reductase activity when injected intraperitoneally. Each of the steroids that inhibited the reductase when injected intraperitoneally failed to inhibit *in vitro,* i.e. when added to incubation mixtures as solutions in propylene glycol or as suspensions in a solution of 10 % ethanol containing 5 % bovine serum albumin.

Along similar lines AVIGAN *et al.* [5] have shown that not only cholesterol but also other sterols can inhibit cholesterol synthesis from acetate, when added as a sonicated suspension with lecithin to cultured human fibroblast cells previously stimulated by delipidated serum. These sterols include some that can give rise to cholesterol by normal metabolic pathways and also some plant sterols that are not converted to cholesterol in mammalian tissues. This work, and that of WATSON [178, 180] with HTC cells, and of SOKOLOFF and ROTHBLAT [157] and others with L cells are the only examples reported of the inhibition of sterol synthesis *in vitro* by added sterols, and this is all with intact cells in culture. No one has yet described an inhibitory effect of cholesterol *per se* on any cell-free system synthesizing cholesterol.

This latter problem, however, may reflect more an inability either to solubilize cholesterol appropriately or to present it to the enzyme system in the correct form, rather than to the lack of direct effect upon the reductase or reductase/membrane complex. For instance, evidence from a number of laboratories [67, 81, 144, 168] suggests that cholesterol ester may be a more important regulator than cholesterol itself. And furthermore, recent work with cells in culture [157, 179] has raised another point which may prove to be particularly important in this context. Apparently sterol synthesis can be influenced significantly not only by sterols in the medium but also by the quantity and quality of exogenous lipoprotein.

Detection of the presence of an actual *intracellular* regulator is another contentious question. Some workers have claimed that both microsomal membranes [144] and soluble cell constituents [27] from cholesterol-fed animals can inhibit microsomal synthesis of cholesterol. Yet others [100, 141, 142], using appropriate mixing experiments and even the addition of cholesterol-rich lipoproteins isolated from rat liver, could not detect the presence of either a soluble or a membrane-bound inhibitor of HMG CoA reductase in the livers of cholesterol-fed animals. The relationship, if any, of this work to that reporting the inhibitory effects of peroxidase isolated from beef liver [32] and more recently to that demonstrating reduced lipogenesis in acatalasemic mice [39] has yet to be determined. And finally, McNamara et al. [105] have isolated from the liver cytosol of suckling rats, and from their mother's milk, an inhibitor of reductase activity. Suckling rats have a low rate of hepatic cholesterol synthesis [34, 105].

Undoubtedly, the two major questions remain unresolved: What is/are the nature and action of intracellular inhibitor(s) responsible for reduced hepatic HMG CoA reductase following cholesterol feeding? – Is this reduction due to lowered activity or lowered amount of the enzyme? It seems to me at this stage that the answer to both these questions may well be found in an examination of the integrity of the relationship between the reductase and its part of the microsomal membrane. As I have already mentioned, cholesterol feeding increases the cholesterol content of the microsomal membrane, and there is a rapidly increasing wealth of evidence that in both model [53, 73, 122] and natural membranes [14, 94] such an increase can profoundly alter membrane properties.

My entire discussion so far presumes of course that HMG CoA reductase is indeed the rate-controlling enzyme of the biosynthetic route to cholesterol, but this may not be so under all circumstances [82, 118]. The evidence presented in favor of the hypothesis is both theoretical and practical, but not necessarily completely convincing.

On theoretical grounds, i.e. by comparison with bacterial systems, control at this point in the pathway would seem logical, in that the reductase can be regarded as the first committed step on the path to cholesterol – mevalonate goes to cholesterol, whereas HMG CoA can give rise also to ketone bodies. But this argument may not hold, for several reasons. As Sugiyama et al. [164] have pointed out recently, cholesterol synthesis from acetyl CoA via HMG CoA is predominantly if not entirely an extramitochondrial process, whereas the production of ketone bodies is from intramitochondrial HMG CoA. These may represent two distinct metabolic pools of HMG CoA that are not readily interconvertible. Furthermore, the enzyme pathway beyond HMG CoA reductase leads not only to cholesterol but also to the isoprene side-chain of ubiquinone, both in liver and in other tissues [118]. Ubiquinone can inhibit cholesterol synthesis [93], and there is considerable experimental evidence for a close relationship between the mechanisms controlling the biosynthesis of both compounds [118].

Practical support for the contention that HMG CoA reductase is rate-limiting comes predominantly from the fact that in virtually all physiological circumstances where cholesterol synthesis is altered, i.e. either increased or decreased, so too is the measured activity of the reductase enzyme, although only recently have these two parameters been measured in the same livers. But, one can recognize other points of control along the enzymatic pathway to cholesterol. Both ELWOOD and MORRIS [50] and SLAKEY et al. [155] have produced evidence that the enzymes beyond mevalonate can be influenced by dietary conditions, although the activity of the reductase could account for the flux along the pathway [155]. Prolonged feeding of cholesterol leads to a reduction of enzyme activity beyond mevalonate [60], but this is probably not of physiological significance. Perhaps of more importance is the fact that both geranyl pyrophosphate and farnesyl pyrophosphate, two further intermediates, can inhibit the enzyme after mevalonate, i.e. mevalonate kinase [44], and more particularly that cholesterol feeding reduces considerably the activity of cytoplasmic HMG CoA synthetase [164, 182]. Also, we have found recently [82] that in at least one experimental situation (animals in continuous light) the rate of cholesterol synthesis, as measured by the incorporation of ^{14}C-acetate into digitonin-precipitable sterols by liver slices, in no way correlates with the activity of HMG CoA reductase measured in the same livers, and indeed RAMASARMA [118] has argued that HMG CoA reductase is unlikely to be rate-controlling at the peak of diurnal rhythm.

An understanding of the physiological regulation of cholesterol synthesis during fasting presents a different set of problems. Present evidence [27, 28, 100, 119, 182] indicates that this control is also mediated mainly via an effect on HMG CoA reductase, but the mechanism of this action is not understood. Both cholesterol synthesis and fatty acid synthesis are drastically curtailed during fasting, and both of these pathways begin with extramitochondrial acetyl CoA. Although this appears to derive predominantly from the action of the citrate cleavage enzyme (CCE), the activity of which is reduced during fasting [91], the action of CCE is not considered to be the critical controlling factor in the lipogenic response to fasting [56]. Nevertheless, the actual supply of extramitochondrial citrate may be limiting, and this question is worth investigating. The recent studies of THORNE and BYGRAVE [171] concerning differential Ca^{++} uptake and release by mitochondria from normal liver and ascites cells may open up fruitful leads in this area. The rate of supply of acetate, possibly a significant source of extramitochondrial acetyl CoA, also needs further consideration [7].

SUBBA RAO and RAMASARMA [163] have suggested recently that the prime change in fasting that leads to reduced lipogenesis is a lack of adenosine triphosphate (ATP), and they have reported a dramatic increase in hepatic cholesterogenesis in fasted rats following the intraperitoneal injection of ATP. If this work is confirmed in other laboratories it will open a new dimension in our consideration of the biological effects of fasting.

There is no doubt that fasting is accompanied by widespread changes in endocrine activity, particularly in circulating levels of insulin and glucagon, and any one or more of these changes could influence sterol biosynthesis, at least indirectly. In 1966 OGILVIE and KAPLAN [111] reported that they had isolated from rat bile a small molecular weight protein that both inhibited cholesterogenesis and increased upon fasting. It seems difficult, however, to assign a major physiological role to this substance, and no further reports of its action have appeared.

Considerable work over the last couple of years has been directed towards the isolation and characterization of the 'sterol carrier protein', and towards an understanding of its role in cholesterol synthesis [120, 139]. While this role is undoubtedly important, especially in the conversion of squalene to cholesterol, the mechanism of action of this protein, particularly with regard to physiological control of the overall rate of sterol biosynthesis, is as yet poorly understood.

B. Hepatic Fatty Acid Synthesis

My concern here is primarily with the mechanisms of the inhibition of fatty acid biosynthesis exerted by fasting and by a high-fat diet, since these are the mechanisms defective in all 10 hepatomas so far examined. Both of these effects are apparently due primarily to inhibition of the enzyme acetyl CoA carboxylase, an enzyme that has been extensively purified, characterized and reviewed over recent years [98, 107, 109, 110, 114, 175].

Evidence to date suggests that there are perhaps three general mechanisms by which the 'activity' of acetyl CoA carboxylase can be affected. First, and probably of primary importance in long-term control, alteration of the rate of synthesis and/or degradation can lead to a change in the level of enzyme protein present in the tissue. This mechanism is certainly important in adjustments made to fasting and to refeeding after fasting [102, 109, 110]. Secondly, the active enzyme is composed (at least *in vitro*) of a large number of subunits, and a wide variety of conditions, including the nature of activating compounds, buffers, ionic strength and pH, can influence the equilibria attained between the monomeric (inactive), protomeric (inactive) and polymeric (active) forms. It seems not unlikely that citrate, which has been known for a long time to activate fatty acid synthesis *in vitro* [21] and perhaps also Mg^{++} [63] may be important physiological regulators in this way. Thirdly, the enzyme can be phosphorylated (by an ATP-dependent kinase) and in this form is considerably less active [33]. Dephosphorylation to the more active form occurs apparently through the action of a Mg^{++}-dependent phosphatase [33, 63].

Moss *et al.* [107] have reported recently that the activity of the enzyme increases with dilution, and they have attributed this to the removal of an inhibitor – just as would occur during purification, which also increases activity. In similar vein SWANSON *et al.* [166] reported several years ago that semi-purified carboxylase from the livers of fasted or even fed rats was partially inhibited, and that this inhibition could be removed by trypsin digestion, aging at $0°C$, or prolonged incubation with citric acid. The nature and mechanisms of these inhibitions have not been determined, although they may involve the phosphorylation/dephosphorylation mechanism described by CARLSON and KIM [33].

It is possible that both a high-fat diet and fasting have an effect on fatty acid synthesis via the same mechanism, particularly since both result in an increased supply to the liver of extrahepatic acyl units. A suggested physiological, *intracellular* effector is long-chain acyl CoA [20, 172, 184], which is known both to increase *in vivo* upon fasting [20, 172] and feeding a high-fat diet [172], and to inhibit acetyl CoA carboxylase *in vitro* [19, 101, 128, 172]. Several authors, however [43, 158, 169], have challenged this hypothesis, mainly on the grounds of the known detergent qualities of long-chain acyl CoA and of its widespread and apparently indiscriminate inhibition *in vitro* of a variety of enzymes. Furthermore, recent work from this laboratory has shown clearly that rates of hepatic fatty acid synthesis in a variety of physiological states are not necessarily correlated with actual tissue levels of acyl CoA [136].

This same work [136] has shown that hepatic lipogenesis is also not necessarily correlated with tissue levels of cAMP, another contender for the role of chief intracellular regulator of lipogenesis [23, 24]. Undoubtedly, under certain conditions *in vitro,* cAMP and its dibutyryl derivative can inhibit fatty acid synthesis, but the physiological importance of this compound in intracellular control of the rate of lipogenesis is still debatable. Plasma levels of cyclic nucleotides do not increase with even prolonged fasting [173]. Perhaps the action of cAMP is via the acetyl CoA carboxylase kinase described above [33], but I am not aware of any reports of such work.

Several workers [35, 36, 114] have suggested, on the basis of good experimental evidence, that malonyl CoA, the major product of the carboxylase reaction, may be in turn an important regulator of the activity of the enzyme. It is certainly a potent competitive inhibitor of the purified enzyme, promoting its dissociation into the inactive protomeric form [64].

The definitive establishment of an *extracellular* regulator of fatty acid synthesis is also debatable. The possible role of polypeptide fragments derived from growth hormone, and detectable in urine, deserves consideration. Extensive work by BORNSTEIN *et al.* [17] has shown that one of these compounds, In-G (now known as Somantin [106]), can inhibit fatty acid synthesis by direct action on acetyl CoA carboxylase. Changes in circulating levels of insulin and glucagon are physiological consequences of fasting and refeeding after fasting and these may be sufficient to initiate the changes seen in lipogenesis [30]. Circulating insulin levels decrease in fasting [26, 31] and in fat-feeding [12] and of course also in diabetes, and all these conditions favor reduced lipogenesis. Furthermore, an acute reduction in insulin level, as for instance induced by anti-insulin serum, results in lowered hepatic fatty acid synthesis [167]. A reciprocal rise in glucagon levels on fasting has been established for man [174] but not so precisely for the rat.

Several reports [3, 4, 9, 108, 133, 156] have favored the hypothesis that the 'feedback' inhibition of hepatic fatty acid synthesis by fat in the diet is due primarily to the linoleic acid content of that fat, but the precise nature of this effect remains to be clarified.

Virtually all the work on regulation of fatty acid biosynthesis described above deals with the enzyme acetyl CoA carboxylase, but the primacy of this as the rate-controlling enzyme has been challenged [114].

There is considerable evidence that in many of the physiological states where fatty acid synthesis is altered there are concomitant changes in the activity and amount of the multi-enzyme complex, fatty acid synthetase [29, 37, 38, 101] and this may have an important regulatory role, particularly with changes occurring in fasting and refeeding after fasting. Recent work by LAKSHMANAN et al. [97] has provided strong evidence that the concentration of this enzyme in liver is under the control of the relative concentrations of insulin and glucagon, possibly acting as coordinated regulators of tissue levels of cAMP. It is my interpretation of the data, however, that all these changes in fatty acid synthetase are perhaps secondary adaptations, i.e. following primary regulatory effects on acetyl CoA carboxylase.

C. Defective Controls in Hepatomas

I shall again treat cholesterol metabolism first, as this has been studied more extensively and defects here are perhaps inherently more important than defects in fatty acid metabolism. But before reviewing the experimental results that have been obtained I should like to examine the theoretical possibilities inherent in the question: Why do hepatomas fail to exhibit dietary feedback control of cholesterol synthesis?

There is a long distance, physiologically, between cholesterol in the gastrointestinal tract and inhibition of the hepatic enzyme, HMG CoA reductase, and this distance spans a lot of unchartered territory. Yet despite an absence of precise knowledge concerning the mechanisms of feedback control in the normal liver, one can nevertheless compile a complete list of the general means by which, at least in theory, the cancerous liver might possibly escape this control; means relating to either the pathway of synthesis, the rate-controlling enzyme or the postulated extracellular and intracellular inhibitors. Thus (a) cholesterol might be synthesized in the hepatoma predominantly by a different metabolic pathway; (b) the rate-controlling enzyme, i.e. probably HMG CoA reductase, might be altered in the tumor, in that although it would perform the same biosynthetic step it might now fail to respond to appropriate intracellular regulation, e.g. through lack of appropriate binding sites or failure to undergo appropriate inhibitory conformational changes following binding; or (c) the extracellular inhibitor, the nature of which is as yet unknown, may fail to elicit effective response from the tumor cell in one of two general ways – if it does not need to enter the cell itself it may fail to stimulate the appropriate intracellular messenger of its action, or this messenger (or the inhibitor itself if entry is necessary) may fail to accumulate in the cell in sufficient quantity at the appropriate site of action, either by failure of entry or intracellular binding, or through increased degradation or passage back out of the cell. These possibilities should all be considered in turn, for although we do not yet know the correct answers we may be able to eliminate some of the incorrect ones.

Even though there has been no detailed study in hepatomas of the complete biosynthetic pathway leading to cholesterol, all available evidence suggests that the tumor synthesizes its sterol requirements predominantly via the same pathway as does the liver, and thus that control breaks down at some point along this route. Indeed, workers in at least three laboratories have now shown that defective dietary feedback control in the hepatoma is localized at the enzyme HMG CoA reductase [57, 85, 154]. Other, less direct, evidence also implicates the same pathway of cholesterogenesis for both tumor and liver, in that MER 29 [25], bile salts [123] and several plant-growth retardants [112], which are known to inhibit cholesterol synthesis at different enzyme steps, can under appropriate conditions inhibit synthesis in the hepatoma as well as in the liver.

In their early papers [144, 149, 150] SIPERSTEIN and his colleagues suggested that the reductase enzyme in the hepatoma might be different from that in the liver, so that it would no longer respond to appropriate intracellular inhibitors. The detailed work of KANDUTSCH and HANCOCK [85], however, would seem to rule out this possibility, in that by the criteria they used, mainly K_m values and heat stability, the enzymes in both tissues are identical.

We come now to a consideration of the 'inhibitor(s)', and its/their action on the tumor. In several papers [123, 126, 128] we have proposed that the extracellular inhibitor, probably cholesterol in some form, fails to reach in the tumor the intracellular site of its action in sufficient quantity to be effective, and we suggested this may be due to impaired uptake from the plasma. Several papers, from this laboratory and elsewhere, have presented experimental evidence in favor of this hypothesis [68, 81, 130]. More recently, however, HORTON et al. [75, 78] have suggested that although uptake of dietary cholesterol from the plasma is impaired in the tumor the extent of this defect – only about 50% reduction in primary hepatomas – is insufficient to account for the almost complete lack of inhibition of synthesis exhibited by the tumor. These authors now believe that the primary defect may be in the inability of the hepatoma to accumulate and store dietary cholesterol [75]. This position is supported by their own results, and more particularly by the earlier work of both HARRY et al. [68] and BRICKER et al. [25] which showed clearly the non-accumulation of cholesterol ester in the hepatoma following cholesterol feeding.

As opposed to this line of reasoning, BRICKER and LEVEY [23] consider that the real defect lies in the tumor's lack of response to cAMP, or in other words an intracellular failure to respond to a secondary inhibitor. SACCONE and SABINE [136] have so far been unable to accept this completely in that, as mentioned earlier, we have found no evidence to correlate rates of hepatic and hepatomal lipogenesis under a variety of conditions with actual tissue levels of cAMP. Of course our results are not definitive either, since 'total' levels of cAMP do not distinguish between possible metabolic pools.

The position at the moment would seem to be that the hepatoma has both an impaired short-term uptake of dietary cholesterol and a long-term failure to accumulate it. But until the mechanism of normal control is revealed we cannot yet say how important these deficiencies are in contributing to defective control, nor indeed even whether this defect is in the control of enzyme activity or of enzyme amount. If my hypothesis earlier in this paper is correct, namely that the answer to the question of regulation in the normal liver lies in the integrity of the relationship between the reductase and its ribosomal membrane, then defective control may be due to the loss of this integrity. Some results may in fact point in this direction. I have already covered the point that whereas the normal liver rapidly accumulates dietary cholesterol, predominantly in its endoplasmic reticulum, the hepatoma does not. Furthermore, SIPERSTEIN and FAGAN [151] have shown that the small amount of 'free' reductase, i,e, enzyme present in the cytoplasm, is in the normal liver not inhibited by dietary cholesterol, and the work of SHEFER et al. [143] indicates that in the intestine, a tissue in which cholesterogenesis does not respond to dietary cholesterol [42], some 50 % of the reductase is not associated with the microsomes. The intracellular distribution of reductase in the hepatoma has yet to be determined.

In considering defective control of fatty acid synthesis in the hepatoma one can build up a picture analogous to that already painted for defective control of cholesterol synthesis. As discussed earlier, the major rate-controlling enzyme of the synthetic pathway leading to long-chain fatty acids is probably acetyl CoA carboxylase, and the defect in the tumor seems localized at this point [41]. The work of MAJERUS *et al.* [101] has shown clearly that this enzyme in the tumor, although unregulated by diet, is essentially identical to that in the normal liver, at least as judged by the following criteria: heat inactivation, affinity for acetyl CoA and ATP, activation and aggregation by citrate, product inhibition by malonyl CoA, pH optima and intrinsic specific activity as determined from biotin content. We had shown a little earlier that this enzyme, in hepatomal as in hepatic cytosol, could be inhibited by long-chain acyl CoA and stimulated by addition of microsomal protein [128].

Changes in the rate of fatty acid synthesis during fasting and refeeding following fasting have been shown to be associated with similar changes in tissue levels of enzyme protein [102, 109, 110]. Thus, for instance, the fall in fatty acid synthesis during fasting is due, at least in considerable measure, to both reduced synthesis and increased degradation of the carboxylase enzyme, although these may be controlled independently [102, 110]. Hence, the failure of the tumor to respond similarly to the liver in these circumstances would seem to be due to a defect in its regulation of specific enzyme synthesis and/or degradation, although, as MAJERUS and KILBURN [102] have noted, the half-life of the hepatic carboxylase, as measured in their studies, varied from 18 to 50 h, and at this rate changes in enzyme protein may be too slow to account for large, short-term alterations in fatty acid synthesis.

Similarly, in the fat-fed animal there is a reduction in the rate of synthesis (but no change in the rate of degradation) of the carboxylase in the liver [102], but presumably not in the hepatoma. Certainly, when the role in lipogenesis of various hormones, particularly insulin, is clarified for the normal liver, we will be more likely to understand the defects in the liver tumor.

In conclusion, it would appear that at present we can go no further than to say that defective control of fatty acid synthesis in the hepatoma is presumably due to either (1) an impaired production, uptake and/or storage of intracellular inhibitor, or (2) lack of an effect of this inhibitor on the mechanism of carboxylase synthesis and/or degradation. I suspect, however, that long-term alterations in the amount of an enzyme protein present in response to various changes in physiological status are a consequence of, and hence subsequent to, changes in enzyme activity and hence it seems to me that the significant defect in the hepatoma will be found relative to the short-term control of enzyme activity.

IV. Defective Control and Carcinogenesis

It is my belief that the bulk of the evidence, as reviewed above, supports the conclusion that defective dietary control of lipid synthesis is an integral feature of the overall carcinogenic process in the liver. This section endeavours to examine how this might be so.

A. Possible Correlative (Causative?) Mechanisms

There are possible only two general mechanisms by which carcinogenesis and defective control of lipogenesis could be correlated. Either (1) well-regulated lipogenesis is essential for the functional integrity of the liver cell and loss of this integrity leads to neoplastic change, or alternatively (2) the defect manifest in lipogenesis is not *per se* particularly important in the carcinogenic process, but is rather a necessary consequence of some other cellular change that is important.

BLOCH [13] has made the point in his Nobel lecture that the sterol molecule is a ubiquitous one, produced or required by all but the most primitive organisms. Moreover, it is not randomly distributed within cells but is primarily associated with intracellular membranes, and indeed the development of membrane-enclosed structures with specialized functions is regarded as a landmark in evolutionary diversification [159]. Subsequent development has given rise to the current situation, where not only is the presence of cholesterol important to the mammalian cell but also its synthesis *in situ* seems essential. In other words, not only does every mammalian cell contain cholesterol but also virtually every cell, with the probable exception of the mature erythrocyte, retains the capacity to synthesize it. This must mean that there is evolutionary significance to the whole organism for each cell to read out from its genetic material the information necessary to synthesize and maintain control over the some 26 enzymes required to convert acetyl CoA to cholesterol. Given this situation it is not difficult to imagine that any impairment of this biosynthetic mechanism and of its control might readily lead to a functional aberration as severe as neoplasia.

Indeed, FENSELAU and WALLIS [54] have suggested a functional advantage such altered cells may hold over their more regular neighbors. From preliminary evidence they suggested that: 'Malignant hepatocytes, as a consequence of their production and release of large quantities of cholesterol, derive metabolic advantage over their neighbors by being able to consume ketone bodies synthesized by normal cells in response to the elevated cholesterol levels.' But this is from work on only one hepatoma and, as the authors themselves admit, more evidence is needed to establish the generality of their conclusion. This is particularly important, since not all hepatomas produce large amounts of cholesterol [152], although they can apparently release significant quantities into the plasma [25]. Also it has yet to be shown that they are all efficient users of ketone bodies.

The preliminary report of KABARA *et al.* [84] is perhaps also important in this context. They have found that some, but not all, hypocholesteremic drugs will decrease the life span of mice inoculated with Ehrlich ascites cells, while Elipten (aminoglutethimide phosphate), which reduces cholesterol utilization by blocking side-chain oxidation, increases the life span.

Obviously this reasoning relating defective cholesterol metabolism to the development of cancer would hold, at least in most species, only for cancer of the liver. Other tissues seem to survive quite effectively with no dietary feedback control of cholesterol synthesis [42, 83]. The guinea pig is an interesting exception, in which a variety of extrahepatic tissues do exhibit feedback inhibition of cholesterol synthesis [165]. I have already mentioned that at least one guinea pig tumor other than a hepatoma, i.e. the leukemic lymphocyte, displays a defective control of cholesterogenesis [147] and it will be interesting to learn whether other tumors do likewise.

The alternative hypothesis, namely that defective control of lipogenesis is not itself critical in initiating or maintaining the cancerous state but merely reflects some other truly cancerous change, seems inherently more tenable. The fact that several controls are missing, controls that appear to operate through different mechanisms, argues that these diverse defects are due to some more fundamental change. For instance, there is and has been for a long time a wealth of literature dealing with qualitative and quantitative changes in cellular membranes, both plasma membranes and intracellular membranes, during carcinogenesis and in the fully developed cancer [1]. In this context the exciting work of RABIN and his colleagues seems particularly relevant. He has suggested [117, 183] that a common mode of action of all chemical carcinogens may be their action on the ribosome/ribosomal membrane complex, as for instance, aflatoxin destroys high-affinity binding sites for steroid hormones [15]. And, as I have already suggested, the control of cholesterol synthesis in the normal liver and its defective control in the cancerous liver are perhaps events primarily related to the functional integrity of the microsomal membrane.

In order to complete this discussion of the possibility of one or perhaps a few common, fundamental cellular alterations giving rise to several diverse manifestations of defective control, I need to mention two other specific instances of defective physiological control over the intracellular metabolism of a cancerous tissue. These are similar to the examples already quoted in this paper, but neither deals directly with lipid metabolism.

Aminolevulinic acid (ALA) synthetase is the rate-controlling enzyme of the biosynthetic pathway leading to heme formation [103]. The end product of this pathway, i.e. heme, exerts a feedback inhibition over the synthesis of this enzyme, in that it can almost completely block its induction by such agents as allylisopropylacetamide (AIA) [46]. We have observed recently [47] that, contrary to previous reports [16], ALA synthetase can be induced by AIA in at least one hepatoma (5123C) but, and this is of considerable importance, feedback control has been lost, i.e. heme fails to inhibit this synthesis of new enzyme. Furthermore, feedback inhibition of ALA synthetase is similarly missing from the liver of the ethionine-fed rat, only four weeks after commencing carcinogen treatment [47]. In this context it is interesting to recall that in an early paper, SIPERSTEIN and FAGAN [150] cited the work of BRESNICK [22] with aspartate transcarbamylase as an example of a feedback mechanism, in that case for pyrimidine synthesis, that had not been lost by the hepatoma. But BRESNICK's work was with a semi-purified enzyme preparation, not with the hepatoma *in vivo*.

In a different approach to this same overall problem, we have reported [134] that the rate of gluconeogenesis in two kidney tumors, MK1 and MK2, fails to increase in fasted animals, a condition known to increase gluconeogenesis in the normal kidney [92]. We have now extended this observation to a third tumor, MK3, which fails to respond not only to fasting but also to alloxan diabetes [176]. This defect in the tumor appears attributable, at least in part, to ineffective control over the enzymes pyruvate carboxylase and phosphoenolpyruvate carboxykinase [176].

The explanation for both these control defects, as with those relating to lipogenesis, must obviously await further work. Nevertheless, we can postulate that a common mechanism could well be operating, in that a variety of extracellular factors may be prohibited from entering or accumulating in the cancerous cell in sufficient concentrations to carry out the regulatory role they perform in a normal cell. And, as a logical extension of this, other potentially more important regulators, such as those governing cell division, may similarly fail to reach their site(s) of action within the tumor cell.

B. Correlation with Other Precancerous Changes

The logic governing studies of carcinogenesis seems to fall into two categories. The vast bulk of such studies, particularly with chemical carcinogens, seems aimed at determining what cellular changes are produced by the carcinogen or its metabolites, and then at attempting to correlate these changes – which are invariably numerous, predominantly toxic, and often lethal – with the subsequent development of neoplasia, a state still uncharacterized in terms of definitive cellular changes. A different philosophy, and the one that we have espoused, aims to seek out those few biochemical parameters that are definitive for a particular tumor or class of tumors, and then to look for these specific changes during various stages of the development of such tumors, particularly the early stages.

As is obvious from the work reviewed in an earlier section of this paper (II.D), this latter approach has been particularly successful with regards to defective control of cholesterol synthesis. The four carcinogens tested – aflatoxin, ethionine, N-2-fluorenylacetamide and 3'-methyl-4-dimethylaminoazobenzene – all produce a significant loss of feedback control of cholesterol synthesis within a few weeks of commencing treatment, i.e. months before the appearance of any histologically discernible tumors. Of course the overwhelming drawback of this approach is that there are as yet very few cellular changes that can be regarded as definitive of the cancerous, and hence precancerous, cell.

STONEHILL and BENDICH [161] have shown that a series of embryonal antigens normally not present in adult tissue are detectable in every one of a complete range of mouse tumors – tumors from 12 tissues and 18 different strains. We have detected these same antigens in livers of cancer-prone mice, C3H-A^{vy}, long before hepatomas appeared [135] and work currently in progress in this laboratory aims to establish a definitive point in time, i.e. during the development prior to hepatoma appearance, at which these antigens first appear and even, with the help of such cell-dispersal techniques such as that of BERRY and FRIEND [10], to localize the particular cells containing antigen. Other precancerous changes, such as those relating to lipid metabolism, would then be correlated with antigen appearance. In this context it is significant that a number of recent papers have reported the early appearance of fetal antigen during chemical carcinogenesis [95, 177].

C. Some Doubts

The question has been raised [8, 78] that since the hepatoma, particularly the transplanted one, has a blood supply that is considerably different from that of the normal liver, this whole discussion of defective controls over metabolic activity may reflect solely this difference. Indeed, the recent work of Horton et al. [75, 78] has shown that the deficiency of the transplanted hepatoma in its short-term uptake of dietary cholesterol may relate more to its altered vasculature than to its neoplasticity.

On the other hand, the findings of Bartley and Abraham [8] are of some importance in this context. They examined some aspects of the dietary control of fatty acid synthesis in the autotransplanted liver. They found that synthesis decreased upon fasting, and from this argued that the lack of sensitivity of the hepatoma could not be due to its impaired, particularly nonportal, blood supply since the autotransplanted liver had a similarly altered supply. On the other hand, their experimental approach did not clearly establish the presence or absence of feedback control of synthesis, although it did appear that the response to a fat-free diet was missing in the autotransplanted liver. This question of the possible explanation for the tumor's insensitivity being its altered blood supply needs further investigation. Another approach would be to examine fatty acid synthesis and its control in the livers of portacaval-shunt animals [89], but I am not aware that this has yet been attempted.

Of perhaps more importance is the work of Watson. In recent publications [178, 180] he has challenged the hypothesis that defective control of cholesterol synthesis is necessarily equated with being a hepatoma. His doubts are based on his observation of the effects of lipid-poor and lipid-rich sera on the rate of lipogenesis by HTC cells – hepatoma cells in tissue culture. As Avigan et al. [5] had shown for fibroblast cells, so too Watson has now shown that the presence of lipid in serum added to growing HTC cells will inhibit lipid synthesis, including cholesterol synthesis.

It seems to me that these findings do not necessarily negate the general hypothesis. HTC cells were derived originally from the Morris hepatoma 7288C, which in the solid form *in vivo* fails to exhibit feedback control of cholesterol synthesis [152]. This tumor was first converted to an ascitic form, and then put into suspension culture [170]. Obviously significant changes must have occurred, at least in or on the external cell wall, to enable the cells first to grow ascitically rather than as a solid, intraperitoneally transplanted tumor, and then to proliferate as individual cells in culture media. Indeed, various workers have failed to detect in HTC cells any adenyl cyclase [62], cAMP [62] or cAMP-binding protein [61], properties certainly different from those of the transplantable hepatoma [2, 52.] It seems not at all unlikely that these or similar changes may now permit again the entry or accumulation of specific metabolic regulators, and all evidence (see above) indicates that if inhibitors can reach their appropriate site of action in the hepatoma they will fulfil the same regulatory role there as in normal liver.

A more serious, indeed fatal, objection to the overall hypothesis would arise if further work, similar to that of KANDUTSCH and HANCOCK [85] showed that not all spontaneous hepatomas had lost feedback control. The validity of the concept depends on there being no exceptions, a condition observed at least until now.

V. Summary

Various aspects of the physiological control of the synthesis of cholesterol and fatty acids have now been examined in 26 different hepatomas, both primary and transplanted, and these controls compared with those operating in normal liver. Every one of the 24 hepatomas tested so far (with possibly one exception), in four different species – rat, mouse, trout, man – displays a defective dietary feedback control of cholesterol synthesis, i.e. synthesis in the liver but not the liver tumor is inhibited by dietary cholesterol. Similarly, fatty acid synthesis in every one of 10 tumors tested, in rats and mice, is insensitive to dietary changes that grossly alter this synthesis in the normal liver.

Control of lipogenesis has also been examined in precancerous liver, i.e. liver from animals either treated with a hepatocarcinogen or of a strain known to be highly prone to spontaneous hepatoma development. With all four chemical carcinogens tested, dietary feedback control of cholesterol synthesis is partially or completely lost within a few weeks of commencing carcinogen treatment, i.e. many weeks or even months before tumors would appear. The response of cholesterol synthesis to fasting and the dietary control of fatty acid synthesis are not necessarily lost at a time, early in carcinogen treatment, when feedback control of cholesterogenesis is missing.

The weight of this evidence supports the hypothesis that defective control of cholesterol synthesis, and to a lesser extent of fatty acid synthesis, is an integral or even essential feature of the carcinogenic process, at least in the liver. As far as I am aware, this is the only specific physiological regulatory mechanism present in a normal tissue but consistently lost both from a large number of tumors derived from that tissue and also from tissue of a carcinogen-treated animal.

This paper reviews current knowledge of the extent of this defective control, suggests the strength of the hypothesis relating defective control to liver carcinogenesis, raises the question of how the two might be correlated, attempts to answer this question in terms of what is known and not known about the control of lipogenesis in normal liver, and suggests appropriate directions future research might take to solve this intriguing aspect of the whole biochemical riddle called 'CANCER'.

VI. References

1 ABERCROMBIE, M. and AMBROSE, E. J.: The surface properties of cancer cells: a review. Cancer Res. 22: 525–548 (1962).
2 ALLEN, D. O.; MUNSHOWER, J.; MORRIS, H. P., and WEBER, G.: Regulation of adenyl cyclase in hepatomas of different growth rates. Cancer Res. 31: 557–560 (1971).
3 ALLMANN, D. W. and GIBSON, D. M.: Fatty acid synthesis during early linoleic acid deficiency in the mouse. J. Lipid Res. 6: 51–62 (1965).
4 ALLMANN, D. W.; HUBBARD, D. D., and GIBSON, D. M.: Fatty acid synthesis during fat-free refeeding of starved rats. J. Lipid Res. 6: 63–74 (1965).
5 AVIGAN, J.; WILLIAMS, C. D., and BLASS, J. P.: Regulation of sterol synthesis in human skin fibroblast cultures. Biochim. biophys. Acta 218: 381–382 (1970).
6 BACK, P.; HAMPRECHT, B., and LYNEN, F.: Regulation of cholesterol biosynthesis in rat liver: diurnal changes of activity and influence of bile acids. Arch. Biochem. Biophys. 133: 11–21 (1969).
7 BALLARD, F. J.: Supply and utilization of acetate in mammals. Amer. J. clin. Nutr. 25: 773–779 (1972).
8 BARTLEY, J. C. and ABRAHAM, S.: Dietary regulation of fatty acid synthesis in rat liver and hepatic autotransplants. Biochim. biophys. Acta 260: 169–177 (1972).
9 BARTLEY, J. C. and ABRAHAM, S.: Hepatic lipogenesis in fasted, re-fed rats and mice: response to dietary fats of differing fatty acid composition. Biochim. biophys. Acta 280: 258–266 (1972).

10 BERRY, M. N. and FRIEND, D. S.: High-yield preparation of isolated rat liver parenchymal cells. J. Cell. Biol. *43:* 506–520 (1969).
11 BISSELL, D. M. and ALPERT, E.: The feedback control of hepatic cholesterol synthesis in Ugandan patients with liver disease. Cancer Res. *32:* 149–152 (1972).
12 BLÁSQUEZ, E. and QUIJADA, C. L.: The effect of a high-fat diet on glucose, insulin sensitivity and plasma insulin in rats. J. Endocrin. *42:* 489–494 (1968).
13 BLOCH, K.: The biological synthesis of cholesterol. Science *150:* 19–28 (1965).
14 BLOJ, B.; MORERO, R. D.; FARIAS, R. N., and TRUCCO, R. E.: Membrane lipid fatty acids and regulation of membrane-bound enzymes. Allosteric behaviour of erythrocyte Mg^{2+}-ATPase, $(Na^+ + K^+)$-ATPase and acetylcholinesterase from rats fed different fat-supplemented diets. Biochim. biophys. Acta *311:* 67–79 (1973).
15 BLYTH, C. A.; FREEDMAN, R. B., and RABIN, B. R.: The effects of aflatoxin B_1 on the sex-specific binding of steroid hormones to microsomal membranes of rat liver. Europ. J. Biochem. *20:* 580–586 (1971).
16 BONKOWSKY, H. L. and TSCHUDY, D. P.: Control of δ-aminolevulinic acid synthetase (ALAS), ALA-dehydrase (ALAD), and tyrosine aminotransferase (TAT) in tumors and livers of tumor-bearing rats. Proc. amer. Ass. Cancer Res. *12:* 8 (1971).
17 BORNSTEIN, J.; ARMSTRONG, J. McD.; NG, F.; PADDLE, B. M., and MISCONI, L.: Structure and synthesis of biologically active peptides derived from pituitary growth hormone. Biochem. biophys. Res. Commun. *42:* 252–258 (1971).
18 BORTZ, W. M.: Noradrenalin-induced increase in hepatic cholesterol synthesis and its blockade by puromycin. Biochim. biophys. Acta *152:* 619–626 (1968).
19 BORTZ, W. M. and LYNEN, F.: The inhibition of acetyl CoA carboxylase by long-chain acyl CoA derivatives. Biochem. Z. *337:* 505–509 (1963).
20 BORTZ, W. M. and LYNEN, F.: Elevation of long-chain acyl CoA derivatives in livers of fasted rats. Biochem. Z. *339:* 77–82 (1963).
21 BRADY, R. O. and GURIN, S.: Biosynthesis of fatty acids by cell-free or water-soluble enzyme systems. J. biol. Chem. *199:* 421–431 (1952).
22 BRESNICK, E.: Feedback inhibition of aspartate transcarbamylase in liver and in hepatoma. Cancer Res. *22:* 1246–1251 (1962).
23 BRICKER, L. A. and LEVEY, G. S.: Autonomous cholesterol and fatty acid synthesis in hepatomas: deletion of the adenosine 3′, 5′-cyclic monophosphate control mechanism of normal liver. Biochem. biophys. Res. Commun. *48:* 362–365 (1972).
24 BRICKER, L. A. and LEVEY, G. S.: Evidence for regulation of cholesterol and fatty acid synthesis in liver by cyclic adenosine 3′, 5′-monophosphate. J. biol. Chem. *247:* 4914–4915 (1972).
25 BRICKER, L. A.; MORRIS, H. P., and SIPERSTEIN, M. D.: Loss of the cholesterol feedback system in the intact hepatoma-bearing rat. J. clin. Invest. *51:* 206–215 (1972).
26 BUCHANAN, K. D.; VANCE, J. E., and WILLIAMS, R. H.: Effect of starvation on insulin and glucagon release from isolated islets of Langerhans of the rat. Metabolism *18:* 155–162 (1969).

27 Bucher, N. L. R.; McGarrahan, K.; Gould, E., and Loud, A. V.: Cholesterol biosynthesis in preparations of liver from normal, fasting, x-irradiated, cholesterol-fed, triton, or Δ^4-cholesten-3-one treated rats. J. biol. Chem. 234: 262–267 (1959).

28 Bucher, N. L. R.; Overath, P., and Lynen, F.: β-hydroxy-β-methylglutaryl coenzyme A reductase, cleavage and condensing enzymes in relation to cholesterol formation in rat liver. Biochim. biophys. Acta 40: 491–501 (1960).

29 Burton, D. N.; Collins, J. M.; Kennan, A. L., and Porter, J. W.: The effects of nutritional and hormonal factors on the fatty acid synthetase level of rat liver. J. biol. Chem. 244: 4510–4516 (1969).

30 Cahill, G. F., jr.: Physiology of insulin in man. Diabetes, N. Y. 20: 785–799 (1971).

31 Cahill, G. F., jr.; Herrera, M. G.; Morgan, A. P.; Soeldner, J. S.; Steinke, J.; Levy, P. L.; Reichard, G. A., jr., and Kipnis, D. M.: Hormone-fuel interrelationships during fasting. J. clin. Invest. 45: 1751–1769 (1966).

32 Caravaca, J. and May, M. D.: The isolation and properties of an active peroxidase from hepatocatalase. Biochem. biophys. Res. Commun. 16: 528–534 (1964).

33 Carlson, C. A. and Kim, K. H.: Regulation of hepatic acetyl coenzyme A carboxylase by phosphorylation and dephosphorylation. J. biol. Chem. 248: 378–380 (1973).

34 Carroll, K. K.: Acetate incorporation into cholesterol and fatty acids by livers of fetal, suckling, and weaned rats. Canad. J. Biochem. 42: 79–86 (1964).

35 Chakrabarty, K. and Leveille, G. A.: Acetyl CoA carboxylase and fatty acid synthetase activities in liver and adipose tissue of meal-fed rats. Proc. Soc. exp. Biol. Med. 131: 1051–1054 (1969).

36 Chang, H. C.; Seidman, I.; Teebor, G., and Lane, M. D.: Liver acetyl CoA carboxylase and fatty acid synthetase: relative activities in the normal state and in hereditary obesity. Biochem. biophys. Res. Commun. 28: 682–686 (1967).

37 Craig, M. C.; Nepokroeff, C. M.; Lakshmanan, M. R., and Porter, J. W.: Effect of dietary changes on the rates of synthesis and degradation of rat liver fatty acid synthetase. Biochem. Soc. Symp. 35: 303–317 (1972).

38 Craig, M. C.; Nepokroeff, C. M.; Lakshmanan, M. R., and Porter, J. W.: Effect of dietary change on the rates of synthesis and degradation of rat liver fatty acid synthetase. Arch. Biochem. Biophys. 152: 619–630 (1972).

39 Cuadrado, R. R. and Bricker, L. A.: An abnormality of hepatic lipogenesis in a mutant strain of acatalasemic mice. Biochim. biophys. Acta 306: 168–172 (1973).

40 Danielsson, H.: Relationship between diurnal variations in biosynthesis of cholesterol and bile acids. Steroids 20: 63–72 (1972).

41 Denton, M. D. and Siperstein, M. D.: Control of fatty acid and cholesterol synthesis in mouse hepatoma. Clin. Res. 12: 265 (1964).

42 Dietschy, J. M. and Siperstein, M. D.: Effect of cholesterol feeding and fasting on sterol synthesis in seventeen tissues of the rat. J. Lipid Res. 8: 97–104 (1967).

43 Dorsey, J. A. and Porter, J. W.: The effect of palmityl coenzyme A on pigeon liver fatty acid synthetase. J. biol. Chem. 243: 3512–3516 (1968).

44 Dorsey, J. K. and Porter, J. W.: The inhibition of mevalonic kinase by geranyl and farnesyl pyrophosphates. J. biol. Chem. 243: 4667–4670 (1968).
45 Dugan, R. E.; Slakey, L. L.; Briedis, A. V., and Porter, J. W.: Factors affecting the diurnal variation in the level of β-hydroxy-β-methylglutaryl coenzyme A reductase and cholesterol-synthesizing activity in rat liver. Arch. Biochem. Biophys. 152: 21–27 (1972).
46 Edwards, A. M. and Elliott, W. H.: Induction of δ-aminolevulinic acid synthetase in perfused rat liver by drugs, steroids, lead and adenosine-3′, 5′-monophosphate; in Pollak and Lee Biochemistry of gene expression in higher organisms, pp. 369–378 (Austr. & N. Z. Book Co., Sydney 1973).
47 Edwards, A. M. and Sabine, J. R.: Regulation of δ-aminolevulinic acid synthetase in Morris hepatomas and livers of ethionine-treated rats (submitted for publication, 1973).
48 Edwards, P. A.: Effect of adrenalectomy and hypophysectomy on the circadian rhythm of β-hydroxy-β-methylglutaryl coenzyme A reductase activity in rat liver. J. biol. Chem. 248: 2912–2917 (1973).
49 Edwards, P. A. and Gould, R. G.: Turnover rate of hepatic 3-hydroxy-3-methylglutaryl coenzyme A reductase as determined by use of cycloheximide. J. biol. Chem. 247: 1520–1524 (1972).
50 Elwood, J. C. and Morris, H. P.: Lack of adaptation in lipogenesis by hepatoma 9121. J. Lipid Res. 9: 337–341 (1968).
51 Emmelot, P. and Bos, C. J.: Differences in the association of two glycolytic enzymes with plasma membranes isolated from rat liver and hepatoma. Biochim. biophys. Acta 121: 434–436 (1966).
52 Emmelot, P. and Bos, C. J.: Studies on plasma membranes. XIV. Adenyl cyclase in plasma membranes isolated from rat and mouse livers and hepatomas, and its hormone sensitivity. Biochim. biophys. Acta 249: 285–292 (1971).
53 Engelman, D. M. and Rothman, J. E.: The planar organization of lecithin-cholesterol bilayers. J. biol. Chem. 247: 3694–3697 (1972).
54 Fenselau, A. and Wallis, K.: Ketone body oxidation by mouse hepatoma BW7756. Life Sci. 12: 185–191 (1973).
55 Fillios, L. C.; Yokono, O.; Pronczuk, A.; Gore, I.; Satoh, T., and Kobayakawa, K.: Synthesis and distribution of cholesterol, and the effect of diet, at the liver endoplasmic reticula and plasma membranes from lean or obese rats. J. Nutr. 98: 105–112 (1969).
56 Foster, D. W. and Srere, P. A.: Citrate cleavage enzyme and fatty acid synthesis. J. biol. Chem. 243: 1926–1930 (1968).
57 Goldfarb, S. and Pitot, H. C.: The regulation of β-hydroxy-β-methylglutaryl coenzyme A reductase in Morris hepatomas 5123C, 7800, and 9618A. Cancer Res. 31: 1879–1882 (1971).
58 Goldfarb, S. and Pitot, H. C.: Stimulatory effect of dietary lipid and cholestyramine on hepatic HMGCoA reductase. J. Lipid Res. 13: 797–801 (1972).
59 Gould, R. G.; Kojola, V. B., and Swyryd, E. A.: Effects of hypophysectomy, adrenalectomy, cholesterol feeding, and puromycin on the radiation-induced increase in hepatic cholesterol biosynthesis in rats. Radiation Res. 41: 57–69 (1970).

60 GOULD, R. G. and SWYRYD, E. A.: Sites of control of hepatic cholesterol biosynthesis. J. Lipid Res. 7: 698–707 (1966).
61 GRANNER, D. K.: Protein kinase: altered regulation in a hepatoma cell line deficient in adenosine 3', 5'-cyclic monophosphate-binding protein. Biochem. biophys. Res. Commun. 46: 1516–1522 (1972).
62 GRANNER, D.; CHASE, L. R.; AURBACH, G. D., and TOMKINS, G. M.: Tyrosine aminotransferase: enzyme induction independent of adenosine 3', 5'-monophosphate. Science 162: 1018–1020 (1968).
63 GREENSPAN, M. and LOWENSTEIN, J. M.: Effect of magnesium ions and adenosine triphosphate on the activity of acetyl co-enzyme A carboxylase. Arch. Biochem. Biophys. 118: 260–263 (1967).
64 GREGOLIN, C.; RYDER, E.; WARNER, R. C.; KLEINSCHMIDT, A. K., and LANE, M. D.: Liver acetyl CoA carboxylase: the dissociation-reassociation process and its relation to catalytic activitiy. Proc. nat. Acad. Sci., Wash. 56: 1751–1758 (1966).
65 GUDER, W.; NOLTE, I., and WIELAND, O.: The influence of thyroid hormones on β-hydroxy-β-methylglutaryl-coenzyme A reductase of rat liver. Europ. J. Biochem. 4: 273–278 (1968).
66 HAMPRECHT, B.; NÜSSLER, C., and LYNEN, F.: Rhythmic changes of hydroxymethylglutaryl coenzyme A reductase activity in livers of fed and fasted rats. FEBS Letters 4: 117–121 (1969).
67 HARRY, D. S.; DINI, M., and MCINTYRE, N.: Effect of cholesterol feeding and biliary obstruction on hepatic cholesterol biosynthesis in the rat. Biochim. biophys. Acta 296: 209–220 (1973).
68 HARRY, D. S.; MORRIS, H. P., and MCINTYRE, N.: Cholesterol biosynthesis in transplantable hepatomas: evidence for impairment of uptake and storage of dietary cholesterol. J. Lipid Res. 12: 313–317 (1971).
69 HELLER, R. A. and GOULD, R. G.: Solubilization and partial purification of hepatic 3-hydroxy-3-methylglutaryl coenzyme A reductase. Biochem. biophys. Res. Commun. 50: 859–865 (1973).
70 HESTON, W. E. and VLAHAKIS, G.: C3H-A^{vy} – a high hepatoma and high mammary tumor strain of mice. J. nat. Cancer Inst. 40: 1161–1166 (1968).
71 HICKMAN, P. E.; HORTON, B. J., and SABINE, J. R.: Effect of adrenalectomy on the diurnal variation of hepatic cholesterogenesis in the rat. J. Lipid Res. 13: 17–22 (1972).
72 HIGGINS, M.; KAWACHI, T., and RUDNEY, H.: The mechanism of the diurnal variation of hepatic HMGCoA reductase activity in the rat. Biochem. biophys. Res. Commun. 45: 138–144 (1971).
73 HINZ, H.-J. and STURTEVANT, J. M.: Calorimetric investigation of the influence of cholesterol on the transition properties of bilayers formed from synthetic L-α-lecithins in aqueous suspension. J. biol. Chem. 247: 3696–3700 (1972).
74 HORTON, B. J.; HICKMAN, P. E., and SABINE, J. R.: The effect of diet on the diurnal variation of cholesterol synthesis in rat liver. Life Sci. 9: 1409–1417 (1970).

75 HORTON, B. J.; HORTON, J. D., and PITOT, H. C.: Abnormal cholesterol uptake, storage, and synthesis in the livers of 2-acetylaminofluorene-fed rats. Cancer Res. 33: 460–464 (1973).

76 HORTON, B. J.; HORTON, J. D., and SABINE, J. R.: Metabolic controls in precancerous liver. II. Loss of feedback control of cholesterol synthesis, measured repeatedly in vivo, during treatment with the carcinogens N-2-fluorenylacetamide and aflatoxin. Europ. J. Cancer 8: 437–443 (1972).

77 HORTON, B. J.; HORTON, J. D., and SABINE, J. R.: Metabolic controls in precancerous liver. V. Loss of control of cholesterol synthesis during feeding of the hepatocarcinogen 3′-methyl-4-dimethylaminoazobenzene. Europ. J. Cancer 9: 513–516 (1973).

78 HORTON, B. J.; MOTT, G. E.; PITOT, H. C., and GOLDFARB, S.: Rapid uptake of dietary cholesterol by hyperplastic liver nodules and primary hepatomas. Cancer Res. 33: 460–464 (1973).

79 HORTON, B. J. and SABINE, J. R.: Metabolic controls in precancerous liver: defective control of cholesterol synthesis in rats fed N-2-fluorenylacetamide. Europ. J. Cancer 7: 459–465 (1971).

80 HORTON, B. J. and SABINE, J. R.: Metabolic controls in precancerous liver. III. Further studies on the control of lipid synthesis during N-2-fluorenylacetamide feeding. Europ. J. Cancer 9: 1–9 (1973).

81 HORTON, B. J. and SABINE, J. R.: Metabolic controls in precancerous liver. IV. Loss of feedback control of cholesterol synthesis and impaired cholesterol uptake in ethionine-fed rats. Europ. J. Cancer 9: 11–17 (1973).

82 JAMES, M. and SABINE, J. R.: Corticosterone and the control of cholesterol synthesis in rat liver (submitted for publication, 1973).

83 JANSEN, G. R.; ZANETTI, M. E., and HUTCHISON, C. F.: Studies on lipogenesis in vivo. Effects of starvation and re-feeding, and studies on cholesterol synthesis. Biochem. J. 99: 333–340 (1966).

84 KABARA, J. J.; CHAPMAN, B. B., and BORIN, B. M.: Effect of hypocholesteremic drugs on tumor-bearing mice. Proc. Soc. exp. Biol. Med. 139: 100–104 (1972).

85 KANDUTSCH, A. A. and HANCOCK, R. L.: Regulation of the rate of sterol synthesis and the level of β-hydroxy-β-methylglutaryl coenzyme A reductase activity in mouse liver and hepatomas. Cancer Res. 31: 1396–1401 (1971).

86 KANDUTSCH, A. A. and PACKIE, R. M.: Comparison of the effects of some C_{27}-, C_{21}-, and C_{19}-steroids upon hepatic sterol synthesis and hydroxymethylglutaryl-coA reductase activity. Arch. Biochem. Biophys. 140: 122–130 (1970).

87 KANDUTSCH, A. A. and SAUCIER, S. E.: Prevention of cyclic and Triton-induced increases in hydroxymethylglutaryl coenzyme A reductase and sterol synthesis by puromycin. J. biol. Chem. 244: 2299–2305 (1969).

88 KAWACHI, T. and RUDNEY, H.: Solubilization and purification of β-hydroxy-β-methylglutaryl coenzyme A reductase from rat liver. Biochemistry 9: 1700–1705 (1970).

89 KENNAN, A. L.: Modifications of the portacaval shunt in the rat. J. appl. Physiol. 20: 1357–1358 (1965).

90 KOPELOVICH, L. and SABINE, J. R.: Control of lipid metabolism in hepatomas: effects of fasting and dietary fat on the activities of several glycolytic and

Krebs-cycle enzymes in mouse liver and hepatoma BW7756. Biochim. biophys. Acta *202:* 269–276 (1970).

91 KORNACKER, M. S. and LOWENSTEIN, J. M.: Citrate and the conversion of carbohydrate into fat. The activities of citrate-cleavage enzyme and acetate thiokinase in livers of starved and re-fed rats. Biochem. J. *94:* 209–215 (1965).

92 KREBS, H. A.; BENNETT, D. A. H.; GASQUET, P. de; GASCOYNE, T., and YOSHIDA, T.: The effect of diet on the gluconeogenic capacity of rat-kidney-cortex slices. Biochem. J. *86:* 22–27 (1963).

93 KRISHNAIAH, K. V. and RAMASARMA, T.: Regulation of hepatic cholesterolgenesis by ubiquinone. Biochim. biophys. Acta *202:* 332–342 (1970).

94 KROES, J.; OSTWALD, R., and KEITH, A.: Erythrocyte membranes – compression of lipid phases by increased cholesterol content. Biochim. biophys. Acta *274:* 71–74 (1972).

95 KROES, R.; WILLIAMS, G. M., and WEISBURGER, J. H.: Early appearance of serum α-fetoprotein during hepatocarcinogenesis as a function of age of rats and extent of treatment with 3′-methyl-4-dimethylaminoazobenzene. Cancer Res. *32:* 1526–1532 (1972).

96 LAKSHMANAN, M. R.; NEPOKROEFF, C. M.; NESS, G. C.; DUGAN, R. E., and PORTER, J. W.: Stimulation by insulin of rat liver β-hydroxy-β-methylglutaryl coenzyme A reductase and cholesterol-synthesizing activities. Biochem. biophys. Res. Commun. *50:* 704–710 (1973).

97 LAKSHMANAN, M. R.; NEPOKROEFF, C. M., and PORTER, J. W.: Control of the synthesis of fatty-acid synthetase in rat liver by insulin, glucagon, and adenosine 3′:5′cyclic monophosphate. Proc. nat. Acad. Sci., Wash. *69:* 3516–3519 (1972).

98 LANE, M. D.; MOSS, J.; RYDER, E., and STOLL, E.: The activation of acetyl CoA carboxylase by tricarboxylic acids. Adv. Enzyme Reg. *9:* 237–251 (1971).

99 LINN, T. C.: The demonstration and solubilization of β-hydroxy-β-methylglutaryl coenzyme A reductase from rat liver microsomes. J. biol. Chem. *242:* 984–989 (1967).

100 LINN, T. C.: The effect of cholesterol feeding and fasting upon β-hydroxy-β-methylglutaryl coenzyme A reductase. J. biol. Chem. *242:* 990–993 (1967).

101 MAJERUS, P. W.; JACOBS, R.; SMITH, M. B., and MORRIS, H. P.: The regulation of fatty acid biosynthesis in rat hepatomas. J. biol. Chem. *243:* 3588–3595 (1968).

102 MAJERUS, P. W. and KILBURN, E.: Acetyl coenzyme A carboxylase. The roles of synthesis and degradation in regulation of enzyme levels in rat liver. J. biol. Chem. *244:* 6254–6262 (1969).

103 MATTEIS, F. de: Disturbances of liver porphyrin metabolism caused by drugs. Pharmacol. Rev. *19:* 523–557 (1967).

104 MCGARRY, J. D. and FOSTER, D. W.: Ketogenesis and cholesterol synthesis in normal and neoplastic tissues of the rat. J. biol. Chem. *244:* 4251–4256 (1969).

105 MCNAMARA, D. J.; QUACKENBUSH, F. W., and RODWELL, V. W.: Regulation of hepatic 3-hydroxy-3-methylglutaryl coenzyme A reductase. J. biol. Chem. *247:* 5805–5810 (1972).

106 Misconi, L.; Bornstein, J.; Armstrong, J. McD., and Waters, M.: Biochemical activities of a synthetic peptide related to the diabetogenic peptide from growth hormone. Proc. austr. biochem. Soc. *4:* 59 (1971).

107 Moss, J.; Yamagishi, M.; Kleinschmidt, A. K., and Lane, M. D.: Acetyl coenzyme A carboxylase. Purification and properties of the bovine adipose tissue enzyme. Biochemistry *11:* 3779–3786 (1972).

108 Muto, Y. and Gibson, D. M.: Selective dampening of lipogenic enzymes of liver by exogenous polyunsaturated fatty acids. Biochem. biophys. Res. Commun. *38:* 9–15 (1970).

109 Numa, S.; Hashimoto, T.; Nakanishi, S., and Okazaki, T.: Regulatory mechanisms for liver acetyl coenzyme A carboxylase. Biochem. Soc. Symp. *35:* 27–39 (1972).

110 Numa, S.; Nakanishi, S.; Hashimoto, T.; Iritani, N., and Okazaki, T.: Role of acetyl coenzyme A carboxylase in the control of fatty acid synthesis. Vitamins Hormones, N. Y. *28:* 213–243 (1970).

111 Ogilvie, J. W. and Kaplan, B. H.: The inhibition of sterol biosynthesis in rat liver homogenates by bile. J. biol. Chem. *241:* 4722–4730 (1966).

112 Paleg, L. and Sabine, J. R.: Inhibition by plant growth retardants of cholesterol biosynthesis in slices of rat liver and hepatoma. Austr. J. biol. Sci. *24:* 1125–1130 (1971).

113 Pitot, H. C.: Some biochemical aspects of malignancy. Annu. Rev. Biochem. *35:* 335–368 (1966).

114 Porter, J. W.; Kumar, S. and Dugan, R. E.: Synthesis of fatty acids by enzymes of avian and mammalian species. Progr. biochem. Pharmacol. *6:* 1–101 (1971).

115 Potter, V. R.: Summary of discussion on neoplasms. Symp. The Developmental Biology of Neoplasia. Cancer Res. *28:* 1901–1907 (1968).

116 Pronczuk, A. and Fillios, L. C.: Changes in cholesterol concentration in rough and smooth endoplasmic reticulum and polysomal profiles in rats fed cholesterol. J. Nutr. *96:* 46–52 (1968).

117 Rabin, B. R.: Chemical carcinogenesis and the interaction of membranes with polysomes. May and Baker Lab. Bull. *10:* 22–26 (1972).

118 Ramasarma, T.: Biogenetic interrelationship of ubiquinone and cholesterol. Biochem. Soc. Symp. *35:* 245–256 (1972).

119 Regen, D.; Riepertinger, C.; Hamprecht, B., and Lynen, F.: The measurement of β-hydroxy-β-methyl-glutaryl-CoA reductase in rat liver; effects of fasting and refeeding. Biochem. Z. *346:* 78–84 (1966).

120 Ritter, M. C. and Dempsey, M. E.: Purification and characterization of a naturally occurring activator of cholesterol biosynthesis from $\Delta^{5,7}$-cholestadienol and other precursors. Biochem. biophys. Res. Commun. *38:* 921–929 (1970).

121 Rodwell, V. W.: Cholesterol regulation of hepatic HMG-CoA reductase. Biochem. Soc. Symp. *35:* 295–301 (1972).

122 Rothman, J. E. and Engelman, D. M.: Molecular mechanism for the interaction of phospholipid with cholesterol. Nature new Biol. *237:* 42–44 (1972).

123 SABINE, J. R.: Control of cholesterol synthesis in hepatomas: the effect of bile salts. Biochim. biophys. Acta 176: 600–604 (1969).
124 SABINE, J. R.: Defective control of cholesterol synthesis and the development of liver cancer: a review; in WOOD Tumor lipids: biochemistry and metabolism, pp. 21–23 (Amer. Oil Chem. Soc. Press, Champaign, Ill. 1973).
125 SABINE, J. R.; ABRAHAM, S., and CHAIKOFF, I. L.: Lack of feedback control of fatty acid synthesis in a transplantable hepatoma. Biochim. biophys. Acta 116: 407–409 (1966).
126 SABINE, J. R.; ABRAHAM, S., and CHAIKOFF, I. L.: Control of lipid metabolism in hepatomas: insensitivity of rate of fatty acid and cholesterol synthesis by mouse hepatoma BW7756 to fasting and to feedback control. Cancer Res. 27: 793–799 (1967).
127 SABINE, J. R.; ABRAHAM, S., and MORRIS, H. P.: Defective dietary control of fatty acid metabolism in four transplantable rat hepatomas: numbers 5123C, 7793, 7795, and 7800. Cancer Res. 28: 46–51 (1968).
128 SABINE, J. R. and CHAIKOFF, I. L.: Control of fatty acid synthesis in homogenate preparations of mouse hepatoma BW 7756. Austr. J. exp. Biol. med. Sci. 45: 541–548 (1967).
129 SABINE, J. R.; HORTON, B. J., and HICKMAN, P. E.: Control of cholesterol synthesis in hepatomas: absence of diurnal rhythm in hepatomas 7794A and 9618A. Europ. J. Cancer 8: 29–32 (1972).
130 SABINE, J. R.; HORTON, B. J., and TAN, C. S.: Control of lipid synthesis in cancerous and precancerous liver. Proc. 10th Int. Cancer Conf., p. 282 (1970).
131 SABINE, J. R.; HORTON, B. J., and WICKS, M. B.: Spontaneous tumors in C3H-A^{vy} and C3H-A^{vy}fB mice: high incidence in the United States and low incidence in Australia. J. nat. Cancer Inst. 50: 1237–1243 (1973).
132 SABINE, J. R.; KOPELOVICH, L.; ABRAHAM, S., and MORRIS, H. P.: Control of lipid metabolism in hepatomas: conversion of glutamate carbon to fatty acid carbon via citrate in several transplantable hepatomas. Biochim. biophys. Acta 296: 493–498 (1973).
133 SABINE, J. R.; MCGRATH, H., and ABRAHAM, S.: Dietary fat and the inhibition of hepatic lipogenesis in the mouse. J. Nutr. 98: 312–318 (1969).
134 SABINE, J. R. and MORRIS, H. P.: Control of glucose metabolism in transplantable kidney tumors: low gluconeogenesis by tumors MK-1 and MK-2. Biochim. biophys. Acta 208: 203–207 (1970).
135 SABINE, J. R. and STONEHILL, E. H.: Appearance of 'tumor-specific' embryonal antigens in pretumorous tissue. Proc. austr. biochem. Soc. 5: 44 (1972).
136 SACCONE, G. T. P. and SABINE, J. R.: The effect of fasting and alloxandiabetes on fatty-acid synthesis and levels of long-chain acyl coenzyme A and cyclic 3'-5'-adenosinemonophosphate in rat liver (submitted for publication, 1973).
137 SAKAKIDA, H.; SHEDIAC, C. C., and SIPERSTEIN, M. D.: Effect of endogenous and exogenous cholesterol on the feedback control of cholesterol synthesis. J. clin. Invest. 42: 1521–1528 (1963).
138 SAUER, F.: Fatty acid, cholesterol, and acetoacetate biosynthesis in liver homogenates from normal and starved guinea pigs. Canad. J. Biochem. Physiol. 38: 635–641 (1960).

139 SCALLEN, T. J.; SCHUSTER, M. W., and DHAR, A. K.: Evidence for a noncatalytic carrier protein in cholesterol biosynthesis. J. biol. Chem. *246:* 224–230 (1971).
140 SCHOTZ, M. C.; RICE, L. I., and ALFIN-SLATER, R. B.: Further studies on cholesterol in liver cell fractions of normal and cholesterol-fed rats. J. biol. Chem. *204:* 19–26 (1953).
141 SHAPIRO, D. J. and RODWELL, V. W.: Diurnal variation and cholesterol regulation of hepatic HMG-CoA reductase activity. Biochem. biophys. Res. Commun. *37:* 867–872 (1969).
142 SHAPIRO, D. J. and RODWELL, V. W.: Regulation of hepatic 3-hydroxy-3-methylglutaryl coenzyme A reductase and cholesterol synthesis. J. biol. Chem. *246:* 3210–3216 (1971).
143 SHEFER, S.; HAUSER, S.; LAPAR, V., and MOSBACH, E. H.: HMG CoA reductase of intestinal mucosa and liver of the rat. J. Lipid Res. *13:* 402–412 (1972).
144 SIPERSTEIN, M. D.: Comparison of the feeback control of cholesterol metabolism in liver and hepatomas; in Developmental and metabolic control mechanisms and neoplasia, pp. 427–451 (Williams & Wilkins, Baltimore 1965).
145 SIPERSTEIN, M. D.: Feedback control of cholesterol synthesis in normal, malignant and premalignant tissues. Proc. canad. Cancer Conf. *7:* 152–162 (1966).
146 SIPERSTEIN, M. D.: Deletion of the cholesterol negative feedback system in precancerous liver. J. clin. Invest. *45:* 1073 (1966).
147 SIPERSTEIN, M. D.: Regulation of cholesterol biosynthesis in normal and malignant tissues. Curr. Top. Cell Reg. *2:* 65–100 (1970).
148 SIPERSTEIN, M. D.: The relationship of cholesterol biosynthesis to cancer. Trans. amer. clin. climat. Ass. *83:* 156–164 (1971).
149 SIPERSTEIN, M. D. and FAGAN, V. M.: Studies on the feedback regulation of cholesterol synthesis. Adv. Enzyme Reg. *2:* 249–264 (1964).
150 SIPERSTEIN, M. D. and FAGAN, V. M.: Deletion of the cholesterol-negative feedback system in liver tumors. Cancer Res. *24:* 1108–1115 (1964).
151 SIPERSTEIN, M. D. and FAGAN, V. M.: Feedback control of mevalonate synthesis by dietary cholesterol. J. biol. Chem. *241:* 602–609 (1966).
152 SIPERSTEIN, M. D.; FAGAN, V. M., and MORRIS, H. P.: Further studies on the deletion of the cholesterol feedback system in hepatomas. Cancer Res. *26:* 7–11 (1966).
153 SIPERSTEIN, M. D. and GUEST, M. J.: Studies on the site of the feedback control of cholesterol synthesis. J. clin. Invest. *39:* 642–652 (1960).
154 SIPERSTEIN, M. D.; GYDE, A. M., and MORRIS, H. P.: Loss of feedback control of hydroxymethylglutaryl coenzyme A reductase in hepatomas. Proc. nat. Acad. Sci., Wash. *68:* 315–317 (1971).
155 SLAKEY, L. L.; CRAIG, M. C.; BEYTIA, E.; BRIEDIS, A.; FELDBRUEGGE, D. H.; DUGAN, R. E.; QURESHI, A. A.; SUBBARAYAN, C., and PORTER, J. W.: The effects of fasting, refeeding, and time of day on the levels of enzymes effecting the conversion of β-hydroxy-β-methylglutaryl-coenzyme A to squalene. J. biol. Chem. *247:* 3014–3022 (1972).
156 SMITH, S. and ABRAHAM, S.: Fatty acid synthesis in developing mouse liver. Arch. Biochem. Biophys. *136:* 112–121 (1970).

157 Sokoloff, L. and Rothblat, G. H.: Regulation of sterol synthesis in L cells: steady-state and transitional responses. Biochim. biophys. Acta *280:* 172–181 (1972).
158 Srere, P. A.: Palmityl-coenzyme A inhibition of the citrate-condensing enzyme. Biochim. biophys. Acta *106:* 445–455 (1965).
159 Stanier, R. Y. and Niel, C. B. van: The concept of a bacterium. Arch. Mikro: biol. *42:* 17–35 (1962).
160 Stein, O.; Stein, Y.; Goodman, D. S., and Fidge, N. H.: The metabolism of chylomicron cholesteryl ester in rat liver. A combined radioautographic-electron microscopic and biochemical study. J. Cell Biol. *43:* 410–431 (1969).
161 Stonehill, E. H. and Bendich, A.: Retrogenetic expression: the reappearance of embryonal antigens in cancer cells. Nature, Lond. *228:* 370–372 (1970).
162 Subba Rao, G. and Ramasarma, T.: Rhythmic activity of biogenesis of cholesterol. Envir. Physiol. *1:* 188–197 (1971).
163 Subba Rao, G. and Ramasarma, T.: ATP in the regulation of cholesterol biosynthesis – a supra-energetic role. Biochem. biophys. Res. Commun. *49:* 225–229 (1972).
164 Sugiyama, T.; Clinkenbeard, K.; Moss, J., and Lane, M. D.: Multiple cytosolic forms of hepatic β-hydroxy-β-methylglutaryl CoA synthetase: possible regulatory role in cholesterol synthesis. Biochem. biophys. Res. Commun. *48:* 255–261 (1972).
165 Swann, A. and Siperstein, M. D.: Distribution of cholesterol feedback control in the guinea pig. J. clin. Invest. *51:* 95a (1972).
166 Swanson, R. F.; Curry, W. M., and Anker, H. S.: The activation of rat liver acetyl-CoA carboxylase by trypsin. Proc. nat. Acad. Sci., Wash. *58:* 1243–1248 (1967).
167 Sweeney, M. J. and Ashmore, J.: Effects of acute insulin insufficiency on liver and adipose tissue fatty acid synthesis. Metabolism *14:* 516–522 (1965).
168 Takasugi, Y. and Imai, Y.: Distribution of lipids in subcellular fractions of fatty livers. I. Lipids in liver cell fractions of rats fed a high fat and cholesterol diet. J. Biochem., Tokyo *60:* 191–196 (1966).
169 Taketa, K. and Pogell, B. M.: The effect of palmityl coenzyme A on glucose 6-phosphate dehydrogenase and other enzymes. J. biol. Chem. *241:* 720–726 (1966).
170 Thompson, E. B.; Tomkins, G. M., and Curran, J. F.: Induction of tyrosine α ketoglutamate transaminase by steroid hormones in a newly established tissue culture cell line. Proc. nat. Acad. Sci., Wash. *56:* 296–303 (1966).
171 Thorne, R. F. W. and Bygrave, F. L.: Interaction of calcium with mitochondria isolated from Ehrlich ascites tumor cells. Biochem. biophys. Res. Commun. *50:* 294–299 (1973).
172 Tubbs, P. K. and Garland, P. B.: Variations in tissue contents of coenzyme A thio esters and possible metabolic implications. Biochem. J. *93:* 550–557 (1964).
173 Turinsky, J.: Study on plasma cyclic nucleotide concentrations in fasting rats. Proc. Soc. exp. Biol. Med. *142:* 1189–1191 (1973).

174 UNGER, R. H.; EISENTRAUT, A. M., and MADISON, L. L.: The effects of total starvation upon the levels of circulating glucagon and insulin in man. J. clin. Invest. *42:* 1031–1039 (1963).
175 VAGELOS, P. R.: Lipid metabolism. Annu. Rev. Biochem. *33:* 139–172 (1964).
176 WALLACE, J. C.; MORRIS, H. P., and SABINE, J. R.: Control of glucose metabolism in transplantable kidney tumours: insensitivity of rate of gluconeogenesis to fasting and diabetes (submitted for publication, 1973).
177 WATABE, H.: Early appearance of embryonic α-globulin in rat serum during carcinogenesis with 4-dimethylaminoazobenzene. Cancer Res. *31:* 1192–1194 (1971).
178 WATSON, J. A.: Regulation of lipid metabolism in *in vitro* cultured minimal deviation hepatoma 7288C. Lipids *7:* 146–155 (1972).
179 WATSON, J. A.: Regulation of 3β-hydroxy sterol synthesis in cultured hepatoma cells. Fed. Proc. *32:* 520a (1973).
180 WATSON, J. A.: Regulation of cholesterol synthesis in HTC cells (minimal deviation hepatoma 7288C); in WOOD Tumor lipids: biochemistry and metabolism, pp. 34–53 (Amer. Oil Chem. Soc. Press, Champaign, Ill. 1973).
181 WEIS, H. J. and DIETSCHY, J. M.: Failure of bile acids to control hepatic cholesterogenesis: evidence for endogenous cholesterol feedback. J. clin. Invest. *48:* 2398–2408 (1969).
182 WHITE, L. W. and RUDNEY, H.: Regulation of 3-hydroxy-3-methylglutarate and mevalonate biosynthesis by rat liver homogenates. Effects of fasting, cholesterol feeding, and Triton administration. Biochemistry *9:* 2725–2731 (1970).
183 WILLIAMS, D. J. and RABIN, B. R.: Disruption by carcinogens of the hormone dependent association of membranes with polysomes. Nature, Lond. *232:* 102–105 (1971).
184 YUGARI, Y.; MATSUDA, T., and SUDA, M.: Control of fatty acid synthesis by long-chain fatty acids in rat liver. Proc. 6th Int. Congr. Biochem. p. 602 (1964).
185 ZUCKERMAN, N. J.; NARDELLA, P.; MORRIS, H. P., and ELWOOD, J. C.: Lack of adaptation in lipogenesis by hepatomas 9098, 7794A, and 9618A. J. nat. Cancer Inst. *44:* 79–83 (1970).

Author's address: JOHN R. SABINE, Department of Animal Physiology, Waite Agricultural Research Institute, University of Adelaide, *Adelaide, S.A.* (Australia)

Dietary Fat in Relation to Tumorigenesis

K. K. Carroll[1] and H. T. Khor

Department of Biochemistry, University of Western Ontario, London, Ont.

Contents

I. Introduction	308
II. Studies with Experimental Animals	309
A. Tumors Enhanced by Dietary Fat	309
1. Skin Tumors in Mice	309
2. Mammary Tumors in Mice and Rats	313
3. Hepatomas in Mice and Rats	318
4. Other Tumors Enhanced by Dietary Fat	319
B. Tumors not Enhanced by Dietary Fat	319
C. Mode of Action of Dietary Fat	322
1. Cocarcinogenesis and the Two-Stage Theory of Tumor Induction	322
2. Effects on the Initiating Agent	323
3. Promoting Action of Dietary Fat	325
III. Epidemiological Data on Humans	329
IV. Discussion	339
V. Summary	344
VI. Acknowlegments	345
VII. References	345

I. Introduction

The role of nutrition in cancer has been a topic of continuing interest for many years and information on the subject has been reviewed on a number of occasions [21, 54, 96, 104, 117, 118, 120, 121, 133]. It is well known that tumor development can be influenced by caloric intake and

---

[1] Medical Research Associate of the Medical Research Council of Canada.

there is also evidence that particular components of the diet can have specific effects which are independent of caloric intake.

This article will be concerned with the role of dietary fat in tumorigenesis and will deal mainly with cocarcinogenic effects of dietary fats and oils. The possible existence of carcinogenic compounds in dietary fat will not be considered in any detail, although lipid-soluble compounds such as polycyclic hydrocarbons and various steroidal compounds are known to have carcinogenic properties [12, 21]. Over the years there has been considerable speculation on whether oxidation or polymerization of fats as a result of heating or other processing in food preparation may result in the formation of carcinogenic compounds, but the evidence from animal experiments is not very convincing [2, 4, 93]. There is likewise little evidence that dietary fat has much effect on tumor growth, but numerous experiments have shown that it can influence the genesis of certain types of neoplasms such as skin tumors and mammary tumors in mice and rats [18, 118, 120, 121].

More recently, there have been indications that dietary fat may also play a role in human carcinogenesis. Several authors have called attention to statistical evidence of a positive correlation between level of dietary fat and age-adjusted mortality from breast cancer in different countries of the world [18, 32, 82, 83, 130, 131]. Possible links between dietary fat and neoplasms of the large bowel [32, 59, 139, 140], pancreas [137] and prostate [138] have also been suggested. These observations lend added interest to the experiments with animals and serve as a stimulus to further work on the effects of dietary fat.

The purpose of this review is to summarize the work on effects of dietary fat on tumorigenesis in animals, to present epidemiological data on dietary fat intake in relation to cancer mortality in humans, and to discuss these findings with regard to possible mechanisms by which dietary fat may exert an effect on tumorigenesis.

II. Studies with Experimental Animals

A. Tumors Enhanced by Dietary Fat

1. Skin Tumors in Mice

The first experiments to show clearly that dietary fat can influence tumorigenesis in animals were carried out by WATSON and MELLANBY [125], who used skin tumors induced in mice by repeated application of coal tar. They reported that addition of 12.5–25 % butter to a basal diet of bread and oats caused a marked increase in tumor incidence. They also observed greater numbers of associated nodules in the lung tissues of the animals on the high fat diet. Subsequent studies by BAUMANN and co-workers [5, 6, 77] showed that dietary fat enhanced the incidence of skin tumors induced by exposure to ultraviolet light or by repeated application of carcinogenic hydrocarbons. Confirmatory evidence was also provided by TANNENBAUM [114]. Results obtained by these workers are summarized in table I. It can be seen that the tumor incidence was consistently higher on the high fat diets although the differential was not as great in TANNENBAUM's experiments as in those carried out by the other workers.

For most of the experiments listed in table I, partially hydrogenated cottonseed oil was used as the main source of fat. JACOBI and BAUMANN [67] also compared the effects of coconut oil, wheat germ oil, Crisco, butter and lard at levels of 15–25 % of the diet and found that addition of any one of these fats decreased the time required for 50 % of the mice to develop tumors after treatment with 3,4-benzpyrene, methylcholanthrene or 1,2,5,6-dibenzanthracene. Subsequently, LAVIK and BAUMANN [78] found that corn oil, coconut oil and lard all gave much the same enhancement of tumor incidence when fed as 10 % of the diet (table II). Similar results were obtained when the oils were first heated to 300° C for 1 h, although in an earlier experiment [77] heated Primex (partially hydrogenated cottonseed oil) appeared to be more effective than the untreated oil. Irradiation or peroxidation of the Primex did not seem to influence its effect on tumor incidence.

LAVIK and BAUMANN [77] also attempted to determine which components of fat were responsible for the enhancement of tumorigenesis. They reported that ethyl laurate was as effective as Primex when fed as 15 % of the diet, while glycerol (5 % of the diet) or unsaponifiable material (2 % of the diet), equivalent to a level of about 30 % Primex, gave a lower tumor incidence. In later experiments [78], they prepared the fatty acids from Primex free of unsaponifiable matter and reconstituted them as triglycerides. These were compared with high and low melting fractions of lard and all three gave similar tumor incidence. Results of these experiments are summarized in table II.

The effect on incidence of skin tumors in mice of feeding high fat diet at different times in relation to treatment with carcinogen was investigated by both LAVIK and BAUMANN [77] and TANNENBAUM [115]. In each case

Table I. Effects of dietary fat on incidence of skin tumors in mice

Strain[1] and sex	Carcinogenic agent	Frequency of treatment	Type of fat	Level in diet, %	Length of experiment	Tumor incidence[2] (%) low fat[3]	Tumor incidence[2] (%) high fat	References
	coal tar	twice weekly for 120 days	butter		330 days		(57)[4]	Watson and Mellanby [125]
		twice weekly for 90 days		12.5–25	240 days	(34)[4]	(82)[4]	
				19–25		(59)[4]		
A ♂	UV light	1 h daily	Primex[6]	30	7.5 months	(55)[5]	(75)[5]	Baumann and Rusch [6]
C	UV light	1 h daily	Crisco[6]	25	4.5 months	5/16 (33)	12/14 (87)	Baumann et al. [5]
	benzpyrene (0.3–0.5% sol. in benzene)	twice weekly	Crisco	25	5.5 months	12/18 (67)	15/16 (94)	
	methylcholanthrene (0.2–0.3% sol. in dioxane)	twice weekly for 2 months	coconut oil	15	4 months	1/15 (6)	10/12 (83)	Lavik and Baumann [77]
			Primex	15	6 months	3/17 (17)	18/21 (86)	
Jax Swiss ♀	benzpyrene (0.3% sol. in benzene)	32 applications over 20 weeks	Kremit[6]	28	42 weeks	22/43 (51)	28/42 (67)	Tannenbaum [114]
C57BL ♂		twice weekly for 13 weeks	Kremit	31	49 weeks	13/49 (27)	17/50 (35)	
DBA ♂		twice weekly for 10 weeks	Kremit	31	56 weeks	34/50 (68)	39/50 (78)	

1 Strain and sex were not specified in some cases.
2 Tumor incidence is expressed in most cases as the percentage of animals with tumors out of the total number alive when the first tumor appeared. The original papers frequently give tumor incidences at various times after the start of treatment with carcinogenic agent.
3 Various natural diets containing not more than 2–3% fat were used. Cod liver oil was given in some cases.
4 Percentage of surviving mice with tumors estimated from charts. 70 mice per group used in the first experiment and 90 per group in the second experiment.
5 From 28 to 37 out of 50 mice were alive at the end of these experiments.
6 Partially hydrogenated cottonseed oil.

Table II. Effects of different components of fats and of fats subjected to various treatments on incidence of skin tumors in mice

Diet	Carcinogenic treatment[1]	Tumor incidence[2]	(%)	References
Experiment 1				
LAVIK and BAUMANN				
Control (low fat)	0.3% methylchol-	0/17	(0)	[77]
10% Primex (untreated)	anthrene in	5/16	(31)	
10% Primex (oxidized)	dioxane	6/21	(28)	
10% Primex (irradiated)		8/20	(40)	
10% Primex (heated)		12/20	(60)	
Experiment 2				
LAVIK and BAUMANN				
Control (low fat)	0.2% methylchol-	6/21	(28)	[78]
10% corn oil	anthrene in	15/21	(71)	
10% corn oil (heated)	dioxane	17/24	(71)	
10% coconut oil (untreated)		16/24	(66)	
10% coconut oil (heated)		16/23	(70)	
10% lard (untreated)		11/18	(61)	
10% lard (heated)		11/16	(69)	
Experiment 3				
LAVIK and BAUMANN				
Control (low fat)	0.3% methylchol-	5/45	(11)	[77]
15% Primex	anthrene in	9/15	(60)	
15% ethyl laurate	dioxane	12/19	(63)	
5% glycerol		8/19	(42)	
2% unsaponifiable from Primex		7/20	(35)	
Experiment 4				
LAVIK and BAUMANN				
Control (low fat)	0.2% methylchol-	1/22	(4)	[78]
10% lard m.p.>37°C	anthrene in	7/23	(30)	
10% lard m.p.<37°C	dioxane	6/24	(25)	
10% reconstituted triglycerides from Primex		6/22	(27)	

1 Carcinogen was applied twice weekly for 2 months.
2 Time of observation was 4 months in each case.

Table III. Effect of time of feeding high fat diet on incidence of skin tumors in mice

Strain and sex	Diet[1]	Duration of carcinogen treatment	Time of high fat feeding	Length of experiment	Tumor incidence[2] (%)
Not specified	control (low fat)	0–2 months	none	6 months	3/17 (17)
	15% Primex	0–2 months	0–2 months	6 months	5/17 (29)
	15% Primex	0–2 months	1.5–3 months	6 months	9/15 (60)
	15% Primex	0–2 months	2–6 months	6 months	9/21 (43)
	15% Primex	0–2 months	0–6 months	6 months	18/21 (86)
C57BL ♂	31% Kremit	13 weeks	0–13 weeks	49 weeks	18/50 (36)
	31% Kremit	13 weeks	13–49 weeks	49 weeks	23/49 (47)
DBA ♂	control (low fat)	10 weeks	none	56 weeks	34/50 (68)
	31% Kremit	10 weeks	0–10 weeks	56 weeks	32/50 (64)
	31% Kremit	10 weeks	10–56 weeks	56 weeks	41/50 (82)
	31% Kremit	10 weeks	0–56 weeks	56 weeks	39/50 (78)

1 The experiments with Primex were carried out by LAVIK and BAUMANN [77] and those with Kremit by TANNENBAUM [115].
2 Results expressed as in table I.

it was found that the high fat diet was more effective when fed after the carcinogen treatment than when it was fed only during treatment (table III).

2. Mammary Tumors in Mice and Rats

TANNENBAUM [114, 116] showed that dietary fat also increased the incidence of mammary tumors developing spontaneously in DBA and C3H mice, and this was later confirmed by SILVERSTONE and TANNENBAUM [106]. More recently, SZEPSENWOL [111] reported that mice of the TM strain showed a much higher incidence of mammary tumors when their diet, which consisted of the Rockland rat diet, was supplemented with an ether extract of whole powdered egg or an alcohol extract of egg yolk. He also found that supplementation with cholesterol and lard gave a higher incidence of mammary adenocarcinomas, whereas supplementation with cholesterol alone did not.

LAVIK and BAUMANN [78] failed to observe any effect of dietary fat on spontaneous mammary tumors in rats as a result of feeding hydrogenated vegetable oil, but later experiments by BENSON *et al.* [7] showed a higher incidence of spontaneous tumors when olive oil was added to the diet. DAVIS et al. [29] also observed a higher incidence of spontaneous tumors in rats on a semisynthetic diet containing 16 % corn oil, compared to rats on a standard laboratory chow diet. In other studies, dietary fat was reported to enhance the incidence of mammary tumors induced in rats by implanting pellets of stilbestrol [33] or by feeding 2-acetylaminofluorene [38], but in each case only small numbers of animals were used. More recent experiments in our laboratory [19, 20, 46] showed that dietary fat increased the incidence of mammary tumors produced in rats by a single oral dose of 7,12-dimethylbenz(a)anthracene (DMBA). Some results from the above experiments are summarized in table IV.

A number of different fats and oils were used for these studies on mammary tumors. The experiments with mice were carried out with partially hydrogenated cottonseed oil, egg lipids or lard, while fats such as lard, olive oil, corn oil and partially hydrogenated cottonseed oil were found to increase the incidence of mammary tumors in rats. On the basis of such results and the findings with skin tumors referred to earlier, it was concluded that enhancement of tumor incidence was related to amount rather than type of dietary fat [54, 67, 118].

On the other hand, experiments carried out in our laboratory with mammary tumors induced in rats with DMBA suggested that type of fat may also be important. In the first experiments (table IV), 20 % corn oil gave a higher tumor incidence than 20 % coconut oil and if one considers the total tumor yield, the difference was even more striking. The 21 tumor-bearing rats on high corn oil diet developed a total of approximately 100 tumors, whereas only about half that number developed on either the low fat diet or the high coconut oil diet. Further studies were subsequently carried out with a series of different fats and oils fed as 20 % of the diet and it was found that the rats on unsaturated fats generally developed more tumors than those on saturated fats (fig. 1). Although a lower dose of DMBA was used for these experiments, the tumor incidence was still 80 % or more in each of the dietary groups and the higher tumor yields with unsaturated fats were largely due to an increased number of tumors per rat. This illustrates the importance of using as many different criteria as possible for assessing effects on tumorigenesis. TANNENBAUM and SILVERSTONE [119], using a large dose of carcinogen which effectively obliterated the difference in incidence of skin tumors, were still able to observe a significant decrease in the length of the latent period in animals on high fat diet as compared to low fat diet.

Table IV. Effects of dietary fat on incidence of mammary tumors in female mice and rats

Species and strain	Carcinogenic agent	Type of fat	Level in diet, %	Length of experiment	Tumor incidence[1] low fat[2]	Tumor incidence[1] high fat	References
Mouse DBA	none	Kremit[3]	12	38 weeks to death	14/44 (32)	24/44 (55)	Tannenbaum [114]
			12	24 weeks to 2 years	16/50 (32)	32/50 (64)	
Rat	none	Primex[3]	30	30 months	3/33 (9)	2/32 (6)	Lavik and Baumann [78]
Rat A × C line 9935	Stilbestrol (4–6 mg implanted s.c. in cholesterol pellets)	Crisco[3]	46	av. survival 383–392 days	9/12 (75)	12/12 (100)[4] 11/12 (92)	Dunning et al. [33]
Rat AES strain	2-acetylamino-fluorene (0.03% in diet)	lard	15	av. survival 28–29 weeks	0/8 (0)	46/62 (74)	Engel and Copeland [38]
			26	31 weeks	1/6 (17)	6/6 (100)	
Rat Sprague-Dawley	none	olive oil	20	10 months	3/25 (12)[5]	5/13 (39)[5]	Benson et al. [7]
Rat Sprague-Dawley	7,12-dimethyl-benz(a)anthracene (DMBA) (10 mg orally)	corn oil	20	4 months	15/21 (71)	21/22 (96)	Gammal et al. [46]
		coconut oil	20			16/21 (76)	

1 Results expressed as in table I.
2 Various natural and purified diets were used. The level of fat in the basal diet varied from 0.5 to 3% except for the experiments of Dunning et al. [33] where the level was 6.5%.
3 Partially hydrogenated cottonseed oil.
4 The first group was pair-fed with the low fat controls and the second group was fed *ad libitum*.
5 Only animals over 18 months of age were included in the evaluation because no tumors developed in younger animals.

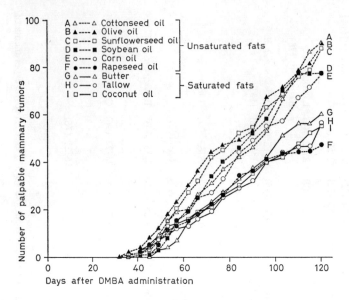

Fig. 1. Effect of different dietary fats and oils fed as 20 % by weight of the diet on cumulative number of palpable mammary tumors in female Sprague-Dawley rats given a single oral dose of 5 mg of DMBA at 50 days of age. 30 rats per group.

The effect of progressively increasing the fat in the diet was investigated by SILVERSTONE and TANNENBAUM [106]. They concluded that the rate of tumor formation was not arithmetically proportional to the fat intake, since increasing the proportion of dietary fat from 2 to 6–8 % gave about the same enhancement as increasing the proportion from 6 to 8 up to 24–26 %. However, the overall difference in response as a result of increasing the dietary fat from 2 to 20 % or more is frequently of borderline significance in any one experiment, and it is therefore difficult to be certain about the linearity of response. In our experiments with induced mammary tumors in rats, a diet containing 20 % corn oil consistently gave a higher tumor yield than diets containing 5 % corn oil or less, but the yields with a 10 % corn oil diet tended to fluctuate between these two extremes.

The relative effectiveness of feeding a high fat diet before or after administration of the carcinogen was also investigated in connection with our studies on rat mammary tumors induced by DMBA [19]. The results indicated that enhancement of the tumor yield only occurred when the

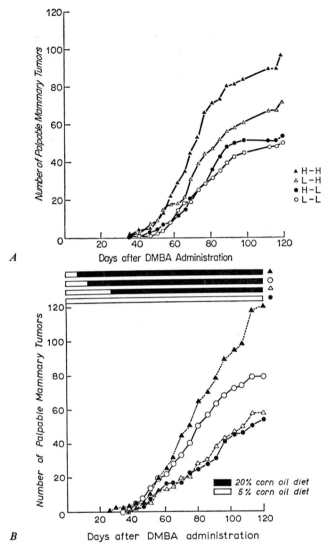

Fig. 2. Cumulative number of palpable mammary tumors in female Sprague-Dawley rats fed low or high corn oil diets before and after administration of DMBA. Diets were fed to groups of 30 rats from the time of weaning and the DMBA was given as a single oral 5 mg dose at 50 days of age. In *A*, H denotes a 20-percent corn oil diet and L a 0.5-percent corn oil diet. The first diet indicated was fed until 2 days before giving the DMBA, the rats were then transferred to commercial feed for 3 days, after which they were fed the second diet indicated. In *B*, the rats were fed a 5-percent corn oil diet until 1, 2 or 4 weeks after the DMBA was given and then transferred to a 20-percent corn oil diet. The control group received a 5-percent corn oil diet throughout the experiment.

high fat diet was fed after the carcinogen had been given (fig. 2A). In subsequent experiments [unpublished] it was found that the tumor yield was still enhanced if rats were transferred to the high corn oil diet one or two weeks after giving the DMBA, but if the transfer was delayed for four weeks little or no enhancement was seen (fig. 2B).

3. Hepatomas in Mice and Rats

Incidental to studies on induced skin tumors in male C3H mice, TANNENBAUM [116] observed a higher incidence of spontaneous hepatomas on high fat compared to low fat diet. In two out of three experiments carried out subsequently by SILVERSTONE and TANNENBAUM [107] the incidence of hepatomas was again higher on high fat diet (table V).

MLLER and MILLER [90] summarized the considerable number of experiments dealing with effects of dietary fat on hepatomas induced by carcinogenic aminoazo dyes. The most clear-cut observation from these experiments was that rats fed a diet containing 0.06 % p-dimethylaminoazobenzene developed a high incidence of hepatomas when the diet contained 5 % corn oil, whereas only a few tumors developed when the diet contained 5 % hydrogenated coconut oil [89]. Increasing the corn oil content from 5 to 20 % of the diet enhanced tumor formation, but diets containing 20 % of Crisco or lard gave results similar to those obtained with 5 % corn oil [73]. Subsequent studies provided evidence that these

Table V. Effect of dietary fat on incidence of spontaneous hepatomas in male C3H mice

Experiment	Tumor incidence (%)				References
	low fat[1]		high fat[1]		
1	3/32	(9)	12/34	(35)	TANNENBAUM [116]
2	22/50	(44)	22/47	(47)	SILVERSTONE and
3	17/44	(39)	24/43	(56)	TANNENBAUM [107]
4	18/50	(36)	24/47	(51)	

1 The basal low fat diets contained approximately 2% fat. From 18 to 21% of partially hydrogenated cottonseed oil was added to the high fat diets. Natural food diets were used in experiments 1 and 2 and purified diets in experiments 3 and 4.

effects were related to hepatic levels of riboflavin, higher levels being associated with a lower tumor incidence [91].

4. Other Tumors Enhanced by Dietary Fat

SILBERBERG and SILBERBERG [105] reported in an abstract that addition of 25 % lard to the diet enhanced the incidence of hypophyseal tumors produced by injection of 400 μCi of ^{131}I. In one experiment with groups of 50 DBA male mice the incidence was increased from 45 to 84 %, and in a second experiment with C57B male mice it increased from 25 to 57 %.

In his experiments on mice of the TM strain, SZEPSENWOL [111] found that supplementation of their Rockland rat diet with egg lipid extracts increased the incidence of lung adenocarcinomas and lymphosarcomas as well as mammary tumors. Supplementation with cholesterol and lard likewise gave a high incidence of lung cancer. Subsequently, he also reported an increased incidence of intracranial tumors in mice on fat enriched diets [112].

B. Tumors not Enhanced by Dietary Fat

In contrast to epitheliomas induced by repeated application of carcinogenic hydrocarbons to the skin, sarcomas induced by subcutaneous injection of these same hydrocarbons seem to be unaffected by dietary fat. The latter finding has been well documented by BAUMANN and co-workers [5, 78] and by TANNENBAUM [114] in experiments with both mice and rats (table VI).

Some of TANNENBAUM's experiments were carried out with strains of mice that normally develop primary lung tumors and these were found in about equal numbers in the low and high fat groups (table VI). In the earlier study by WATSON and MELLANBY [125], it was reported that lung tumors occur more frequently on high fat diet, but only mice bearing skin tumors were examined and the lung tumors were of both metastatic and primary origin. The incidence reported by TANNENBAUM, on the other hand, was based on lung tumors in mice free of other tumors and thus presumed to be primary rather than metastatic.

LAWRASON and KIRSCHBAUM [79] investigated the effect of dietary fat on spontaneous and induced leukemia in mice (table VI). Spontaneous

Table VI. Tumors not influenced by dietary fat

Type of tumor	Species and strain	Carcinogenic agent	Method of administration	Type of fat	Level in diet, %	Length of experiment	Tumor incidence[1] (%) low fat[2]	Tumor incidence[1] (%) high fat	Reference
Sarcoma	mouse strain C	benzpyrene	0.5 mg s.c. in corn oil	Crisco	25	5 months	(85)[3]	(80)[3]	BAUMANN et al. [5]
		methylcholanthrene	0.5 mg s.c. in corn oil	Crisco	25	5 months	(90)	(92)	
		dibenzanthracene	1.25 mg s.c. in corn oil	Crisco	25	8 months	(50)	(75)	
Sarcoma	mouse JAX Swiss ♀	3,4-benzpyrene	0.15 mg s.c. in lard	Kremit	28	52 weeks	19/39 (49)	12/40 (30)	TANNENBAUM [114]
	JAX ABC ♀	3,4-benzpyrene	0.1 mg s.c. in lard	Kremit	28	51 weeks	7/40 (18)	6/37 (16)	
Sarcoma	mouse	methylcholanthrene	0.02 mg s.c. in corn oil	Primex	15	7 months	4/23 (17)	5/30 (16)	LAVIK and BAUMANN [78]
	rat	methylcholanthrene	1 mg s.c.	Primex	30	7 months	5/8 (62)	5/8 (62)	
	rat	methylcholanthrene	2 mg s.c.	Primex	30	10 months	5/6 (83)	6/7 (85)	
Lung	mouse JAX Swiss ♀	3,4-benzpyrene[4]	0.5 mg in benzene 32 applications over 20 weeks	Kremit	28	52 weeks[5]	8/15 (52)	4/9 (44)	TANNENBAUM [114]
	JAX Swiss ♀	3,4-benzpyrene[4]	0.15 mg s.c. in lard	Kremit	28	62 weeks[5]	4/15 (27)	4/16 (25)	
	JAX ABC ♀	3,4-benzpyrene[4]	0.1 mg s.c. in lard	Kremit	28	60 weeks[5]	10/29 (34)	13/36 (36)	

Tumor	Animal	Carcinogen	Administration	Fat	% fat	Time	High fat	Low fat	Reference
Leukemia	mouse strain F DBA subline 12 ♂	none	—	Kremit	32	700 days	29/48 (60)	27/48 (56)	Lawrason and Kirschbaum [79]
		methylcholanthrene	twice weekly percutaneously as 0.5% sol. in benzene	Kremit	32		21/33 (64)	16/22 (73)	
	♀	methylcholanthrene	twice weekly percutaneously as 0.5% sol. in benzene	Kremit	32		12/15 (80)	14/27 (52)	
Submaxillary gland	rat	methylcholanthrene	1 mg s.c.	Primex	30	10 months	2/5 (40)	2/5 (40)	Lavik and Baumann [78]
Ear duct	rat	2-acetylaminofluorene	0.03% in diet	lard	15	28–29 weeks[6]	5/8 (63)	29/62 (47)	Engel and Copeland [38]
Ear duct	rat	2-acetylaminofluorene	0.03% in diet	lard	26	31 weeks[6]	6/6 (100)	5/6 (83)	

1 Results expressed as in table I.
2 The low fat diets contained no more than 3% fat.
3 20 mice per group at the start. Many of the animals given dibenzanthracene died before tumors developed.
4 Primary lung tumors develop spontaneously in these strains of mice.
5 Mean time of appearance of tumors.
6 Average survival time.

leukemia in strain F mice tended to appear somewhat earlier in the high fat group, but the overall incidence was no greater. In this experiment the growth was more rapid and the caloric intake higher on the high fat diet. With induced leukemia in DBA mice there was again no difference in overall incidence in the two groups, but the onset was significantly delayed in the high fat group. Food intake and rate of growth were similar in both groups.

LAVIK and BAUMANN [78] observed no effect of dietary fat on submaxillary gland tumors induced in rats with methylcholanthrene, but only a small number of animals were used. In experiments by ENGEL and COPELAND [38] the incidence of tumors of the ocular orbit induced by 2-acetylaminofluorene appeared to decrease when the content of fat in the diet was increased. The animals grew much better on the high fat diet even when they were pair-fed. A low fat diet in which most of the protein was supplied by alcohol-extracted peanut meal actually gave the highest incidence of tumors (6 out of 11) but no corresponding high fat diet was fed. The rats used in these experiments also developed tumors of the ear duct, and the incidence of this type of tumor appeared to be unaffected by the level of dietary fat (table VI).

C. Mode of Action of Dietary Fat

1. Cocarcinogenesis and the Two-Stage Theory of Tumor Induction
The results of many experiments with skin tumors have given rise to the concept that tumor induction is a discontinuous process involving two or more discrete stages [8, 10, 13, 34, 99]. The first stage, initiation, involves an interaction with a carcinogenic agent, in which normal tissue cells are converted into potentially neoplastic cells. Initiation is considered to be a rapid process, since it can occur after only one treatment with a carcinogen; and to be essentially irreversible, since initiated cells can be stimulated to develop into tumors long after the initial contact with a carcinogenic agent [8, 100]. The second or developmental stage covers the period from the initial transformation until the cells divide and multiply to form a discrete tumor. It is at this stage that the process of tumor induction appears to be most susceptible to environmental influences [97]. This period may, in fact, involve more than one stage in the development of tumor cells [13]. There may also be a progressive selection of cell populations for characteristics such as resistance to cytotoxicity, autono-

mous growth and malignancy [39–41] and once a tumor has formed, it may progress to more malignant types by further permanent, irreversible changes in one or more of the characters of its cells [43, 44].

The initial stimulus in tumorigenesis is generally referred to as the initiating action and any subsequent stimulus which leads to tumor formation is referred to as promoting action [45]. Carcinogens may act as promotors as well as initiators, but promoting agents may be incapable of producing tumors in the absence of an initial carcinogenic stimulus. These concepts have been derived primarily from work with skin tumors, but there is a certain amount of evidence that more than one stage is also involved in the development of other types of tumor [100].

BERENBLUM [9] has defined cocarcinogenic action as any form of augmentation of tumor induction brought about by some added factor other than the carcinogen and has summarized the different types of modifying influence in cocarcinogenesis as follows: An *additive or synergistic action,* when the modifying agent possesses carcinogenic activity of its own; an *'incomplete' carcinogenic action,* when the modifying agent is responsible for only one phase of carcinogenesis, initiation or promotion; a *preparative action,* which renders the target organ more responsive to the carcinogenic action; a *permissive influence,* affecting the scope of action of the carcinogen by altering its absorption, distribution or metabolism; an *influence on viral action* involved in the carcinogenic process, even when the initial stimulus is a nonviral carcinogen; or a *conditional influence on the tumor* concerned with factors such as hormonal dependence of the tumor or immunological resistance of the host.

As noted earlier, there is little evidence to date that dietary fats are themselves carcinogenic or that they render target organs more responsive to the initiating action of a carcinogen. Their possible influence on viral action does not appear to have been investigated. The following discussion will therefore be mainly concerned with possible effects of dietary fats on the carcinogenic agent involved and with ways in which dietary fats may exert a promoting effect on the process of tumorigenesis.

2. Effects on the Initiating Agent

In the early experiments with skin tumors, WATSON and MELLANBY [125] observed that mice on high fat diets had a characteristic fatty or oily coat. They suggested that this might enhance tumor incidence by facilitating absorption of the carcinogenic tar, since they observed that enhancement could also be obtained by local application of fats or oils to the skin, whereas the opposite effect was observed when the skin was pretreated with petroleum ether to remove natural fats. Further evidence for a local effect of dietary fat on skin tumorigenesis was obtained by BAUMANN and co-workers [67, 78].

Although it seemed that fat could act locally to facilitate passage of a carcinogen into the epidermis or prevent its loss from the skin by mechanical means, other findings indicated that this could not account entirely for the effect of dietary fat on skin tumorigenesis. LAVIK and BAUMANN [78] were able to increase tumor incidence by adding as little as 7% fat to the diet or by emulsifying fat in the drinking water; conditions under which no appreciable greasiness of the skin was observed. In experiments carried out by TANNENBAUM [115], the mice did not have an oily appearance even when high levels of fat were fed, possibly because the fat was better absorbed into the dietary mixture. In this case the fat did not appear to have as great an augmenting effect but there was nevertheless a definite enhancement of tumor yield and reduction of the latent period.

The effects of dietary fat on incidence of hepatomas induced by aminoazo dyes appear to be related to liver levels of riboflavin, which in turn may be involved in metabolism of these dyes [91]. The possibility that dietary fat may exert an effect on yields of DMBA-induced mammary tumors by altering the distribution or metabolism of the carcinogen has also been considered [46]. However, when the concentration of DMBA was measured in mammary tissue of rats on different diets at various times after giving the carcinogen, the small differences observed did not seem adequate to explain the differences in tumor yield [47]. It was also found in later studies [CARROLL and KHOR, unpublished experiments] that high corn oil diet effectively increased the yield of mammary tumors even when it was instituted one to two weeks after giving the carcinogen, by which time the concentration of DMBA had decreased to very low levels.

It might be argued that the measurements of tissue levels of carcinogen do not constitute valid evidence since they were based on concentration of DMBA in the tissue as a whole. Most of this would be present in the adipose tissue and the values may not accurately reflect the level in epithelial cells from which the tumors arise. Recent studies by JANSS *et al.* [68] have shown that tritiated DMBA associates with DNA in mammary parenchymal cells and can persist in that form for 2 weeks or more after it is administered to rats. Thus, the possibility that dietary fat exerts a direct effect on the carcinogenic agent cannot be ruled out completely.

3. Promoting Action of Dietary Fat

Promoting action is defined as an augmentation of tumor induction which occurs when an added factor is administered after completion of the initiating action but not when the sequence is reversed [9]. Since high fat diets appear to exert an effect on tumorigenesis when administered after the carcinogenic agent (table III, fig. 2), dietary fat may be considered to exert a promoting action on skin tumors and mammary tumors. Both LAVIK and BAUMANN [78] and TANNENBAUM [114, 115] considered this possibility, which they referred to as a systemic or cocarcinogenic action exerted during the developmental stage of tumorigenesis.

The above definition does not specify the mechanisms involved in promoting action and these presumably can vary from one situation to another. The experiments of BERENBLUM and SHUBIK [11] on promotion by croton oil of skin tumors induced by DMBA, indicated that the croton oil could still promote tumors as long as 43 weeks after treatment with the carcinogen. In our experiments with mammary cancer, on the other hand, dietary fat appeared to be ineffective unless it was fed within the first few weeks after giving the DMBA. Possibly this is related in some way to the stage of development of the animal, which may be more important in the case of the mammary gland. It is known that the ability of 3-methylcholanthrene (3-MC) to induce mammary tumors in rats decreases quite rapidly as they grow older [63].

In postulating a cocarcinogenic action of dietary fat, TANNENBAUM [114, 115, 118] suggested metabolic stimulation of potential tumor cells as a possible mechanism of action, and the report of DUNNING et al. [33] provides some evidence that dietary fat can exert a stimulatory effect on the mammary gland. In rats implanted with pellets containing stilbestrol, they observed that the mammary glands from rats on high fat diet showed histological evidence of more hypertrophy and secretory activity than those from rats on diets of lower fat content. The tumor incidence was also higher in the rats on high fat diet (table IV).

The development of normal mammary glands and mammary tumors are both strongly influenced by hormonal effects. Estrogens and prolactin seem to be most important in this regard [88, 92, 108] although tumorigenesis may also be influenced by other hormones such as progesterone [66]. An attempt to investigate possible effects of dietary fat on estrogen metabolism was carried out in our laboratory by injecting tritiated estradiol into rats on diets containing either 5 or 20 % corn oil and determining the distribution of radioactivity in different tissues. The results of

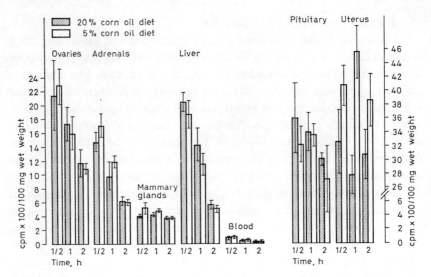

Fig. 3. Effects of dietary fat on distribution of radioactivity in female Sprague-Dawley rats following intravenous injection of tritiated estradiol-17β (0.2 μg containing 2×10^6 cpm in 1 ml of saline containing 5% ethanol). The diets were fed from 50 days of age and the injections were given between 90 and 100 days of age with the rats in diestrus. Each value represents the mean $\pm$ SEM for 5 rats.

a typical experiment are illustrated in figure 3. The trend toward lower levels of activity in the adrenals and mammary glands of rats on the 20% corn oil diet was also seen in two other experiments, but the only statistically significant decrease ($p < 0.05$) was seen in the uterus 1 h after injection of the labeled estradiol.

The general pattern of distribution of estradiol in different tissues is well-known from studies in other laboratories [69-72] and administration of unlabeled estradiol was found to decrease the accumulation of radioactivity in tissues such as the uterus and mammary gland [36, 37, 101]. The lower levels of radioactivity seen in our experiments in rats on the 20% corn oil diet might therefore be simply an indication of higher levels of endogenous estradiol.

Although hormones appear to be essential for the production of mammary tumors, hormonal stimulation can also effectively inhibit tumor formation [25, 26]. Pregnancy ensuing soon after a single dose of 3-methylcholanthrene caused a marked reduction in tumor yield [27]. Induction of mammary tumors by DMBA was inhibited by administration of estradiol and progesterone 15 days after carcinogen treatment and by stimulation of the ovary with gonadotrophin to increase ovarian steroid hormone production [64]. The work of MEITES and co-workers [22, 88, 128] indicated that the stimulated mammary gland is relatively refractory to carcinogenic action. Thus, if dietary fat acts by stimulating the mammary gland, one might expect some inhibition of tumor incidence as a result of feeding high fat diet prior to giving the carcinogen, but such does not seem to be the case (fig. 2A) [19].

In considering possible ways in which dietary fat could influence developing tumor cells, it must be kept in mind that only certain types of tumor respond to this stimulus. The two that have been studied most extensively are skin tumors and mammary tumors. Diets high in fat have consistently been shown to enhance the yields of these two types of tumor, but it is a matter of conjecture whether the same mechanisms are involved in each case. The experiments with mammary tumors have of necessity been carried out with female rats and mice, whereas male mice were used in many of the experiments on skin tumors. This does not rule out the possibility that dietary fat may act by altering the hormonal environment but it seems unlikely that the same hormones would be involved in both cases.

In the mammary gland, the associated adipose tissue appears to have a significant influence on the growth and development of the glandular epithelium. Successful transplantation and subsequent growth seem to be dependent on the presence of adipose tissue, and development occurs only within the framework of the fat pads [61]. It is therefore conceivable that tumorigenesis may be influenced by alterations in the composition or metabolism of adipose tissue. The fatty acid composition of mammary tissue can be altered considerably by the nature of the dietary fat [46], and it is also known that diets high in fat have an inhibitory effect on fatty acid biosynthesis in adipose tissue [53]. However, corn oil and coconut oil were found to be equally effective at inhibiting fatty acid biosynthesis in lactating mammary gland [23], whereas in our experiments, corn oil appeared to have a greater effect on mammary tumorigenesis (table IV, fig. 1).

High fat diets have a high caloric content and since it is well-known that tumorigenesis can be inhibited by restriction of caloric intake, careful consideration has been given to the possibility that observed effects of dietary fat may be related to differences in caloric intake [118]. However, paired feeding experiments have shown in a number of cases that dietary

Table VII. Effect of caloric intake versus fat intake on incidence of skin tumors and mammary tumors in mice

Caloric intake, cal/day	Type of fat	Fat intake, % of diet		Tumor incidence (%)		References
		low	high	low fat	high fat	
Skin tumors						
12	corn oil	2[1]	8.5[1]	29/54 (54)	33/50 (66)	LAVIK and
8	corn oil	2.9[1]	12.9[1]	0/29 (0)	8/28 (28)	BAUMANN [78]
9.7	corn oil	1.6	12.3	42/48 (87)	30/48 (63)	RUSCH et al.
6.7	corn oil	2	17.1	2/29 (7)	9/38 (24)	[97]
14.3–14.4	Kremit	2	18	43/50 (86)	46/50 (92)	TANNENBAUM
10.0–10.1	Kremit	2	18	18/49 (37)	27/49 (55)	[116]
8.5	Kremit	2	18	9/45 (20)	14/48 (29)	
10	corn oil (low fat)	2	27.2	28/44 (64)	34/47 (72)	BOUTWELL et al. [14]
9.5	Primex (high fat)	2	60.5	25/39 (64)	36/46 (78)	
6	Primex (high fat)	3.4	29.1	8/44 (18)	14/42 (33)	
Mammary tumors						
11.7–11.9	Kremit	2	18	35/60 (58)	27/30 (90)	TANNENBAUM
10.0–10.1	Kremit	2	18	12/30 (40)	20/30 (67)	[116]
9.2	Kremit	2	18	4/28 (14)	18/30 (60)	
8.5	Kremit	2	18	0/29 (0)	5/29 (17)	
7.8	Kremit	2	18	0/28 (0)	1/28 (4)	

[1] Mice received 60 and 240 mg of corn oil per day on the low and high fat diets, respectively.

fat has an effect on both skin tumors and mammary tumors in mice over and above that due to differences in caloric intake (table VII).

In our studies with rats treated with DMBA, groups on high fat diets were usually slightly heavier at the end of the experiment, but there appeared to be little difference in caloric intake between the different groups [46]. Analysis of tumor incidence within any particular dietary group also failed to indicate a tendency toward increased incidence in the larger animals of the group. BOUTWELL et al. [14] attempted to explain the effect of dietary fat on the basis that efficiency of utilization of food energy increases with increasing fat content of the diet, but the validity of this argument was challenged by SILVERSTONE and TANNENBAUM [106].

Other considerations make it seem unlikely that the observed effects of dietary fat are simply a result of increased caloric intake. Only certain types of tumors seem to respond to dietary fat, whereas caloric restriction appears to have a general inhibitory effect on tumorigenesis, although it affects some tumors more than others [118, 129]. In our experiments, rats fed unsaturated fats developed more tumors than those on saturated fats (fig. 1) and it seems unlikely that this could be explained on the basis of differences in caloric intake.

Lipid peroxidation may be a factor in tumor enhancement by polyunsaturated fats [103, 126], but in our experiments olive oil gave a relatively high tumor yield in spite of its rather low content of polyunsaturated fatty acids. Rapeseed oil, on the other hand, failed to stimulate tumorigenesis although it contains a higher percentage of linoleic acid than olive oil [20]. In this case the results may be influenced by the high content of eicosenoic acid (20:1) and erucic acid (22:1) in rapeseed oil.

The experiments with skin tumors were carried out with a number of different types of dietary fat (table I). There was no indication that one type of fat was more effective than another, but this was not investigated systematically. LAVIK and BAUMANN [77, 78] provided evidence that the effect was mediated by fatty acids rather than the nonsaponifiable fraction of the fat. They also considered the possibility that products of incomplete metabolism of fat might be the agents responsible for the increased tendency toward tumor formation, but feeding trials with ketone bodies provided no evidence to support this hypothesis. Increasing the riboflavin or protein content of the diet also failed to counteract the tumor-promoting action of dietary fat on skin tumors.

III. Epidemiological Data on Humans

The evidence from animal experiments that underfeeding reduces tumor incidence led TANNENBAUM [113] to review studies on human insurance statistics dealing with body weight in relation to cancer mortality. In 6 of 7 surveys covered by his review, the figures indicated an increase in cancer mortality with increasing body weight. For the most part these surveys dealt with male policy holders and gave no breakdown of cancer at different sites. In one survey of women based on 'standard experience', the relationship between cancer mortality rate and increasing weight was most apparent in the older age group, especially for cancer of the female genital organs and of the intestines. There was no distinct trend according to weight for cancer of the breast, stomach, liver and gallbladder.

TANNENBAUM pointed out the many complicating factors in the collection and assessment of this kind of data. In particular, the weight was measured only at the time the policy was issued and many years might elapse before death ensued. Other factors such as variations in age and hereditary characteristics, medical selection and lack of adequate confirmation of the cause of death all combine to lend a high degree of uncertainty to any conclusions based on such data. The general conclusion that overweight individuals have an increased incidence of cancer does not seem warranted [133]. There does, however, seem to be an association between obesity and certain types of cancer such as cancer of the endometrium [132] and cancer of the gallbladder [136]. Obesity has also been reported to be slightly more frequent in breast cancer patients than in controls [122, 136].

Although body weight may be influenced by caloric intake, it is not necessarily related to level of fat in the diet and recent evidence of a correlation between fat intake and mortality from certain types of tumor is more relevant to the subject of the present review. LEA [80] observed a correlation between geographical latitude and mortality from breast cancer, and in looking for an explanation of this observation, noted first a correlation with environmental temperature and subsequently a correlation ($p < 0.001$) between dietary fat intake and mortality from breast cancer [81, 82]. He also reported a significant positive correlation between fat consumption and death rates from neoplasms of the pancreas, ovary, prostate, bladder, intestines, rectum, leukemia in the age group over 55, Hodgkin's disease and lymphosarcoma [82, 83]. For many of these tumors the death rates were also positively correlated with consumption of sugar, animal protein, eggs and milk and with incidence of coronary disease. On the other hand, another group of tumors, which included neoplasms of the stomach, liver, and body of the uterus, showed a negative correlation with consumption of fats and oils, with other dietary components and with coronary disease [83].

During extensive studies of epidemiological data on cancer in humans, WYNDER et al. [134] observed a correlation between breast cancer and

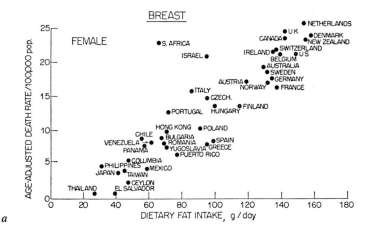

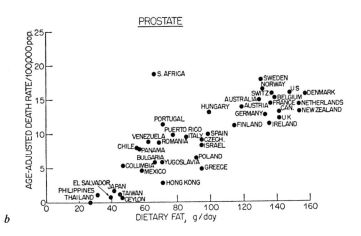

Fig. 4. Positive correlation between per caput consumption of dietary fat [42] and age-adjusted mortality from cancer at certain sites [102]. The values for dietary fat are averages for 1964–66 and those for cancer mortality are for 1964–65, except in a few cases where data were only available for 1960–61 or 1962–63. The cancer mortality for the United States was estimated from values given for US white and US non-white and that for the UK from values given for England and Wales, Scotland and Northern Ireland. According to SEGI *et al.* [102], the cancer mortality for South Africa is based on the European population only and that for Israel on the Jewish population only.

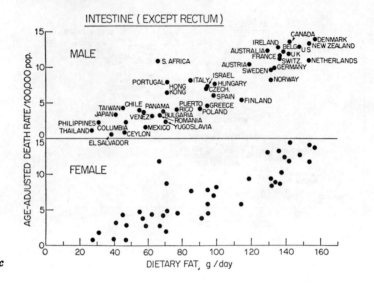

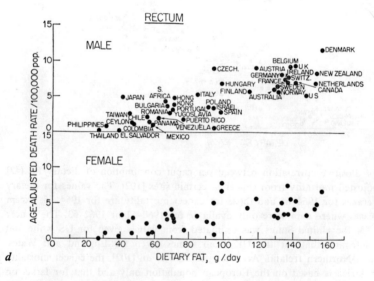

Fig. 4. Continuation

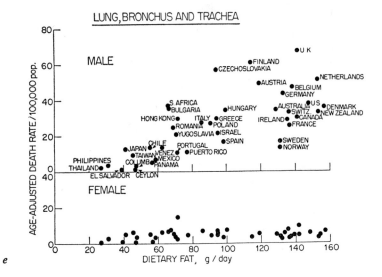

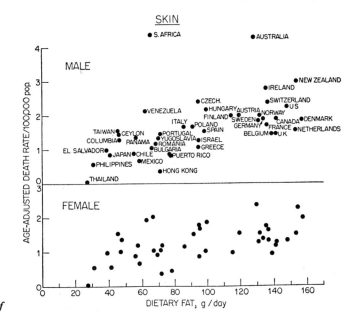

Fig. 4. Continuation

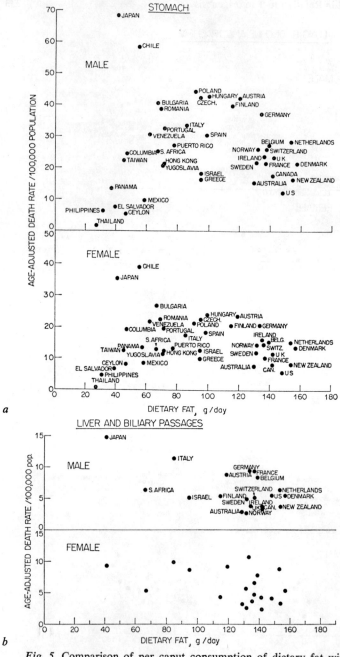

Fig. 5. Comparison of per caput consumption of dietary fat with age-adjusted death rate from malignant neoplasm of the stomach and of the liver and biliary passages. See caption to figure 4 for details.

Table VIII. Correlation coefficients between per capita consumption of dietary fat in various countries of the world and age-adjusted mortality in those countries from malignant neoplasms at different sites

Site	Sex	Correlation coefficient	Site	Sex	Correlation coefficient
Breast	F	+0.935	bladder and other urinary organs	F	+0.345[1]
Intestine (except rectum)	M	+0.928	buccal cavity and pharynx	M	+0.260
Intestine (except rectum)	F	+0.911	lung, bronchus and trachea	F	+0.252
Prostate		+0.892	larynx	M	+0.170
Leukemia and aleukemia	M	+0.857	esophagus	M	+0.088
Leukemia and aleukemia	F	+0.838	thyroid	F	+0.029
Rectum	M	+0.834	thyroid	M	+0.014
Rectum	F	+0.786	stomach	M	+0.010
Ovary, fallopian tube and broad ligament		+0.726[1]	esophagus	F	−0.014
			uterus (all parts)		−0.064
Lung, bronchus and trachea	M	+0.710	stomach	F	−0.112
			buccal cavity and pharynx	F	−0.245
Pancreas	M	+0.666[1]	larynx	F	−0.276
Skin	M	+0.634	liver and biliary passages	F	−0.487[1]
Skin	F	+0.550			
Bladder and other urinary organs	M	+0.523[1]	liver and biliary passages	M	−0.676[1]
Pancreas	F	+0.501			

1 These values are based on data for 17–19 countries. All other values are based on data from 40 countries.

cancer of the colon and suggested that both might be influenced by dietary factors, perhaps in terms of fat intake and/or related socioeconomic variables. In more recent publications, WYNDER and co-workers have again drawn attention to the correlation between level of dietary fat and incidence of breast cancer [130, 131] and cancer of the colon and rectum [135, 139] and have suggested that 'overnutrition' probably has an adverse effect on a variety of different types of cancer [136]. The role of

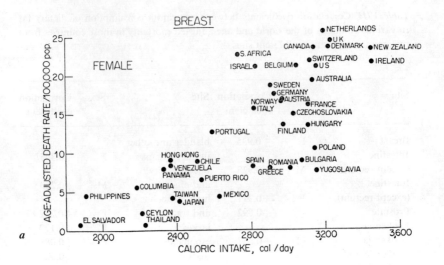

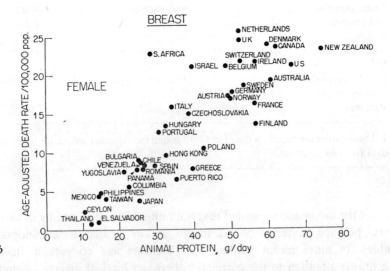

Fig. 6. Plots of age-adjusted death rate from malignant neoplasm of female breast against per caput consumption of calories, protein and animal protein. The values for caloric intake and protein intake are those given by FAO [42]. See caption to figure 4 for details of the cancer mortality data.

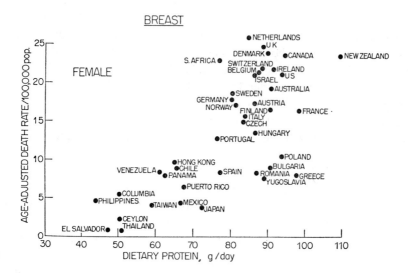

dietary fat in breast cancer has been reviewed by DE WAARD [122], and HEMS [55] has carried out statistical analyses which suggest that the influence of fat intake is more important in breast cancer late in life.

Our own studies on the effect of dietary fat on incidence of mammary tumors induced in rats with DMBA led us independently to consider the role of dietary fat in human breast cancer [18]. When the data of SEGI et al. [102] for mortality from breast cancer in different countries are plotted against the fat intake for these countries published by the Food and Agriculture Organization of the United Nations [42], there is a strong positive correlation (r = +0.935), as shown in figure 4. Only two countries, South Africa and Israel, deviate appreciably from the general trend and these results are biased by the fact that the cancer mortality for South Africa is based on the European population only and that for Israel on the Jewish population only, whereas the fat intake is calculated for each country as a whole. The age-adjusted incidence of breast cancer in the white population of South Africa is much greater than in the non-white and Jews in Israel show a much higher incidence than non-Jews [30].

There is also a strong positive correlation with cancer of the prostate and with cancer of the intestine and of the rectum in both males and females (fig. 4). Some other types, such as lung cancer in males and skin cancer in both males and females, also show a positive correlation, while cancer at a number of other sites shows only a random scatter. Liver cancer shows a tendency towards a negative correlation (table VIII). Plots for stomach cancer and liver cancer are shown in figure 5. Leukemia and aleukemia combined show a strong positive correlation with dietary fat intake for both males and females (table VIII), but when this is broken down into age groups (under 15, 15–54 and above 55 years of age) only the group over age 55 shows the strong positive correlation. The cancer incidence for the different age groups, however, is based on data for only slightly more than half the total number of countries. These observations, based on more recent data, are in general agreement with those reported earlier by Lea [83].

Breast cancer mortality also shows a strong positive correlation with caloric intake (fig. 6), but the correlation is not as strong ($r = +0.796$) as with fat intake. There is even less correlation with total protein intake ($r = +0.737$), but the correlation with animal protein is strongly positive ($r = +0.898$). These results are indicative of positive correlations between caloric intake, fat intake and protein intake and the degree of correlation for 40 of the countries from which Segi's data were collected is shown in table IX. It is interesting to note the lack of correlation of total fat with either vegetable fat or carbohydrates and the negative correlation between total fat and vegetable protein.

Interest in the possible role of dietary fat in human carcinogenesis has been stimulated by a recent report of Pearce and Dayton [94]. On analysis of their data from an 8-year controlled clinical trial of a diet high in polyunsaturated vegetable fat as a means of preventing complications of atherosclerosis, they observed more deaths due to cancer in the experimental group than in a control group on normal diet. This study was carried out with a total of 846 men in an older age group and the increased number of carcinomas was not localized to any particular site. These findings stimulated a reinvestigation of data from other long-term dietary trials involving the feeding of polyunsaturated fat, but no correlation with cancer mortality was observed [35].

Other existing data also fail to show evidence of an association between cancer mortality and intake of unsaturated fat. The data in table IX show a strong positive correlation between total fat intake and animal fat intake but there is no significant correlation with vegetable fat, which is presumably more unsaturated than animal fat. Furthermore, it has been estimated that Americans and Japanese have similar intakes of dietary linoleic acid. Adipose tissue from Japanese actually contains a higher proportion of linoleic acid than that from Americans [65],

Table IX. Correlation coefficients between dietary variables[1]

	Total fat	Animal protein	Animal fat	Total calories	Total protein	Vegetable fat	Carbohydrate
Animal protein	+0.946 $p<0.01$						
Animal fat	+0.939 $p<0.01$	+0.938 $p<0.01$					
Total calories	+0.844 $p<0.01$	+0.791 $p<0.01$	+0.811 $p<0.01$				
Total protein	+0.792 $p<0.01$	+0.796 $p<0.01$	+0.691 $p<0.01$	+0.943 $p<0.01$			
Vegetable fat	+0.188 n.s.	+0.032 n.s.	−0.141 n.s.	+0.102 n.s.	+0.243 n.s.		
Carbohydrates	−0.263 n.s.	−0.258 n.s.	−0.184 n.s.	+0.256 n.s.	+0.248 n.s.	−0.224 n.s.	
Vegetable protein	−0.329 $p<0.05$	−0.410 $p<0.01$	−0.403 $p<0.01$	+0.149 n.s.	+0.226 n.s.	+0.144 n.s.	+0.787 $p<0.01$

1 Based on Food Balance Sheets [42] for 40 countries (excluding South Africa) from which cancer mortality data were collected by SEGI et al. [102].

although the death rate from breast cancer, prostatic cancer and intestinal cancer is much higher in Americans than in Japanese (fig. 4) [102].

IV. Discussion

The epidemiological data on humans show a positive correlation between dietary fat intake and mortality from cancer at about half the sites reported (table VIII), and the correlation is particularly strong for cancer of the breast, intestinal cancer and cancer of the prostate. However, great caution must be exercised in drawing conclusions from such data, since there are many possible sources of error in epidemiological studies on humans.

In the present instance, the figures on food consumption in different countries are based on figures for production, gross exports, gross imports and changes in stocks, with allowance for factors such as use for animal feed and for seed, and losses due to processing and waste. These figures, which are themselves subject to error, only give an estimate of the food available and this cannot be equated with the net food consumed. The average per caput consumption will also be influenced by factors such as the age distribution and eating habits of diverse groups within the country, and these may bias the results considerably.

Similarly, the data for cancer mortality, although adjusted for age, may be influenced by factors such as standards of health care and the frequency of other disease processes. The heterogeneous nature of the population in many countries increases the difficulty of interpretation and data on cancer mortality do not necessarily give a good indication of cancer incidence. In view of these many uncertainties, it is difficult to say whether observed correlations are actually relevant to the etiology of cancer in humans, but it may be significant that dietary fat intake appears to be selectively correlated with cancer incidence at certain sites.

Although there are obvious limitations to the cancer mortality data, these are perhaps less serious for cancer of the breast than for cancer at other sites. Breast cancer can be readily recognized as a cause of death and mortality is probably a reasonably good indication of morbidity. The strong positive correlation with breast cancer and the observed enhancement of mammary tumors in animals by high fat diets provide some indication that results obtained with the animal models may be relevant to the etiology of breast cancer in humans [18].

Skin cancer has also been clearly shown to be influenced by dietary fat in animals and epidemiological data on skin cancer in humans show a positive, though less striking, correlation with fat intake (fig. 4). In this case, however, the limitations of the epidemiological data are more apparent. The geographical distribution of skin cancer is influenced by factors such as exposure to sunlight [84] and to coal tar products [21], and because of the high rate of success in treating skin cancer, mortality is less likely to be correlated with morbidity. The age-adjusted values for incidence of skin cancer in different countries compiled by DOLL et al. [30] are generally 10–20 times higher than the age-adjusted figures for mortality due to skin cancer according to data collected by SEGI et al. [102], whereas the age-adjusted values for breast cancer incidence are only 2–3 times higher than those for breast cancer mortality.

Cancer of the prostate, like breast cancer, arises in a secondary sex organ responsive to steroid hormones. The selectivity of the correlation for these tissues might therefore be due to some effect of dietary fat on the hormonal environment [86, 138]. It was suggested by WYNDER [130] that dietary fat may affect hormone production or the retention of hormones in adipose tissue. He also proposed that the high incidence of cancer of the colon on high fat diets may be related to increased amounts of fat in the intestine. A similar argument might be used to explain the positive correlation between dietary fat and cancer of the rectum.

HILL et al. [59] observed that people in 'Western' countries living on diets high in fat and animal protein excreted greater concentrations of fecal steroids than people from African and Eastern countries. The steroids were also more degraded on the Western-type diets. Reducing the fat content in the diet of volunteer subjects caused a marked reduction in the excretion of acid steroids and a more gradual decrease in neutral steroid excretion [57]. These workers postulated that intestinal bacteria might be able to produce carcinogens from dietary fats or bile steroids and that geographical variations in carcinoma of the colon might depend partly on differences in composition of the intestinal bacterial flora resulting from differences in diet [31, 59].

The importance attached to the high fat content of the diet was questioned on epidemiological grounds by WALKER [123, 124], who also pointed out that ingestion of high-fat low-residue diets does not necessarily lead to increased excretion of fecal steroids. If carcinogenic substances are present in the intestinal contents, the potential hazard may be increased by the slower transit of intestinal contents observed with refined diets of low fiber content [15–17]. However, DRASAR and IRVING [32] analyzed data from 37 different countries and found no correlation between dietary fiber and either cancer of the colon or breast cancer, whereas dietary fat and animal protein each showed a strong correlation.

Recently, REDDY and WYNDER [95] have reported that Americans consuming a mixed diet excrete more neutral steroids and bile acids than American vegetarians, Seventh-day Adventists or recent immigrants from Japan, Taiwan and Hong Kong. The fecal cholesterol and bile acids were also more extensively degraded in Americans on the Western-type diet than in the other groups. There results and earlier epidemiological data [135, 140] were taken as evidence that diet can affect the risk of colon cancer.

Other investigators have shown that several bile acids can induce sarcomas in animals at the site of injection [24, 75, 76] and intestinal microflora may be capable of transforming bile acids into carcinogens [59, 95]. It has been shown that intestinal bacteria are capable of carrying out at least some of the reactions leading to aromatization of the steroid nucleus [3, 48, 58].

Hill and co-workers [31, 60] have also proposed that gut bacteria may be involved in the aetiology of breast cancer. This suggestion is based on evidence that intestinal bacteria are capable of converting bile acids and cholesterol derivatives to steroidal estrogens. Since the intestinal flora can be altered by diet this offers a possible explanation for the observed correlations between diet and incidence of breast cancer.

MacMahon and Cole [85] and MacMahon et al. [86, 87] suggested that differences in urinary estrogen profiles between Asian and North American women may be related to the observed differences in breast cancer in these geographical areas. The ratio of estriol to estrone and estradiol was found to be higher in the Asian population, and the difference was particularly marked among young women. In view of the positive correlation between breast cancer and dietary fat intake, it would be of interest to see whether the estrogen profile can be altered by dietary fat.

Although the apparent enhancement of certain types of tumor by dietary fat may be linked to affects on steroidal sex hormones, there is less reason to think that this may be so for other tumors. It is interesting to note that the correlations with fat intake are rather similar in males and females for tumors at most of the sites common to both sexes (table VIII). Lung cancer is the most notable exception (fig. 4), and in this case the much higher incidence in males may tend to emphasize a trend toward positive correlation. Hormonal influences may not be limited, however, to tumors arising in primary or secondary sex organs. For example, possible adverse effects of estrogen on malignant melanoma are discussed in a recent paper by Sadoff et al. [98].

Wynder et al. [137] have recently reviewed the epidemiology of cancer of the pancreas and have called attention to suggestions that it may be associated with dietary factors, particularly dietary fat. As a working hypothesis, they proposed that bile may contain carcinogens, possibly originating in the diet, or from tobacco, or the occupational environment, and that pancreatic cancer may be caused by this bile refluxing into the pancreatic duct.

Higginson [56] concluded that, apart from a few exceptional situations, genetic factors are of little significance to the overall incidence of cancer in man. Studies on migrant populations offer one approach to determining the relative importance of genetic and environmental influences on cancer. Kmet [74] has recently reviewed studies along these lines and has pointed out some of the difficulties of gathering and interpreting such data. For cancer at some sites, such as stomach and breast, the cancer incidence in emigrants seems to conform more closely to that in their country of origin [50, 52, 109, 110] while for other sites, such as colon and rectum, the pattern is more like that of the country to which they immigrate [51, 109].

In view of the many environmental factors which can influence cancer incidence [1, 21, 28, 56] it is all the more remarkable that certain types of cancer show such a strong positive correlation with dietary fat intake. Such a relationship can of course not be regarded as necessarily causative, since other environmental influences which parallel fat intake may be involved. Attention has been drawn to the positive correlations between fat intake, caloric intake and protein intake (table IX), and GREGOR et al. [49] noted a correlation between gastrointestinal cancer and animal protein intake. It may prove difficult to dissociate these different environmental influences in human studies and this emphasizes the potential value of further experiments with animals.

It would be desirable, for example, to have more information on the effects of dietary fat on additional types of tumors in animals. This approach is somewhat limited by the availability of suitable methods for producing tumors at particular sites, but further experiments along these lines are possible. It is known, for example, that older dogs have a tendency to develop cancer of the prostate [62] and it would be of interest to know whether this tendency is accentuated by increasing the fat content of the diet. Methods are also available for producing intestinal tumors in experimental animals [127], and the effects of low and high fat diets of animal or vegetable origin on the incidence of large bowel cancer produced by such compounds is currently under investigation [139]. Such experiments have their limitations, since tumors at particular sites in experimental animals may not share the characteristics of their counterparts in humans and may not be influenced in the same way by environmental factors. However, by building up a body of comparative data, it may be possible to gain a better idea of the applicability of results with animals to the etiology of human cancer.

It is apparent that one approach to the prevention of cancer is by preventing the initiating event from occurring. This may be feasible in cases where the carcinogen can be identified and where it is practicable to remove it from the environment or otherwise avoid contact with it, but in practice this is often not possible. Some known carcinogenic agents are difficult to avoid entirely and other potential carcinogens may be formed in the body as a result of normal metabolic processes.

Assuming that cancer develops in stages, another approach might be to avoid conditions which favor development of a tumor after an initiating event has taken place. There is normally a long and variable latent period between exposure to a carcinogen and development of a tumor and this is taken as an indication of the susceptibility of tumorigenesis to environmental influences during the early stages of tumor formation. Experimental and epidemiological evidence presented in this review indicate that dietary fat may exert such an influence on the formation of certain types of tumor. It seems possible that in man, as in animals, the incidence of some kinds of cancer may be altered by modifying either the quantity or type of fat in the diet. A better understanding of the mechanisms by which dietary fat influences tumorigenesis in animals might also lead to new concepts which could be applied to cancer prevention in humans.

V. Summary

Experiments with mice and rats have shown that increasing the level of dietary fat enhances the yield of certain types of tumors, such as mammary tumors and skin tumors. Yields of other tumors such as sarcomas, lung tumors and leukemia do not seem to be affected. Pair-feeding experiments and data on growth rates indicate that the enhancement by dietary fat cannot be explained by differences in caloric intake. With hepatomas induced by aminoazo dyes and mammary tumors induced by DMBA, unsaturated fats appear to have a greater effect than saturated fats, but this distinction has not been apparent in studies with other tumors.

Enhancement by dietary fat has been observed both with spontaneous tumors and with tumors induced in various ways. With induced tumors, the high fat diet is generally more effective when fed after exposure to the carcinogen, rather than before. In some of the early studies with skin tumors, the animals on high fat diet had oily fur and it appeared that this may have facilitated absorption of carcinogens painted on the skin. Otherwise, there is little direct evidence that dietary fat influences the distribution or metabolism of the inducing agent. It seems possible that dietary fat may enhance mammary tumor yields by affecting the production, distribution or metabolism of hormones involved in growth and development of the mammary gland.

Obesity does not seem to be an important predisposing factor in human cancer, but epidemiological evidence shows a strong positive correlation between dietary fat intake and mortality from certain kinds of cancer such as breast cancer, prostatic cancer and cancer of the colon. It has been suggested that excess fat in the intestine may alter sterol and bile acid metabolism, possibly through changes in the bacterial flora, and that this may lead to production of potential carcinogens and may also lead to prolonged contact of carcinogens with the intestinal wall, as well as increasing the possibility of their absorption.

In considering the possible significance of observed correlations between cancer mortality and dietary fat intake, the difficulties of collecting the epidemiological data should be kept in mind. Even if a positive correlation can be established, the association is not necessarily causative. Caloric intake and protein intake both tend to increase along with fat intake and it is very possible that a number of environmental variables are associated with differences in dietary fat intake. However, the fact that only certain kinds of cancer show a positive correlation suggests that this is a productive field for further research.

VI. Acknowledgments

Studies by the authors described in this review were supported by the National Cancer Institute of Canada and the National Live Stock and Meat Board, Chicago, Ill. The technical assistance of R. RASMUSSEN and GAIL SHEEHAN in these studies is gratefully acknowledged. Dr. KATHLEEN STAVRAKY assisted with statistical analysis of the human data and both she and Dr. E. B. GAMMAL provided valuable commentaries on the review.

VII. References

1 ACKERMAN, L. V.: Some thoughts on food and cancer. Nutr. Today 7: 2–9 (1972).
2 ARFFMANN, E.: Heated fats and allied compounds as carcinogens. A critical review of experimental results. J. nat. Cancer Inst. 25: 893–926 (1960).
3 ARIES, V. C.; GODDARD, P., and HILL, M. J.: Degradation of steroids by intestinal bacteria. III. 3-Oxo-5β-steroid Δ^1-dehydrogenase and 3-oxo-5β-steroid Δ^4-dehydrogenase. Biochim. biophys. Acta 248: 482–488 (1971).
4 ARTMAN, N. R.: The chemical and biological properties of heated and oxidized fats; in PAOLETTI and KRITCHEVSKY Adv. Lipid Res., vol. 7, pp. 245–330 (Academic Press, New York 1969).
5 BAUMANN, C. A.; JACOBI, H. P., and RUSCH, H. P.: The effect of diet on experimental tumor production. Amer. J. Hyg. 30A: 1–6 (1939).

6 BAUMANN, C. A. and RUSCH, H. P.: Effect of diet on tumors induced by ultraviolet light. Amer. J. Cancer 35: 213–221 (1939).
7 BENSON, J.; LEV, M., and GRAND, C. G.: Enhancement of mammary fibroadenomas in the female rat by a high fat diet. Cancer Res. 16: 135–137 (1956).
8 BERENBLUM, I.: Carcinogenesis and tumor pathogenesis. Adv. Cancer Res. 2: 129–175 (1954).
9 BERENBLUM, I.: A re-evaluation of the concept of cocarcinogenesis; in HOMBURGER Prog. exp. Tum. Res., vol. 11, pp. 21–30 (Karger, Basel 1969).
10 BERENBLUM, I. and SHUBIK, P.: A new, quantitative, approach to the study of the stages of chemical carcinogenesis in the mouse's skin. Brit. J. Cancer 1: 383–391 (1947).
11 BERENBLUM, I. and SHUBIK, P.: The persistence of latent tumour cells induced in the mouse's skin by a single application of 9:10-dimethyl-1:2-benzanthracene. Brit. J. Cancer 3: 384–386 (1949).
12 BISCHOFF, F.: Carcinogenic effects of steroids; in PAOLETTI and KRITCHEVSKY, Adv. Lipid Res., vol. 7, pp. 165–244 (Academic Press, New York 1969).
13 BOUTWELL, R. K.: Some biological aspects of skin carcinogenesis; in HOMBURGER Progr. exp. Tum. Res., vol. 4, pp. 207–250 (Karger, Basel 1964).
14 BOUTWELL, R. K.; BRUSH, M. K., and RUSCH, H. P.: The stimulating effect of dietary fat on carcinogenesis. Cancer Res. 9: 741–746 (1949).
15 BURKITT, D. P.: Epidemiology of cancer of the colon and rectum. Cancer, Philad. 28: 3–13 (1971).
16 BURKITT, D. P.: Cancer of the colon and rectum. Epidemiology and possible causative factors. Minnesota Med. 55: 779–783 (1972).
17 BURKITT, D. P.; WALKER, A. R. P., and PAINTER, N. S.: Effect of dietary fibre on stools and transit-times, and its role in the causation of disease. Lancet ii: 1408–1412 (1972).
18 CARROLL, K. K.; GAMMAL, E. B., and PLUNKETT, E. R.: Dietary fat and mammary cancer. Canad. med. Ass. J. 98: 590–594 (1968).
19 CARROLL, K. K. and KHOR, H. T.: Effects of dietary fat and dose level of 7,12-dimethylbenz(α)anthracene on mammary tumor incidence in rats. Cancer Res. 30: 2260–2264 (1970).
20 CARROLL, K. K. and KHOR, H. T.: Effects of level and type of dietary fat on incidence of mammary tumors induced in female Sprague-Dawley rats by 7,12-dimethylbenz(α)anthracene. Lipids 6: 415–420 (1971).
21 CLAYSON, D. B.: Chemical carcinogenesis (Churchill, London 1962).
22 CLEMENS, J. A.; WELSCH, C. W., and MEITES, J.: Effects of hypothalamic lesions on incidence and growth of mammary tumors in carcinogen-treated rats. Proc. Soc. exp. Biol. Med. 127: 969–972 (1968).
23 CONIGLIO, J. G. and BRIDGES, R.: The effect of dietary fat on fatty acid synthesis in cell-free preparation of lactating mammary gland. Lipids 1: 76–80 (1966).
24 COOK, J. W.; KENNAWAY, E. L., and KENNAWAY, N. M.: Production of tumours in mice by deoxycholic acid. Nature, Lond. 145: 627 (1940).

25 Dao, T. L.: Studies on mechanism of carcinogenesis in the mammary gland; in Homburger Progr. exp. Tum. Res., vol. 11, pp. 235–261 (Karger, Basel 1969).
26 Dao, T. L.: Inhibition of tumor induction in chemical carcinogenesis in the mammary gland; in Homburger, Van Duuren and Rubin Progr. exp. Tum. Res., vol. 14, pp. 59–88 (Karger, Basel 1971).
27 Dao, T. L.; Bock, F. G., and Greiner, M. J.: Mammary carcinogenesis by 3-methylcholanthrene. II. Inhibitory effect of pregnancy and lactation on tumor induction. J. nat. Cancer Inst. 25: 991–1003 (1960).
28 Davies, J. N. P.: Cancer and the internal and external environments; in Scholefield Canad. Cancer Conf., vol. 9, pp. 10–22 (Univ. of Toronto Press, Toronto 1972).
29 Davis, R. K.; Stevenson, G. T., and Busch, K. A.: Tumor incidence in normal Sprague-Dawley female rats. Cancer Res. 16: 194–197 (1956).
30 Doll, R.; Muir, C., and Waterhouse, J.: Cancer incidence in five continents, vol. 2 (Springer, Berlin 1970).
31 Drasar, B. S. and Hill, M. J.: Intestinal bacteria and cancer. Amer. J. clin. Nutr. 25: 1399–1404 (1972).
32 Drasar, B. S. and Irving, D.: Environmental factors and cancer of the colon and breast. Brit. J. Cancer 27: 167–172 (1973).
33 Dunning, W. F.; Curtis, M. R., and Maun, M. E.: The effect of dietary fat and carbohydrate on diethylstilbestrol-induced mammary cancer in rats. Cancer Res. 9: 354–361 (1949).
34 Duuren, B. L. van: Tumor-promoting agents in two-stage carcinogenesis; in Homburger Progr. exp. Tum. Res., vol. 11, pp. 31–68 (Karger, Basel 1969).
35 Ederer, F.; Leren, P.; Turpeinen, O., and Frantz, I. D., jr.: Cancer among men on cholesterol-lowering diets. Experience from five clinical trials. Lancet ii: 203–206 (1971).
36 Eisenfeld, A. J. and Axelrod, J.: Selectivity of estrogen distribution in tissues. J. Pharmacol. exp. Ther. 150: 469–475 (1965).
37 Eisenfeld, A. J. and Axelrod, J.: Effect of steroid hormones, ovariectomy, estrogen pretreatment, sex and immaturity on the distribution of ^{3}H-estradiol. Endocrinology 79: 38–42 (1966).
38 Engel, R. W. and Copeland, D. H.: Influence of diet on the relative incidence of eye, mammary, ear-duct, and liver tumors in rats fed 2-acetylaminofluorene. Cancer Res. 11: 180–183 (1951).
39 Farber, E.: Studies on the molecular mechanisms of carcinogenesis; in Whelan and Schultz Homologies in enzymes and metabolic pathways. Metabolic alterations in cancer, pp. 314–334 (North-Holland, Amsterdam 1970).
40 Farber, E.: Chemical carcinogenesis; in Anfinsen, Potter and Schechter Current research in oncology, 1972, pp. 95–123 (Academic Press, New York 1973).
41 Farber, E.: Hyperplastic liver nodules; in Busch Methods in cancer research, vol. 7, pp. 345–375 (Academic Press, New York 1973).
42 Food Balance Sheets, 1964–66 Average (Food and Agriculture Organization of the United Nations, Rome 1971).

43 Foulds, L.: The experimental study of tumor progression. A review. Cancer Res. *14:* 327–339 (1954).
44 Foulds, L.: Neoplastic development, vol. 1 (Academic Press, New York 1969).
45 Friedewald, W. R. and Rous, P.: The initiating and promoting elements in tumor production. An analysis of the effects of tar, benzpyrene, and methylcholanthrene on rabbit skin. J. exp. Med. *80:* 101–126 (1944).
46 Gammal, E. B.; Carroll, K. K., and Plunkett, E. R.: Effects of dietary fat on mammary carcinogenesis by 7,12-dimethylbenz(α)anthracene in rats. Cancer Res. *27:* 1737–1742 (1967).
47 Gammal, E. B.; Carroll, K. K., and Plunkett, E. R.: Effects of dietary fat on the uptake and clearance of 7,12-dimethylbenz(α)anthracene by rat mammary tissue. Cancer Res. *28:* 384–385 (1968).
48 Goddard, P. and Hill, M. J.: Degradation of steroids by intestinal bacteria. IV. The aromatisation of ring A. Biochim. biophys. Acta *280:* 336–342 (1972).
49 Gregor, O.; Toman, R., and Prusová, F.: Gastrointestinal cancer and nutrition. Gut *10:* 1031–1034 (1969).
50 Haenszel, W.: Cancer mortality among the foreign-born in the United States. J. nat. Cancer Inst. *26:* 37–132 (1961).
51 Haenszel, W. and Dawson, E. A.: A note on mortality from cancer of the colon and rectum in the United States. Cancer, Philad. *18:* 265–272 (1965).
52 Haenszel, W. and Kurihara, M.: Studies of Japanese migrants. I. Mortality from cancer and other diseases among Japanese in the United States. J. nat. Cancer Inst. *40:* 43–68 (1968).
53 Hausberger, F. X. and Milstein, S. W.: Dietary effects on lipogenesis in adipose tissue, J. biol. Chem. *214:* 483–488 (1955).
54 Haven, F. L. and Bloor, W. R.: Lipids in cancer. Adv. Cancer Res. *4:* 237–314 (1956).
55 Hems, G.: Epidemiological characteristics of breast cancer in middle and late age. Brit. J. Cancer *24:* 226–234 (1970).
56 Higginson, J.: Present trends in cancer epidemiology; in Morgan Canad. Cancer Conf., vol. 8, pp. 40–75 (Pergamon, Oxford 1969).
57 Hill, M. J.: The effects of some factors on the faecal concentration of acid steroids, neutral steroids and urobilins. J. Path. *104:* 239–245 (1971).
58 Hill, M. J.: Gut bacteria, steroids and cancer of the large bowel; in Williams and Briggs Some implications of steroid hormones in cancer, pp. 94–106 (Heinemann, London 1971).
59 Hill, M. J.; Drasar, B. S.; Aries, V.; Crowther, J. S.; Hawksworth, G., and Williams, R. E. O.: Bacteria and aetiology of cancer of the large bowel. Lancet *i:* 95–100 (1971).
60 Hill, M. J.; Goddard, P., and Williams, R. E. O.: Gut bacteria and aetiology of cancer of the breast. Lancet *ii:* 472–473 (1971).
61 Hoshino, K.: Morphogenesis and growth potentiality of mammary glands in mice. I. Transplantability and growth potentiality of mammary tissue of virgin mice. J. nat. Cancer Inst. *29:* 835–851 (1962).
62 Huggins, C.: Endocrine-induced regression of cancers. Science *156:* 1050–1054 (1967).

63 HUGGINS, C.; GRAND, L. C., and BRILLANTES, F. P.: Mammary cancer induced by a single feeding of polynuclear hydrocarbons, and its suppression. Nature, Lond. *189:* 204–207 (1961).
64 HUGGINS, C.; MOON, R. C., and MORII, S.: Extinction of experimental mammary cancer. I. Estradiol-17β and progesterone. Proc. nat. Acad. Sci., Wash. *48:* 379–386 (1962).
65 INSULL, W. jr.; LANG, P. D.; HSI, B. P., and YOSHIMURA, S.: Studies of arteriosclerosis in Japanese and American men. I. Comparison of fatty acid composition of adipose tissue. J. clin. Invest. *48:* 1313–1327 (1969).
66 JABARA, A. G.; TOYNE, P. H., and HARCOURT, A. G.: Effects of time and duration of progesterone administration on mammary tumors induced by 7,12-dimethylbenz(a)anthracene in Sprague-Dawley rats. Brit. J. Cancer *27:* 63–71 (1973).
67 JACOBI, H. P. and BAUMANN, C. A.: The effect of fat on tumor formation. Amer. J. Cancer *39:* 338–342 (1940).
68 JANSS, D. H.; MOON, R. C., and IRVING, C. C.; The binding of 7,12-dimethylbenz(a)anthracene to mammary parenchyma DNA and protein *in vivo*. Cancer Res. *32:* 254–258 (1972).
69 JENSEN, E. V.; BLOCK, G. E.; SMITH, S.; KYSER, K., and DESOMBRE, E. R.: Estrogen receptors and breast cancer response to adrenalectomy. Nat. Cancer Inst. Monogr. *34:* 55–70 (1971).
70 JENSEN, E. V. and DESOMBRE, E. R.: Mechanism of action of the female sex hormones. Annu. Rev. Biochem. *41:* 203–230 (1972).
71 JENSEN, E. V. and JACOBSON, H. I.: Basic guides to the mechanism of estrogen action. Recent Progr. Hormone Res. *18:* 387–414 (1962).
72 JENSEN, E. V.; NUMATA, M.; BRECHER, P. I., and DESOMBRE, E. R.: Hormone-receptor interaction as a guide to biochemical mechanism. Biochem. Soc. Symp. *32:* 133–159 (1971).
73 KLINE, B. E.; MILLER, J. A.; RUSCH, H. P., and BAUMANN, C. A.: Certain effects of dietary fats on the production of liver tumors in rats fed *p*-dimethylaminoazobenzene. Cancer Res. *6:* 5–7 (1946).
74 KMET, J.: The role of migrant population in studies of selected cancer sites. A review. J. chron. Dis. *23:* 305–324 (1970).
75 LACASSAGNE, A.; BUU-HOÏ, N. P., and ZAJDELA, F.: Carcinogenic activity of apocholic acid. Nature, Lond. *190:* 1007–1008 (1961).
76 LACASSAGNE, A.; BUU-HOÏ, N. P., and ZAJDELA, F.: Carcinogenic activity *in situ* of further steroid compounds. Nature, Lond. *209:* 1026–1027 (1966).
77 LAVIK, P. S. and BAUMANN, C. A.: Dietary fat and tumor formation. Cancer Res. *1:* 181–187 (1941).
78 LAVIK, P. S. and BAUMANN, C. A.: Further studies on the tumor-promoting action of fat. Cancer Res. *3:* 749–756 (1943).
79 LAWRASON, F. D. and KIRSCHBAUM, A.: Dietary fat with reference to the spontaneous appearance and induction of leukemia in mice. Proc. Soc. exp. Biol. Med. *56:* 6–7 (1944).
80 LEA, A. J.: New observations on distribution of neoplasms of female breast in certain European countries. Brit. med. J. *i:* 488–490 (1965).

81 LEA, A. J.: Relationship between environmental temperature and the death rate of women from neoplasms of the breast. Nature, Lond. *209:* 57–59 (1966).
82 LEA, A. J.: Dietary factors associated with death-rates from certain neoplasms in man. Lancet *ii:* 332–333 (1966).
83 LEA, A. J.: Neoplasms and environmental factors. Ann. roy. Coll. Surg. Engl. *41:* 432–438 (1967).
84 LEE, J. A. H. and ISSENBERG, H. J.: A comparison between England and Wales and Sweden in the incidence and mortality of malignant skin tumors. Brit. J. Cancer *26:* 59–66 (1972).
85 MACMAHON, B. and COLE, P.: The ovarian etiology of human breast cancer; in GRUNDMANN and TULINIUS Recent results in cancer research. Current problems in the epidemiology of cancer and lymphomas, vol. 39, pp. 185–192 (Heinemann, London; Springer, Berlin 1972).
86 MACMAHON, B.; COLE, P., and BROWN, J.: Etiology of human breast cancer. A review. J. nat. Cancer Inst. *50:* 21–42 (1973).
87 MACMAHON, B.; COLE, P.; BROWN, J. B.; AOKI, K.; LIN, T. M.; MORGAN, R. W., and WOO, N.-C.: Oestrogen profiles of Asian and North American women. Lancet *ii:* 900–902 (1971).
88 MEITES, J.: Relation of prolactin and estrogen to mammary tumorigenesis in the rat. J. nat. Cancer Inst. *48:* 1217–1224 (1972).
89 MILLER, J. A.; KLINE, B. E.; RUSCH, H. P., and BAUMANN, C. A.: The carcinogenicity of *p*-dimethylaminoazobenzene in diets containing hydrogenated coconut oil. Cancer Res. *4:* 153–158 (1944).
90 MILLER, J. A. and MILLER, E. C.: The carcinogenic aminoazo dyes. Adv. Cancer Res. *1:* 339–396 (1953).
91 MILLER, E. C.; MILLER, J. A.; KLINE, B. E., and RUSCH, H. P.: Correlation of the level of hepatic riboflavin with the appearance of liver tumors in rats fed aminoazo dyes. J. exp. Med. *88:* 89–98 (1948).
92 MÜHLBOCK, O.: Role of hormones in the etiology of breast cancer. J. nat. Cancer Inst. *48:* 1213–1216 (1972).
93 O'GARA, R. W.; STEWART, L.; BROWN, J., and HUEPER, W. C.: Carcinogenicity of heated fats and fat fractions. J. nat. Cancer Inst. *42:* 275–287 (1969).
94 PEARCE, M. L. and DAYTON, S.: Incidence of cancer in men on a diet high in polyunsaturated fat. Lancet *i:* 464–467 (1971).
95 REDDY, B. S. and WYNDER, E. L.: Large-bowel carcinogenesis. Fecal constituents of populations with diverse incidence rates of colon cancer. J. nat. Cancer Inst. *50:* 1437–1442 (1973).
96 RUSCH, H. P.: Extrinsic factors that influence carcinogenesis. Physiol. Rev. *24:* 177–204 (1944).
97 RUSCH, H. P.; KLINE, B. E., and BAUMANN, C. A.: The influence of caloric restriction and of dietary fat on tumor formation with ultraviolet radiation. Cancer Res. *5:* 431–435 (1945).
98 SADOFF, L.; WINKLEY, J., and TYSON, S.: Is malignant melanoma an endocrine-dependent tumor? The possible adverse effects of estrogen. Oncology *27:* 244–257 (1973).

99 SAFFIOTTI, U. and SHUBIK, P.: Studies on promoting action in skin carcinogenesis. Nat. Cancer Inst. Monogr. *10:* 489–507 (1963).
100 SALAMAN, M. H. and ROE, F. J. C.: Cocarcinogenesis. Brit. med. Bull. *20:* 139–144 (1964).
101 SANDER, S.: The uptake of 17β-oestradiol in breast tissue of female rats. Acta endocrin., Kbh. *58:* 49–56 (1968).
102 SEGI, M.; KURIHARA, M., and MATSUYAMA, T.: Cancer mortality for selected sites in 24 countries, No. 5 (1964–1965) (Department of Public Health, Tohoku University School of Medicine, Sendai 1969).
103 SHAMBERGER, R. J.: Increase of peroxidation in carcinogenesis. J. nat. Cancer Inst. *48:* 1491–1497 (1972).
104 SHILS, M. E.: Nutritional and dietary factors in neoplastic development. CA *21:* 399–406 (1971).
105 SILBERBERG, R. and SILBERBERG, M.: Hypophyseal tumors produced by radioactive iodine (I^{131}) in mice of various strains fed a high fat diet. Proc. amer. Ass. Cancer Res. *1:* 52–53 (1953).
106 SILVERSTONE, H. and TANNENBAUM, A.: The effect of the proportion of dietary fat on the rate of formation of mammary carcinoma in mice. Cancer Res. *10:* 448–453 (1950).
107 SILVERSTONE, H. and TANNENBAUM, A.: The influence of dietary fat and riboflavin on the formation of spontaneous hepatomas in the mouse. Cancer Res. *11:* 200–203 (1951).
108 SINHA, D.; COOPER, D., and DAO, T. L.: The nature of estrogen and prolactin effect on mammary tumorigenesis. Cancer Res. *33:* 411–414 (1973).
109 STASZEWSKI, J.: Migrant studies in alimentary tract cancer; in GRUNDMANN and TULINIUS Recent results in cancer research. Current problems in the epidemiology of cancer and lymphomas, vol. 39, pp. 85–97 (Heinemann, London; Springer, Berlin 1972).
110 STASZEWSKI, J. and HAENSZEL, W.: Cancer mortality among the Polish-born in the United States. J. nat. Cancer Inst. *35:* 291–297 (1965).
111 SZEPSENWOL, J.: Carcinogenic effect of ether extract of whole egg, alcohol extract of egg yolk and powdered egg free of the ether extractable part in mice. Proc. Soc. exp. Biol. Med. *116:* 1136–1139 (1964).
112 SZEPSENWOL, J.: Intracranial tumors in mice of two different strains maintained on fat enriched diets. Europ. J. Cancer *7:* 529–532 (1971).
113 TANNENBAUM, A.: Relationship of body weight to cancer incidence. Arch. Path. *30:* 509–517 (1940).
114 TANNENBAUM, A.: The genesis and growth of tumors. III. Effects of a high-fat diet. Cancer Res. *2:* 468–475 (1942).
115 TANNENBAUM, A.: The dependence of the genesis of induced skin tumors on the fat content of the diet during different stages of carcinogenesis. Cancer Res. *4:* 683–687 (1944).
116 TANNENBAUM, A.: The dependence of tumor formation on the composition of the calorie-restricted diet as well as on the degree of restriction. Cancer Res. *5:* 616–625 (1945).

117 TANNENBAUM, A.: The role of nutrition in the origin and growth of tumors; in MOULTON Approaches to tumor chemotherapy, pp. 96–127 (American Association for the Advancement of Science, Washington 1947).
118 TANNENBAUM, A.: Nutrition and cancer; in HOMBURGER The physiopathology of cancer; 2nd ed., pp. 517–562 (Hoeber-Harper, New York 1959).
119 TANNENBAUM, A. and SILVERSTONE, H.: Dosage of carcinogen as a modifying factor in evaluating experimental procedures expected to influence formation of skin tumors. Cancer Res. 7: 567–574 (1947).
120 TANNENBAUM, A. and SILVERSTONE, H.: Nutrition in relation to cancer. Adv. Cancer Res. 1: 451–501 (1953).
121 TANNENBAUM, A. and SILVERSTONE, H.: Nutrition and the genesis of tumours; in RAVEN Cancer, vol. 1, pp. 306–334 (Butterworth, London 1957).
122 WAARD, F. DE: The epidemiology of breast cancer; review and prospects. Int. J. Cancer 4: 577–586 (1969).
123 WALKER, A. R. P.: Diet and cancer of the colon. Lancet i: 593 (1971).
124 WALKER, A. R. P.: Diet, bowel motility, faeces composition and colonic cancer. Sth. afr. med. J. 45: 377–379 (1971).
125 WATSON, A. F. and MELLANBY, E.: Tar cancer in mice. II. The condition of the skin when modified by external treatment or diet, as a factor in influencing the cancerous reaction. Brit. J. exp. Path. 11: 311–322 (1930).
126 WATTENBERG, L. W.: Inhibition of carcinogenic and toxic effects of polycyclic hydrocarbons by phenolic antioxidants and ethoxyquin. J. nat. Cancer Inst. 48: 1425–1430 (1972).
127 WEISBURGER, J. H.: Colon carcinogens. Their metabolism and mode of action. Cancer, Philad. 28: 60–70 (1971).
128 WELSCH, C. W.; CLEMENS, J. A., and MEITES, J.: Effects of multiple pituitary homografts or progesterone on 7,12-dimethylbenz(α)anthracene-induced mammary tumors in rats. J. nat. Cancer Inst. 41: 465–471 (1968).
129 WHITE, F. R.: The relationship between underfeeding and tumor formation, transplantation, and growth in rats and mice. Cancer Res. 21: 281–290 (1961).
130 WYNDER, E. L.: Current concepts of the aetiology of breast cancer; in FORREST and KUNKLER Prognostic factors in breast cancer. Proc. 1st Tenovus Symp., pp. 32–49 (Livingstone, Edinburgh 1968).
131 WYNDER, E. L.: Identification of women at high risk from breast cancer. Cancer, Philad. 24: 1235–1240 (1969).
132 WYNDER, E. L.; ESCHER, G. C., and MANTEL, N.: An epidemiological investigation of cancer of the endometrium. Cancer, Philad. 19: 489–520 (1966).
133 WYNDER, E. L. and HOFFMAN, D.: Nutrition and cancer; in RAVEN and ROE The prevention of cancer, pp. 11–18 (Butterworth, London 1967).
134 WYNDER, E. L.; HYAMS, L., and SHIGEMATSU, T.: Correlations of international cancer death rates. An epidemiological exercise. Cancer, Philad. 20: 113–126 (1967).
135 WYNDER, E. L.; KAJITANI, T.; ISHIKAWA, S.; DODO, H., and TAKANO, A.: Environmental factors of cancer of the colon and rectum. II. Japanese epidemiological data. Cancer, Philad. 23: 1210–1220 (1969).

136 WYNDER, E. L. and MABUCHI, K.: Etiological and preventive aspects of human cancer. Prev. Med. *1:* 300–334 (1972).
137 WYNDER, E. L.; MABUCHI, K.; MARUCHI, N., and FORTNER, J. G.: Epidemiology of cancer of the pancreas. J. nat. Cancer Inst. *50:* 645–667 (1973).
138 WYNDER, E. L.; MABUCHI, K., and WHITMORE, W. F.: Epidemiology of cancer of the prostate. Cancer, Philad. *28:* 344–360 (1971).
139 WYNDER, E. L. and REDDY, B.: Guest editorial. Studies of large-bowel cancer: human leads to experimental application. J. nat. Cancer Inst. *50:* 1099–1106 (1973).
140 WYNDER, E. L. and SHIGEMATSU, T.: Environmental factors of cancer of the colon and rectum. Cancer, Philad. *20:* 1520–1561 (1967).

Authors' addresses: K. K. CARROLL, Health Sciences Centre, Department of Biochemistry, The University of Western Ontario, *London, Ontario, N6A 3K7* (Canada); H. T. KHOR, Department of Biochemistry, University of Malaya, *Kuala Lumpur* (Malaysia)

Author Index

Aaes-Jorgensen, E. 28, 32
Abdel-Rahman, Y. M. 81, 86, 89, 109
Abe, S. 181, 185, 194
Abeles, R. H. 255, 261
Abercrombie, H. N. 151, 157
Abercrombie, M. 291, 296
Abraham, S. 28, 36, 45, 46, 48, 69, 73, 148, 164, 271, 273, 274, 275, 285, 287, 294, 296, 304, 305
Abrahamson, S. 206, 222
Abramovich, D. A. 234, 262
Ackerman, L. V. 343, 345
Ackermann, H. 4, 5, 13, 36
Adams, R. A. 151, 159
Adelsberger, L. 170, 194
Adler, C. R. 113, 119, 121, 122, 123, 124, 133
Aexel, R. T. 232, 233, 265
Agranoff, B. W. 17, 28, 31, 34, 60, 69, 144, 147, 157
Ahrens, E. H. Jr. 116, 124, 134
Albutt, E. C. 256, 261
Alfin-Slater, R. B. 279, 305
Allen, A. 50, 58, 75
Allen, D. O. 295, 296
Allende, C. C. 202, 203, 222
Allende, J. E. 202, 203, 222
Allmann, D. W. 285, 296
Almeida, J. O. 170, 194
Alonzo, N. 13, 36
Alonzo, N. F. 128, 133, 153, 163, 243, 265

Alper, S. 10, 23, 32
Alpert, E. 277, 297
Altman, P. L. 95, 107
Alvarez, R. R. de 94, 97, 107
Ambrose, E. J. 291, 296
Anderer, F. A. 147, 159
Andersen, H. 230, 242, 261
Anderson, B. 188, 191
Anderson, R. E. 7, 30, 31, 139, 143, 144, 158
Andrews, E. P. 179, 194
Andrews, H. D. 128, 132, 168, 170, 171, 178, 185, 186, 192
Angervall, L. 5, 14, 31
Anker, H. S. 284, 306
Ansell, G. B. 25, 26, 31
Aoki, K. 342, 350
Apffel, C. A. 155, 158
Apostolescu, I. 5, 13, 35
Archibald, F. M. 77, 79, 111, 113, 115, 118, 119, 121, 122, 123, 124, 125, 133
Arffmann, E. 309, 345
Arias, I. M. 67, 72
Aries, V. 309, 341, 348
Aries, V. C. 341, 345
Armstrong, J. McD. 285, 297, 303
Artman, N. R. 309, 345
Ashbrook, J. D. 56, 57, 73, 74
Ashmore, J. 45, 75, 285, 306
Ata, A. A. 81, 86, 89, 109
Aurbach, G. D. 295, 300

Author Index

Avigan, J. 145, 150, 151, 158, 166, 236, 263, 280, 294, 296
Axelrod, J. 326, 347
Azarnoff, D. L. 241, 261
Azarnoff, R. S. 254, 266

Back, P. 278, 296
Baer, E. 17, 31
Bailey, J. M. 27, 34, 46, 47, 48, 50, 62, 63, 68, 69, 70, 138, 143, 144, 145, 147, 148, 149, 150, 155, 156, 158, 159, 161
Bailey, P. J. 139, 158
Baker, N. 44, 54, 55, 56, 64, 66, 67, 69, 72
Ballard, F. J. 283, 296
Bandi, Z. L. 28, 32
Bandyopadhyay, A. K. 204, 205, 222
Banik, N. L. 234, 238, 261
Barclay, M. 52, 63, 69, 77, 79, 80, 81, 86, 87, 89, 91, 93, 94, 95, 96, 99, 101, 102, 104, 106, 107, 108, 109, 111, 113, 115, 116, 117, 118, 119, 121, 122, 123, 124, 125, 131, 133, 134, 152, 159, 219, 222
Barclay, R. K. 63, 69, 77, 79, 80, 81, 86, 89, 95, 104, 108, 111
Barker, J. R. 155, 158
Barker, M. 246, 252, 254, 256, 261, 264
Barley, F. W. 255, 261
Barnes, B. A. 86, 89, 100, 108
Bartley, J. C. 285, 294, 296
Bassin, R. 154, 160
Basu, M. 168, 185, 190
Battaner, E. 214, 226
Baumann, C. A. 310, 311, 312, 313, 314, 315, 318, 319, 320, 321, 322, 324, 325, 328, 329, 345, 346, 349, 350
Baumann, W. J. 8, 36
Baxter, C. F. 230, 231, 266
Bazzano, G. 30, 36
Bednařík, T. 94, 99, 110, 111
Beeler, M. F. 217, 222
Begemann, P. H. 8, 36
Begg, R. W. 51, 71, 74, 100, 108, 109, 111
Bell, C. 4, 34
Bell, O. E. Jr. 239, 261
Benda, P. 252, 254, 261
Bendich, A. 293, 306

Bennett, A. 143, 160
Bennett, D. A. H. 292, 302
Bennett, G. 170, 190
Bennett, K. 248, 254, 256, 258, 259, 261, 267
Ben-Porat, T. 145, 147, 158
Benson, J. 313, 315, 346
Berenblum, I. 322, 323, 325, 346
Berg, G. 94, 108
Bergelson, L. D. 14, 32, 33, 35, 127, 129, 131
Bergeron, J. J. M. 147, 163
Berliner, D. L. 145, 158
Bermek, E. 209, 210, 211, 213, 216, 220, 224
Bernheimer, H. 245, 261
Berry, M. N. 293, 297
Berwald, Y. 136, 158
Beytia, E. 278, 282, 305
Bielka, H. 217, 226
Bierring, F. 241, 242, 243, 244, 261
Bilheimer, D. W. 99, 109
Biner, J. 99, 111
Bingas, B. 239, 264
Bischoff, F. 309, 346
Bissell, D. M. 277, 297
Bjorntorp, P. 5, 14, 31
Black, D. D. 220, 223
Black, P. H. 168, 195
Blair, C. D. 145, 147, 158
Blanchette-Mackie, E. J. 63, 73
Blank, M. L. 4, 7, 8, 10, 14, 16, 17, 18, 19, 23, 27, 29, 30, 32, 37, 38, 40, 217, 225
Blásquez, E. 285, 297
Blass, J. P. 145, 150, 151, 158, 280, 294, 296
Bloch, H. 170, 171, 172, 185, 186, 187, 192
Bloch, K. 290, 297
Bloch-Frankenthal, L. 59, 70
Block, G. E. 326, 349
Bloj, B. 281, 297
Bloor, W. R. 3, 34, 51, 71, 100, 109, 308, 314, 348
Blyth, C. A. 291, 297
Bock, F. G. 326, 347
Bodley, J. W. 216, 225

Bole, G. G. 139, 158
Bollinger, J. N. 7, 32
Bonkowsky, H. L. 292, 297
Borek, C. 136, 137, 158
Borin, B. M. 291, 301
Bornstein, J. 285, 297, 303
Borrone, C. 170, 171, 191
Bortz, W. M. 278, 284, 297
Bos, C. J. 59, 70, 274, 295, 299
Bosch, H. van den 24, 26, 27, 32
Bosch, L. 45, 70
Boulos, B. M. 254, 266
Boutwell, R. K. 322, 328, 329, 346
Boxer, G. E. 28, 32, 60, 70
Boyd, E. M. 217, 222
Boyd, J. E. 232, 233, 262
Boyd, R. 138, 144, 145, 148, 164
Boyle, J. J. 54, 70
Bradley, R. M. 128, 132, 154, 162, 168, 194
Brady, R. O. 97, 98, 108, 128, 131, 132, 154, 158, 159, 160, 162, 168, 185, 190, 194, 284, 297
Brante, G. 239, 261
Braun, A. C. 137, 152, 158
Brecher, P. I. 326, 349
Bremer, J. 23, 32
Brennan, P. J. 145, 147, 158, 168, 169, 171, 187, 195
Brenneman, D. E. 44, 49, 53, 58, 60, 61, 62, 63, 67, 68, 70, 73, 74, 155, 156, 165
Bresnick, E. 292, 297
Bricker, L. 12, 39, 126, 134
Bricker, L. A. 271, 272, 273, 281, 285, 287, 288, 291, 297, 298
Bridges, R. 327, 346
Briedis, A. 278, 282, 305
Briedis, A. V. 278, 279, 299
Brillantes, F. P. 325, 349
Bronnestam, R. 245, 267
Brown, G. W. Jr. 59, 70
Brown, J. 309, 340, 342, 350
Brown, J. B. 342, 350
Brown, R. K. 86, 89, 100, 108
Brush, M. K. 328, 329, 346
Bryant, J. C. 47, 50, 70, 144, 160
Buchanan, K. D. 285, 297

Bucher, N. L. R. 278, 279, 280, 282, 298
Buchheit, F. 128, 132, 243, 244, 264
Buchko, M. K. 145, 164
Buck, C. A. 154, 166, 168, 190, 195
Bugiani, O. 170, 171, 191
Burger, M. M. 151, 159, 168, 190
Burkitt, D. P. 341, 346
Burns, C. H. 256, 266
Burns, E. R. 57, 65, 70
Burton, D. N. 286, 298
Busch, K. A. 314, 347
Buskirk, H. H. 127, 132
Butel, J. S. 136, 137, 159
Butler, J. D. 149, 150, 159, 161
Butler, R. E. 95, 96, 110, 229, 264
Buu-Hoï, N. P. 341, 349
Bygrave, F. L. 283, 306

Caffier, H. 208, 225
Cahill, G. F. Jr. 285, 298
Cain, C. E. 239, 261
Calathes, D. N. 86, 91, 94, 95, 96, 108
Cantero, A. 11, 40, 129, 132
Cappuccino, J. G. 52, 69, 101, 102, 108
Caravaca, J. 281, 298
Carbonara, A. O. 95, 110
Carlson, C. A. 284, 285, 298
Carrel, A. 136, 159
Carroll, K. K. 281, 298, 309, 314, 315, 316, 324, 327, 328, 329, 337, 340, 346, 348
Carruthers, C. 5, 13, 32, 217, 222
Cass, L. 170, 171, 175, 194
Castles, J. J. 216, 225
Castor, C. W. 139, 158
Cattabeni, F. 232, 240, 262
Cederqvist, L. L. 139, 160
Celis, J. 202, 203, 222
Chae, K. 20, 23, 27, 32
Chaikoff, I. L. 45, 48, 59, 69, 70, 73, 148, 164, 271, 273, 284, 287, 289, 304
Chakrabarty, K. 285, 298
Chang, H. C. 285, 298
Chang, R. S. 139, 159
Chao, F. 141, 142, 159
Chapman, B. B. 291, 301

Author Index

Chase, L. R. 295, 300
Chatterjee, S. 169, 190
Cheema, P. 168, 190
Cheng, S. 6, 32, 168, 190
Christensen Lou, H. O. 11, 32, 230, 241, 242, 243, 244, 245, 261
Christian, J. C. 13, 33
Christie, W. H. 18, 38
Chu, F. 116, 124, 134
Ciaccio, E. I. 60, 70
Clark, M. 28, 29, 38
Clausen, J. 11, 32, 230, 241, 242, 243, 244, 261
Clayson, D. B. 308, 309, 340, 343, 346
Clemens, J. A. 327, 346, 352
Clevenger, M. 4, 38
Clinkenbeard, K. 281, 282, 306
Cogin, G. E. 86, 91, 94, 95, 96, 108
Cohen, L. 88, 102, 111
Cohn, E. J. 86, 89, 100, 108
Cole, P. 340, 342, 350
Collins, J. M. 286, 298
Condie, R. M. 238, 266
Coniglio, J. G. 327, 346
Conner, R. L. 256, 266
Connor, W. E. 232, 261
Cook, J. W. 341, 346
Cooper, D. 325, 351
Copeland, D. H. 314, 315, 321, 322, 347
Cornatzer, W. E. 140, 147, 164, 166
Costa, G. 51, 70
Craig, J. M. 156, 164
Craig, M. C. 278, 282, 286, 298, 305
Cravioto, H. 234, 235, 248, 249, 251, 252, 254, 256, 258, 259, 261, 267, 268
Creinin, H. L. 51, 70, 105, 109
Cress, E. A. 6, 7, 38
Crim, J. A. 127, 132
Cristofalo, V. J. 137, 138, 146, 159, 162
Critchley, D. R. 168, 190
Crookston, J. H. 183, 194
Crookston, M. C. 183, 194
Crowther, J. S. 309, 341, 348
Cuadrado, R. R. 281, 298
Culling, C. A. 137, 159
Cumar, F. A. 154, 159, 168, 185, 190
Cumings, J. N. 118, 132, 228, 268

Cumming, R. B. 7, 17, 18, 30, 31, 38, 40, 139, 143, 144, 158
Cunningham, D. D. 30, 33, 152, 153, 159
Curran, G. L. 241, 261
Curran, J. F. 295, 306
Curtis, M. R. 314, 315, 325, 347
Curry, W. M. 284, 306
Czernobilsky, B. 47, 70

Dabelsteen, E. 182, 183, 190
Daehnfeldt, J. L. 46, 50, 72
Danes, B. S. 139, 160
Danielsson, H. 278, 298
Dao, T. L. 325, 326, 347, 351
Dargeon, H. W. 77, 80, 87, 96, 108
Davidsohn, I. 182, 183, 190
Davie, E. W. 204, 222
Davies, J. N. P. 343, 347
Davis, R. K. 314, 347
Davison, A. N. 234, 238, 251, 261, 267
Dawson, E. A. 343, 348
Day, E. A. 217, 222
Dayton, S. 338, 350
Deal, C. 138, 144, 145, 148, 164
Dean, P. D. G. 234, 262
Deenen, L. L. M. van 10, 11, 12, 24, 26, 27, 32, 33, 37, 39
Defendi, V. 151, 159
Della, C. 170, 171, 191
De Medio, G. E. 19, 23, 33
Dempsey, M. E. 283, 303
Den, H. 168, 185, 190
Dencker, S. J. 245, 262, 267
Dennick, R. G. 234, 262
Denton, M. D. 271, 273, 289, 298
De Paul, L. 238, 267
Derouaux, G. 86, 89, 100, 108
Desmond, W. 138, 143, 145, 161
Desnuelle, P. 23, 26, 27, 33
DeSombre, E. R. 326, 349
Deuel, H. J. Jr. 141, 159
Deutscher, M. P. 204, 205, 222
Dhar, A. K. 283, 305
Dick, A. R. 60, 73
Dickens, F. 64, 70
Dickinson, T. E. 100, 108
Diehl, V. 137, 162

Dietschy, J. M. 278, 288, 291, 298, 307
DiLorenzo, J. C. 86, 91, 94, 95, 96, 108
Dini, M. 280, 300
DiPaolo, J. A. 5, 33
Diringer, H. 147, 159, 168, 191, 193
Dirksen, T. R. 145, 159
Disait' Agnese, P. 169, 191
Dische, Z. 169, 191
Dittmann, J. 217, 226
Dittmer, J. C. 4, 11, 36, 228, 241, 242, 266
DiTullio, N. W. 233, 234, 236, 237, 264
Djordjevich, J. 88, 102, 111
Dodo, H. 335, 341, 352
Dods, R. F. 152, 159
Dohr, H. 11, 33
Dolejš, L. 199, 224
Doležal, A. 99, 111
Doll, R. 337, 340, 347
Donisch, V. 145, 159
Dorsey, J. A. 284, 298
Dorsey, J. K. 282, 299
Dougherty, T. F. 145, 158
Downie, R. 252, 254, 261
Drasar, B. S. 309, 341, 342, 347, 348
Druckrey, H. 249, 250, 262
Dubin, I. N. 47, 70
Dugan, R. E. 278, 279, 282, 283, 285, 299, 302, 303, 305
Dunbar, L. M. 46, 47, 48, 50, 62, 68, 69, 70, 138, 143, 158, 159
Dulak, A. C. 138, 165
Dunn, W. L. 137, 159
Dunning, W. F. 314, 315, 325, 347
Durand, P. 170, 171, 191
Dušek, Z. 200, 201, 202, 203, 204, 205, 208, 209, 210, 211, 212, 213, 214, 215, 216, 220, 224
Dutch, P. H. 4, 34
Duuren, B. L. van 322, 347
Dyatlovitskaya, E. V. 14, 32, 33, 35, 127, 131

Eagle, H. 151, 159
Earl, D. C. N. 207, 222
Earle, W. R. 148, 159
Ederer, F. 338, 347
Edmonson, J. H. 52, 70

Edwards, A. M. 292, 299
Edwards, P. A. 278, 279, 299
Eichberg, J. 8, 33, 146, 160, 254, 255, 257, 262, 263
Eilser, M. 170, 191
Eisenberg, S. 99, 109
Eisenfeld, A. J. 326, 347
Eisentraut, A. M. 285, 307
Eliason, J. 252, 254, 261
Elkins, W. L. 80, 108
Elliott, W. H. 292, 299
Elwood, J. C. 45, 48, 49, 70, 271, 273, 274, 282, 299, 307
Embree, L. J. 252, 253, 262
Emmelot, P. 45, 59, 70, 274, 295, 299
Eng, L. F. 141, 142, 159
Engel, R. W. 314, 315, 321, 322, 347
Engelman, D. M. 281, 299, 303
Erbland, J. F. 22, 33
Erf, L. 91, 94, 95, 110
Erwin, G. V. 29, 39
Escher, G. C. 77, 80, 86, 91, 94, 95, 96, 108, 109, 124, 131, 330, 352
Essner, E. 152, 159
Eto, Y. 237, 262
Euler, H. von 101, 109
Evans, A. J. 63, 73
Evans, J. S. 127, 132
Evans, V. J. 47, 50, 70, 144, 160

Fabiani, A. 249, 250, 251, 263, 268
Fagan, V. M. 148, 165, 271, 277, 278, 279, 287, 288, 291, 292, 295, 305
Fallani, A. 4, 9, 12, 36
Farber, E. 323, 347
Farias, R. N. 281, 297
Farr, A. L. 115, 132
Feizi, T. 188, 191
Feldbruegge, D. H. 278, 282, 305
Feldman, G. L. 116, 124, 134
Fenselau, A. 290, 299
Ferber, E. 146, 164
Ferrell, W. J. 29, 33
Fetzer, V. A. 77, 79, 111
Fidge, N. H. 98, 109, 279, 306
Fiecchi, A. 232, 233, 240, 262, 265
Fieser, L. F. 233, 262

Figard, P. H. 12, 23, 33, 45, 71, 141, 145, 160, 168, 191
Fillerup, D. L. 50, 54, 57, 71
Fillios, L. C. 279, 299, 303
Fine, A. W. 119, 121, 122, 123, 133
Fischer, H. O. L. 17, 31
Fischman, P. 168, 190
Fish, W. A. 232, 233, 262
Fishman, P. H. 154, 160
Fletcher, J. E. 56, 57, 73, 74
Folch-Pi, J. 113, 131, 132, 134
Foley, G. E. 151, 159
Forstner, G. G. 170, 171, 176, 191
Fortner, J. G. 309, 342, 353
Foster, D. W. 275, 283, 299, 302
Fothergill, J. 182, 194
Foulds, L. 323, 348
Foxman, C. J. 98, 109
Frantz, I. D. Jr. 338, 347
Frederick, G. L. 51, 71, 100, 109
Fredrickson, D. S. 97, 98, 109, 110
Freedman, R. B. 291, 297
Freedman, S. 186, 187, 188, 191
Friedewald, W. R. 323, 348
Friedricks, B. 147, 163
Friend, D. S. 293, 297
Fry, G. L. 58, 74
Fuhrer, J. P. 154, 166, 168, 195
Fujita, T. 106, 110
Fulling, J. H. 182, 183, 190
Fumagalli, R. 229, 231, 233, 234, 236, 237, 238, 240, 241, 246, 262, 265
Furukawa, K. 181, 182, 185, 193

Gahmberg, C. G. 139, 143, 152, 163, 164, 169, 191
Gaiti, A. 19, 23, 33
Galli, C. 142, 166, 231, 234, 236, 262
Galli, G. 231, 232, 233, 234, 240, 262, 263, 265, 268
Galli-Kienle, M. 232, 240, 262
Gammal, E. B. 309, 314, 315, 324, 327, 328, 337, 340, 346, 348
Gardas, A. 170, 171, 172, 173, 175, 178, 179, 191, 193
Garfinkel, E. 63, 69
Gargiulo, A. W. 182, 194

Garland, P. B. 284, 306
Gascoyne, T. 292, 302
Gasquet, P. de 292, 302
Gatica, M. 202, 203, 222
Gatmaitan, Z. 67, 72
Gatt, S. 25, 26, 41
Gehrke, C. W. 115, 133
Gerhardt, U. 217, 226
Gerschenson, L. E. 47, 54, 70, 138, 143, 145, 160, 161
Gerstl, B. 11, 33
Gey, G. O. 155, 156, 158
Gey, M. K. 155, 156, 158
Geyer, R. P. 54, 71, 143, 144, 155, 156, 160, 162
Ghanta, V. 171, 176, 193
Gibbs, C. C. 4, 34
Gibson, D. M. 285, 296, 303
Gielen, W. 128, 134, 243, 267
Giessen, G. J. van 127, 132
Gigg, J. 27, 33
Gigg, R. 27, 33
Gilbertson, J. R. 8, 29, 33, 34
Gilden, R. V. 151, 161
Gillespie, J. M. 86, 89, 100, 108
Ginsburg, V. 169, 194
Girgis, G. R. 216, 223
Glick, M. C. 168, 190
Glover, J. 245, 263
Goddard, P. 341, 342, 345, 348
Goebel, H. H. 249, 252, 261
Gofman, J. W. 89, 109
Gold, J. 187, 191
Gold, P. 186, 187, 188, 191
Golde, L. M. G. van 10, 24, 26, 27, 32, 33
Goldfarb, S. 148, 160, 253, 263, 271, 272, 278, 287, 294, 299, 301
Good, J. J. 77, 79, 111, 116, 124, 133
Goodell, B. W. 94, 97, 107
Goodman, D. S. 56, 71, 279, 306
Gopal, K. 4, 11, 34, 239, 240, 253, 263
Goracci, G. 19, 23, 33
Gordon, G. S. 246, 256, 264
Gore, I. 279, 299
Gor'Kova, N. P. 14, 32, 33
Gorniak, H. 170, 171, 172, 173, 175, 178, 179, 193

Goswitz, F. A. 4, 38
Gottfried, E. L. 13, 34, 139, 140, 160
Gould, E. 278, 280, 282, 298
Gould, R. G. 278, 279, 282, 299, 300
Graf, L. 128, 133, 153, 163, 243, 265
Grafnetter, D. 237, 262, 263
Grand, C. G. 313, 315, 346
Grand, L. C. 325, 349
Granner, D. 295, 300
Granner, D. K. 295, 300
Gray, G. M. 5, 9, 11, 12, 16, 34
Green, C. 57, 75
Green, H. 137, 166
Greenberg, D. M. 12, 23, 32, 33
Greene, E. M. 77, 80, 87, 89, 91, 94, 95, 96, 108, 124, 131
Greenspan, M. 284, 300
Gregolin, C. 285, 300
Gregor, O. 343, 348
Greiner, M. J. 326, 347
Griffin, A. 141, 142, 159
Griffin, A. C. 220, 223
Griffiths, J. B. 151, 160
Grimes, W. J. 168, 185, 191
Gromek, A. 46, 50, 72
Gronroos, J. 187, 191
Gross, S. K. 153, 164
Grossi, E. 4, 11, 34, 234, 236, 237, 239, 240, 253, 262, 263, 265
Grossi Paoletti, E. 229, 231, 232, 233, 234, 236, 240, 241, 249, 250, 251, 253, 259, 262, 263, 265, 268
Guarino, M. J. 254, 266
Guder, W. 278, 300
Guest, M. J. 278, 305
Guia, M. de 63, 69
Gupta, R. M. 89, 109
Gurcay, O. 252, 254, 261
Gurd, F. R. N. 86, 89, 100, 108
Gurin, S. 284, 297
Gyde, A. M. 287, 305

Haas, G. H. de 23, 27, 34, 37
Haenszel, W. 343, 348, 351
Haggerty, D. F. Jr. 47, 54, 71, 138, 143, 145, 160, 161, 255, 264
Hajra, A. K. 17, 18, 23, 28, 31, 34, 35, 60, 69, 144, 147, 157
Hakkinen, I. 186, 187, 191
Hakomori, S. 128, 129, 131, 132, 133, 134, 153, 154, 161, 168, 169, 170, 171, 172, 173, 175, 177, 178, 179, 180, 183, 185, 186, 187, 188, 189, 191, 192, 193, 194, 195, 196, 243, 263
Hallauer, C. 170, 192
Halpenny, G. W. 87, 88, 102, 110
Ham, R. G. 47, 71, 138, 161
Hamasato, Y. 170, 192, 193
Hamberger, A. 228, 230, 242, 263
Hamilton, J. G. 30, 36
Hamosh, M. 63, 73
Hamprecht, B. 278, 282, 296, 300, 303
Hanafusa, H. 151, 161
Hancock, R. L. 271, 272, 275, 276, 277, 287, 295, 301
Handa, S. 172, 193
Harary, H. 138, 160
Harary, I. 47, 54, 71, 138, 143, 145, 161
Harcourt, A. G. 325, 349
Hardesty, B. 209, 225
Harry, D. S. 148, 161, 271, 279, 280, 287, 288, 300
Hartzell, R. W. 145, 156, 164
Hasegawa, S. 14, 39
Hashimoto, T. 283, 284, 289, 303
Hasselquist, H. 101, 109
Hatanka, M. 151, 161
Hatch, F. T. 79, 86, 89, 95, 109, 110
Hausberger, F. X. 327, 348
Hauser, G. 146, 160, 254, 255, 257, 262, 263
Hauser, S. 288, 305
Haven, F. L. 3, 34, 51, 71, 100, 109, 308, 314, 348
Hawksworth, G. 309, 341, 348
Hay, R. J. 136, 161
Hayflick, L. 136, 161
Hayman, R. B. 11, 33
Healy, K. 17, 40
Heaysman, J. E. M. 151, 157
Hecht, L. I. 200, 223
Hehl, J. L. 29, 36
Heining, A. 5, 33
Heller, G. 214, 226

Heller, R. A. 279, 300
Helmich, O. 221, 223
Helms, J. 138, 145, 161, 259, 263
Helper, A. 86, 91, 94, 95, 96, 108
Hemminki, K. 228, 263
Hems, G. 337, 348
Henderson, J. F. 50, 68, 71
Henle, G. 137, 162
Henle, W. 137, 162
Hepp, D. 50, 71
Herbst, B. 47, 70
Heremans, J. F. 95, 110
Herrera, M. G. 285, 298
Hess, H. H. 115, 124, 132, 252, 253, 262
Heston, W. E. 277, 300
Heubner, R. T. 151, 161
Heyn, G. 29, 39
Hickman, P. E. 271, 272, 278, 279, 300, 304
Higashino, S. 137, 158
Higazi, A. M. 81, 86, 89, 109
Higgins, M. 278, 279, 300
Higginson, J. 342, 343, 348
Hijmans, J. C. 155, 156, 162
Hildebrand, J. 168, 193
Hilf, R. 4, 34
Hill, M. J. 309, 341, 342, 345, 347, 348
Hindley, S. T. 207, 222
Hinse, C. H. 236, 263
Hinz, H.-J. 281, 300
Hiramoto, R. 171, 176, 193
Hironaga, K. 106, 110
Ho, W. K. L. 67, 72
Hoagland, M. B. 200, 223
Hoak, J. C. 58, 74
Hoffman, D. 308, 330, 352
Hogan, E. L. 243, 244, 263
Hokin, L. E. 23, 34
Hokin, M. R. 23, 34
Holasek, A. 19, 35
Holland, J. F. 51, 70
Holley, R. W. 138, 165
Holmer, G. 28, 32
Holmes, R. 138, 145, 161, 259, 263
Holmes, W. L. 233, 234, 236, 237, 264
Horhammer, L. 119, 134
Horlick, L. 236, 263

Horowitz, M. I. 171, 177, 179, 194
Horton, B. J. 253, 263, 271, 272, 273, 276, 277, 278, 279, 280, 287, 288, 294, 300, 301, 304
Horton, J. D. 276, 279, 287, 288, 294, 301
Hosaka, K. 17, 18, 20, 41
Hoshino, K. 327, 348
Hoshino, T. 252, 254, 261
Howard, B. V. 27, 34, 62, 63, 68, 69, 71, 140, 141, 142, 143, 144, 145, 147, 148, 149, 154, 155, 156, 158, 161, 168, 193
Hradec, J. 197, 198, 199, 200, 201, 202, 203, 204, 205, 206, 207, 208, 209, 210, 211, 212, 213, 214, 215, 216, 217, 218, 220, 221, 223, 224, 225
Hsi, B. P. 338, 349
Hubbard, D. D. 285, 296
Hueper, W. C. 309, 350
Huggins, C. 325, 327, 343, 348, 349
Hutchison, C. F. 291, 301
Hyams, L. 330, 352

Ibuki, F. 209, 224
Iida, T. 170, 171, 196
Ikoda, T. 127, 132
Ikuta, F. 248, 263
Illingworth, D. R. 245, 263
Imai, Y. 279, 280, 306
Inbar, M. 151, 161, 168, 193
Ingenito, E. F. 156, 164
Insull, W. Jr. 338, 349
Irie, R. 5, 12, 40, 139, 143, 166, 170, 171, 172, 196
Iritani, N. 283, 284, 289, 303
Irving, C. C. 324, 349
Irving, D. 309, 341, 347
Iseki, S. 181, 182, 185, 193
Ishihara, K. 181, 182, 185, 193
Ishikawa, S. 335, 341, 352
Issenberg, H. J. 340, 350
Itoh, K. 4, 7, 34
Ivankovic, S. 249, 250, 262
Iwanaga, M. 170, 196

Jabara, A. G. 325, 349
Jacob, F. 208, 224
Jacobi, H. P. 310, 311, 314, 319, 320, 324, 345, 349

Jacobs, R. 46, 48, 71, 151, 162, 273, 274, 284, 286, 289, 302
Jacobs, R. A. 144, 145, 151, 162
Jacobson, H. I. 326, 349
James, M. 281, 282, 301
Jandl, J. H. 123, 132
Jansen, G. R. 291, 301
Janss, D. H. 324, 349
Jarvi, O. 187, 191
Jayaraman, J. 241, 263
Jeanloz, R. W. 128, 132, 170, 171, 172, 178, 185, 186, 187, 192
Jenkin, H. M. 7. 30, 39
Jensen, E. V. 326, 349
Jensen, L. C. 79, 86, 89, 110
John, K. 56, 74
Johnson, G. S. 151, 163
Johnson, M. 5, 34
Johnson, R. C. 29, 34
Johnson, R. M. 4, 34
Johnston, R. 232, 261
Jones, J. P. 232, 233, 265

Kaariainen, L. 139, 143, 152, 163, 164
Kabara, J. J. 231, 237, 264, 291, 301
Kabat, E. A. 188, 191
Kahnt, F. W. 86, 89, 100, 108
Kajitani, T. 335, 341, 352
Kalinina, E. V. 94, 109
Kameneva, T. I. 94, 109
Kamimura, M. 170, 171, 172, 196
Kandutsch, A. A. 236, 264, 271, 272, 275, 276, 277, 278, 280, 287, 295, 301
Kanfer, J. 23, 36
Kanfer, J. N. 245, 264
Kang, K. W. 13, 33
Kaplan, A. S. 145, 147, 158
Kaplan, B. H. 283, 303
Karnovsky, M. L. 8, 33
Kasama, K. 4, 7, 14, 34, 39
Kates, M. 3, 34
Katsuta, H. 50, 74, 143, 165
Katz, J. 59, 70
Katz, M. 151, 165
Kaufman, R. J. 77, 80, 86, 89, 91, 94, 95, 96, 108, 109, 124, 131
Kaufmann, R. 208, 224

Kawachi, T. 278, 279, 300, 301
Kawalek, J. C. 29, 34
Kawanami, J. 4, 9, 11, 35, 128, 132
Kawasaki, H. 180, 181, 182, 185, 193, 194
Kay, H. E. H. 182, 193
Kayden, H. J. 234, 246, 247, 248, 268
Keith, A. 281, 302
Kellen, J. 89, 109
Keller, D. 139, 158
Keller, D. L. 60, 70
Keller, E. B. 200, 223
Kennan, A. L. 286, 294, 298, 301
Kennaway, E. L. 341, 346
Kennaway, N. M. 341, 346
Keene, W. R. 123, 132
Kennedy, E. P. 10, 19, 23, 25, 27, 35, 37, 39
Kerr, H. A. 47, 50, 70, 144, 160
Kessler, R. J. 29, 33
Khor, H. T. 314, 316, 327, 329, 346
Kidder, E. D. 80, 86, 91, 94, 95, 96, 108, 109
Kiernan, J. A. 138, 165
Kijimoto, S. 128, 129, 132, 168, 185, 192, 193
Kilburn, E. 284, 289, 302
Kim, K. H. 284, 285, 298
Kimura, Y. 45, 46, 50, 71
Kipnis, D. M. 285, 298
Kirschbaum, A. 319, 321, 349
Kiyasu, J. Y. 10, 23, 35, 40
Klc, G. M. 28, 35
Kleinschmidt, A. K. 283, 284, 285, 300, 303
Kletzien, R. 152, 163
Kline, B. E. 318, 319, 322, 324, 328, 349, 350
Klintworth, G. K. 155, 156, 162
Kmet, J. 342, 349
Kobayakawa, K. 279, 299
Koch, M. A. 147, 159, 168, 191, 193
Kohn, G. 137, 162
Kojola, V. B. 279, 299
Kolesnichenko, T. S. 198, 225
Kolodny, E. H. 154, 159, 168, 185, 190

Komárková, E. 207, 208, 209, 215, 220, 224, 225
Komeiji, T. 45, 46, 50, 71
Kopelovich, L. 274, 301, 304
Koprowski, H. 151, 159
Kornacker, M. S. 282, 302
Kornberg, A. 17, 20, 35
Kosaki, T. 127, 132
Koscielak, J. 170, 171, 172, 173, 175, 178, 179, 185, 186, 187, 191, 192, 193
Kossjakow, P. N. 170, 171, 193, 195
Koster, J. 8, 36
Kostić, D. 128, 132, 243, 244, 264
Kotani, Y. 127, 132
Kovarik, S. 182, 183, 190
Kraemer, P. M. 151, 154, 162
Krebs, H. A. 292, 302
Kreuzer, T. 208, 225
Krishnaiah, K. V. 282, 302
Kritchevsky, D. 30, 36, 43, 62, 71, 140, 141, 142, 143, 144, 145, 148, 154, 156, 161, 164, 168, 193, 233, 234, 236, 237, 264
Kritchevsky, G. 230, 231, 266
Kroes, J. 281, 302
Kroes, R. 294, 302
Kruml, J. 198, 224
Kuhl, W. E. 62, 63, 71
Kulas, H. P. 168, 193
Kumanishi, T. 248, 263
Kumar, S. 283, 285, 303
Kummerow, F. A. 101, 105, 106, 110
Kunkel, H. G. 86, 89, 109
Kurihara, M. 331, 337, 339, 340, 343, 348, 351
Kurland, C. G. 216, 225
Kurokawa, T. 168, 195
Kusin, Z. 170, 171, 195
Kuz'mina, S. N. 14, 35
Kyser, K. 326, 349

LaBelle, E. F. Jr. 23, 35
Lacassagne, A. 341, 349
Lakshmanan, M. R. 278, 286, 298, 302
Lalla, O. de 89, 109
Landrey, J. R. 256, 266
Lands, W. E. M. 20, 25, 35, 40

Landsteiner, K. 170, 193
Lane, M. D. 281, 282, 283, 284, 285, 298, 300, 302, 303, 306
Lang, P. D. 338, 349
Langan, J. 59, 70
Lapar, V. 288, 305
Lavik, P. S. 310, 311, 312, 313, 315, 319, 320, 321, 322, 324, 325, 328, 329, 349
Law, M. D. 201, 225
Lawrason, F. D. 319, 321, 349
Lazarus, H. 151, 159
Lea, A. J. 309, 330, 338, 349, 350
Leader, D. P. 216, 225
Leblond, C. P. 170, 190
Ledeen, R. W. 116, 134
Lee, C. L. 182, 183, 190
Lee, J. A. H. 340, 350
Lee, P. 10, 39
Lee, T.-C. 8, 36
Lees, M. 113, 131
Lees, R. S. 43, 71, 95, 97, 98, 109, 110
Legakis, N. J. 49, 72
Lehrer, G. M. 228, 264
LeKim, D. 29, 39
Lengle, E. 148, 151, 162
Lennard-Jones, J. E. 87, 102, 110
Lepage, G. A. 50, 68, 71
Leren, P. 338, 347
Letnansky, K. 28, 35
Lev, M. 313, 315, 346
Leveille, G. A. 285, 298
Lever, W. F. 86, 89, 100, 108
Levey, G. S. 272, 273, 285, 288, 297
Levine, A. S. 45, 71, 141, 145, 160, 168, 191
Levine, E. M. 151, 159
Levis, G. 60, 71
Levy, P. L. 285, 298
Levy, R. I. 97, 98, 99, 109, 110
Lewin, E. 115, 124, 132
Lewis, G. M. 49, 72
Lewis, L. A. 87, 110
Lewis, M. R. 136, 162
Lewis, W. H. 136, 162
Limnell, I. 101, 109
Lin, D. S. 232, 261
Lin, T. M. 342, 350

Lindberg, M. 25, 39
Lindgren, F. T. 79, 86, 89, 110
Lindlar, F. 13, 35, 239, 264
Linn, T. C. 278, 279, 281, 282, 302
Lis, H. 168, 193
Littman, M. L. 52, 68, 71
Liu, C. H. 86, 89, 100, 108
Lockhard, W. L. 47, 73
Loftfield, R. B. 205, 225
Long, H. W. 47, 73
Losticky, C. 94, 110
Loud, A. V. 278, 280, 282, 298
Love, W. C. 45, 75
Lowenstein, J. M. 282, 284, 300, 302
Lowenstein, W. R. 137, 158
Lowenthal, A. 245, 246, 264
Lowry, O. H. 115, 132
Lubich, W. 246, 256, 264
Lucas, D. O. 146, 162
Ludwig, E. H. 54, 70
Lumb, R. H. 26, 30, 35
Lynch, R. D. 155, 162
Lynch, T. P. Jr. 115, 118, 119, 125, 133
Lynen, F. 278, 279, 282, 284, 296, 297, 298, 300, 303

Mabuchi, K. 309, 330, 335, 340, 342, 353
Mach, O. 212, 214, 215, 220, 224
MacKenzie, C. G. 143, 144, 154, 155, 156, 162
MacKenzie, J. B. 143, 144, 154, 155, 156, 162
MacMahon, B. 340, 342, 350
Macpherson, I. 153, 164, 168, 190
Madison, L. L. 285, 307
Mailhe, H. 145, 156, 164
Majerus, P. W. 46, 48, 71, 144, 145, 151, 162, 273, 274, 284, 286, 289, 302
Makajeva, Z. 170, 171, 195
Maker, H. S. 228, 264
Makita, A. 171, 175, 176, 195
Malamos, B. 60, 71
Malataner, A. M. 170, 194
Malcolm, G. T. 217, 222
Malek, A. 81, 86, 89, 109
Malins, D. C. 117, 132
Malone, B. 4, 7, 8, 10, 16, 18, 19, 23, 25, 27, 29, 37, 38, 40
Mancini, G. 95, 110
Mangold, H. K. 7, 8, 28, 32, 36, 117, 132
Manning, J. A. 67, 72
Mansour, K. 81, 86, 89, 109
Manteca, A. 11, 33
Mantel, N. 330, 352
Mantozos, J. 60, 71
Maragoudakis, M. E. 43, 71
Marchesi, V. T. 179, 194
Marcus, D. 170, 171, 175, 194
Marggraf, W. D. 147, 159, 168, 193
Marinetti, G. V. 22, 33, 145, 159
Marsch, W. L. 188, 191
Martinell, M. 156, 164
Martinez-Polomo, A. 151, 162
Marton, L. J. 246, 256, 264
Maruchi, N. 309, 342, 353
Marwan, G. 94, 108
Mary, G. E. S. 101, 105, 106, 110
Masamichi, H. 239, 240, 241, 242, 264
Masamune, H. 180, 181, 185, 194
Maspes, P. E. 229, 264
Masukawa, A. 180, 194
Matamala, M. 202, 203, 222
Matsuda, T. 284, 307
Matsumoto, M. 113, 132, 170, 196
Matsumoto, R. 113, 132
Matsuyama, T. 331, 337, 339, 340, 351
Matteis, F. de 292, 302
Matthaei, H. 208, 209, 210, 211, 213, 216, 220, 224, 225
Matuda, Y. 180, 194
Matzke, J. 246, 261
Maun, M. E. 314, 315, 325, 347
May, M. D. 281, 298
Mays, E. T. 52, 72
McCluer, R. H. 243, 266
McEntegart, H. 182, 194
McEwen, H. D. 217, 222
McFarland, V. 154, 160
McFarland, V. W. 128, 132, 154, 159, 160, 162, 168, 185, 190, 194
McGarrahan, K. 278, 280, 282, 298
McGarry, J. D. 275, 302
McGrath, H. 285, 304
McIntosh, R. 145, 147, 162

McIntyre, N. 148, 161, 271, 279, 280, 287, 288, 300
McKeehan, W. L. 209, 225
McKibbin, J. M. 171, 175, 176, 193, 194, 195
McMorris, F. A. 255, 268
McNamara, D. J. 281, 302
McPherson, I. 136, 165
Mead, J. F. 47, 50, 54, 57, 71, 138, 143, 145, 161, 255, 264
Mead, J. G. 138, 160
Medes, G. 44, 50, 52, 54, 57, 59, 64, 65, 72
Meezan, E. 168, 195
Mehler, A. H. 204, 225
Meites, J. 325, 327, 346, 350, 352
Mellanby, E. 309, 311, 319, 323, 352
Mellman, W. J. 137, 138, 162
Melnik, J. L. 136, 137, 159, 168, 169, 171, 187, 195
Menkes, J. H. 145, 162
Mennel, H. D. 249, 250, 262
Menon, I. A. 208, 225
Menšík, P. 198, 201, 206, 224
Merance, D. R. 182, 195
Mercer, G. 138, 145, 161, 259, 263
Merker, P. L. 52, 69, 101, 102, 108
Merler, E. 146, 162
Mermier, P. 44, 54, 55, 56, 64, 66, 67, 69, 72
Messer, J. 252, 254, 261
Meyer, G. 151, 162
Michaels, M. A. 183, 194
Michel, I. 4, 34
Micu, D. 5, 13, 35
Mider G. B. 51, 72
Miettinen, M. 94, 110
Migliori, J. C. 50, 54, 57, 71
Mihailescu, E. 5, 13, 35
Miller, B. J. 91, 94, 95, 110
Miller, E. C. 318, 319, 324, 350
Miller, J. A. 318, 319, 324, 349, 350
Miller, K. 152, 163
Millington, R. H. 50, 58, 75
Milstein, S. W. 327, 348
Miras, C. 60, 71
Miras, C. J. 49, 72

Misconi, L. 285, 297, 303
Mishkin, S. 67, 72
Mittleman, D. 86, 89, 100, 108
Moehl, A. 8, 35
Moldave, K. 209, 212, 224, 225
Monod, J. 208, 224
Monro, E. R. 214, 226
Montagnier, L. 137, 138, 162
Moon, R. C. 324, 327, 349
Moore, J. H. 237, 264
Moorhead, P. S. 136, 161
Mora, G. 202, 203, 222
Mora, P. T. 5, 34, 128, 131, 132, 154, 158, 159, 160, 162, 168, 185, 190, 194
Morero, R. D. 281, 297
Morgan, A. P. 285, 298
Morgan, R. W. 342, 350
Morganti, P. 237, 262, 263
Morii, S. 327, 349
Moritsch, P. 170, 191
Morris, H. P. 7, 10, 14, 27, 34, 38, 45, 46, 48, 49, 51, 59, 67, 70, 71, 72, 73, 75, 101, 102, 105, 106, 110, 148, 151, 161, 162, 165, 168, 190, 217, 219, 225, 271, 272, 273, 274, 275, 278, 279, 282, 284, 286, 287, 288, 289, 291, 292, 295, 296, 297, 299, 300, 302, 304, 305, 307
Morton, H. S. 87, 88, 102, 110
Morton, J. J. 51, 72
Mosbach, E. H. 52, 68, 71, 288, 305
Moskowitz. M. S. 155, 163
Moss, J. 281, 282, 283, 284, 302, 303, 306
Mott, G. E. 253, 263, 287, 294, 301
Moulton, M. J. 241, 266
Mouton, R. F. 86, 89, 100, 108
Mreana, G. 99, 111
Mueller, C. H. 80, 108
Mueller, P. S. 52, 72, 97, 110
Mühlbock, O. 325, 350
Muir, C. 337, 340, 347
Mulder, I. 11, 12, 39
Müldner, H. G. 118, 132
Munshower, J. 295, 296
Murakami, W. T. 128, 132, 153, 161, 168, 188, 189, 192, 243, 263
Murray, R. K. 128, 134, 154, 166, 168, 190, 255, 268

Musser, E. A. 127, 132
Muto, Y. 285, 303

Nagai, Y. 245, 264
Nairn, R. C. 182, 194
Nakagawa, H. 95, 96, 110
Nakagawa, S. 127, 132
Nakanishi, S. 283, 284, 289, 303
Nanava, I. G. 88, 110
Narayan, K. A. 51, 67, 70, 72, 79, 101, 102, 105, 106, 109, 110, 219, 225
Nardella, P. 273, 274, 307
Nayyar, S. N. 11, 35
Neacsu, C. 99, 111
Neimark, J. M. 155, 156, 160
Nepokroeff, C. M. 278, 286, 298, 302
Ness, G. C. 278, 302
Neth, R. 214, 226
Nettesheim, P. 16, 37
Neufeld, A. H. 87, 88, 102, 110
Neuman, R. E. 256, 264
Neville, H. E. 238, 266
Newell, L. R. 243, 244, 263
Ng, F. 285, 297
Ni, Y. 182, 183, 190
Nicholas, H. J. 230, 231, 232, 233, 234, 264, 265
Nicholls, D. M. 216, 223
Nichols, A. V. 63, 72
Nicolson, G. L. 168, 194
Niel, C. B. van 290, 306
Nielsen, S. L. 252, 254, 261
Niemiro, R. 233, 238, 262
Nigam, V. N. 129, 132
Nishimura, S. 170, 171, 172, 196
Nishioka, M. 106, 110
Niwa, T. 45, 46, 50, 71
Noble, R. C. 237, 264
Nojima, T. 151, 166
Nolte, I. 278, 300
Nordheim, W. 217, 226
Norton, W. T. 228, 242, 264, 265
Numa, S. 17, 18, 20, 41, 283, 284, 289, 303
Numata, M. 326, 349
Nüssler, C. 278, 300
Nydegger, U. E. 95, 96, 110, 229, 264

Ockner, R. K. 67, 72
O'Gara, R. W. 309, 350
Ogilvie, J. W. 283, 303
Oh-Uti, K. 180, 185, 194
Ohya, A. 94, 111
Okazaki, T. 283, 284, 289, 303
Olson, J. A. 233, 264
Onodera, Y. 239, 240, 241, 242, 264
Opalka, E. 4, 9, 39, 240, 252, 253, 267
Osawa, T. 168, 195
Ostwald, R. 281, 302
Otsuka, H. 4, 9, 11, 35
Overath, P. 278, 279, 282, 298
Oyama, K. 181, 185, 194

Packie, R. M. 280, 301
Paddle, B. M. 285, 297
Paden, G. 44, 50, 52, 64, 72
Page, I. H. 87, 110
Painter, N. S. 341, 346
Paleg, L. 272, 273, 287, 303
Palekar, L. 234, 251, 252, 254, 261, 268
Pallavicini, C. 169, 191
Paltauf, F. 4, 19, 35
Pangborn, M. C. 170, 194
Paoletti, P. 4, 11, 34, 229, 234, 236, 239, 240, 241, 246, 249, 250, 251, 253, 259, 262, 263, 264, 265, 268
Paoletti, R. 142, 166, 229, 231, 232, 233, 234, 236, 238, 240, 246, 253, 259, 262, 263, 265
Papas, T. S. 204, 225
Pardee, A. B. 151, 152, 159, 163
Parker, R. F. 136, 165
Parmeggiani, A. 208, 225
Pastan, I. 151, 163
Pasternak, C. A. 147, 163
Payne, S. 145, 147, 162
Pearce, M. L. 338, 350
Pearson, M. L. 255, 268
Peck, F. 240, 267
Peck, W. A. 145, 159
Pedersen, B. N. 46, 50, 72
Peery, C. V. 151, 163
Perdue, J. F. 152, 163
Perlmann, P. 188, 195
Peškova, D. 99, 111

Petering, H. G. 127, 132
Petermann, M. L. 80, 86, 91, 94, 95, 96, 108, 109
Peterson, J. A. 62, 72, 155, 156, 163
Peterson, R. F. 118, 133, 134
Pezzotta, S. 249, 250, 251, 263
Pfleger, R. C. 25, 35
Philippart, M. 170, 171, 191, 194
Piantadosi, C. 6, 10, 17, 18, 20, 23, 25, 27, 28, 29, 32, 35, 38, 39, 40, 168, 190
Piasek, A. 170, 171, 172, 173, 175, 178, 179, 193
Pierson, R. W. Jr. 138, 165
Pitot, H. C. 148, 160, 253, 263, 269, 271, 272, 278, 279, 287, 288, 294, 299, 301, 303
Plotkin, H. R. 182, 195
Plunkett, E. R. 309, 314, 315, 324, 327, 328, 337, 340, 346, 348
Poduslo, S. E. 228, 242, 264, 265
Pogell, B. M. 284, 306
Polheim, D. 4, 35
Poon, Y. C. 13, 33
Popova, G. M. 243, 265
Poppenhausen, R. B. 67, 72
Porcellati, G. 19, 23, 33
Porter, J. W. 278, 279, 282, 283, 284, 285, 286, 298, 299, 302, 303, 305
Potop, I. 99, 111
Potter, V. R. 273, 303
Predmore, G. 152, 163
Prendergast, R. C. 182, 194
Preussmann, R. 249, 250, 262
Pricer, W. E. Jr. 17, 20, 35
Promyslov, M. S. 243, 265
Pronczuk, A. 279, 299, 303
Prusóvá, F. 343, 348
Prusse, E. 50, 71
Puck, T. T. 136, 166

Quackenbush, F. W. 281, 302
Quigley, J. P. 141, 142, 146, 163
Quijada, C. L. 285, 297
Qureshi, A. A. 278, 282, 305

Rabin, B. R. 291, 297, 303, 307
Rabinowitz, Z. 137, 163
Radin, N. S. 171, 194

Raff, R. A. 50, 72
Raghavan, S. 23, 36
Rainey, W. T. Jr. 18, 38
Ramasarma, T. 272, 278, 279, 281, 282, 283, 302, 303, 306
Ramsey, R. B. 230, 231, 232, 233, 265
Randall, C. 51, 71
Randall, R. J. 115, 132
Ransohoff, J. 234, 235, 246, 247, 248, 249, 251, 252, 256, 258, 259, 261, 267, 268
Rao, G. A. 28, 36
Rapp, F. 151, 165
Rapport, M. M. 13, 36, 128, 133, 153, 163, 228, 243, 265
Rawlins, F. A. 231, 238, 265
Ray, T. K. 5, 9, 36
Re Cecconi, D. 234, 236, 262
Reddy, B. 309, 335, 343, 353
Reddy, B. S. 341, 350
Reed, C. F. 62, 72
Reed, F. 155, 163
Rees, E. D. 4, 5, 13, 36
Regen, D. 278, 282, 303
Reggio, R. B. 116, 124, 133
Reich, E. 141, 142, 146, 163
Reichard, G. A. Jr. 285, 298
Reid, D. E. 137, 159
Reisner, R. M. 30, 36
Reiss, O. K. 143, 144, 154, 155, 156, 162
Rejnek, J. 94, 99, 110, 111
Renkonen, O. 10, 36, 139, 143, 152, 163, 164
Řeřábková, E. 99, 111
Resch, K. 146, 164
Rhoads, D. 23, 36
Rice, L. I. 279, 305
Richman, N. 216, 225
Riepertinger, C. 278, 282, 303
Rifkin, D. P. 141, 142, 146, 163
Ritter, M. C. 283, 303
Roach, D. 115, 133
Robbins, P. W. 128, 133, 153, 164, 168, 195
Rodriguez, G. A. 254, 266
Rodwell, V. W. 272, 278, 279, 281, 302, 303, 305

Author Index

Roe, F. J. C. 322, 323, 351
Roger, J. 23, 34
Rohr, A. 143, 160
Rosebrough, N. J. 115, 132
Roseman, S. 168, 185, 190
Rosenberg, R. N. 254, 266
Rosenblum, I. 4, 9, 39, 252, 253, 267
Rossiter, R. J. 145, 159
Rothblat, G. H. 30, 36, 138, 144, 145, 148, 149, 155, 156, 164, 254, 256, 257, 259, 266, 280, 306
Rothman, J. E. 281, 299, 303
Rous, P. 323, 348
Rouser, G. 116, 124, 134, 230, 231, 266
Roy, S. C. 5, 9, 36
Rubin, H. 62, 72, 136, 155, 156, 163, 165
Ruddle, F. H. 255, 268
Rudney, H. 278, 279, 282, 300, 301, 307
Ruggieri, S. 4, 9, 12, 36
Rusch, H. P. 308, 310, 311, 318, 319, 320, 322, 324, 328, 329, 345, 346, 349, 350
Russell, W. C. 145, 147, 162
Rutstein, D. D. 156, 164
Ryder, E. 283, 285, 300, 302
Rytter, D. J. 147, 164

Sabine, J. R. 46, 48, 73, 148, 164, 270, 271, 272, 273, 274, 275, 276, 277, 278, 279, 280, 281, 282, 284, 285, 287, 288, 289, 292, 293, 299, 300, 301, 303, 304, 307
Saccone, G. T. P. 285, 288, 304
Sachs, L. 136, 137, 151, 158, 161, 163, 164, 168, 193
Sacktor, B. 60, 73
Sadoff, L. 342, 350
Saffiotti, U. 322, 351
Saito, T. 153, 161, 168, 192
Saka, T. 127, 132
Sakakida, H. 279, 304
Sakamoto, Y. 95, 96, 111
Sakiyama, H. 128, 133, 153, 164
Sakurai, Y. 168, 195
Salaman, M. H. 322, 323, 351
Sambrook, J. 136, 164
Samuelsson, K. 245, 266

Sand, D. M. 29, 36
Sander G. 208, 225
Sander, S. 326, 351
Sanders, J. 118, 134
Sansone-Bazzano, G. 30, 36
Santos, E. C. 57, 73
Sanwal, B. D. 255, 268
Sarda, L. 23, 34
Sato, G. H. 255, 261
Sato, S. 50, 59, 73
Satoh, T. 279, 299
Saucier, S. E. 236, 264, 278, 301
Sauer, F. 278, 304
Savary, P. 23, 26, 27, 33
Savchuck, W. B. 47, 73
Savluchinskaya, L. A. 198, 225
Scala, A. 233, 265
Scallen, T. J. 238, 266, 283, 305
Scanu, A. 89, 111
Schachter, H. 183, 194
Scheer, J. van der 170, 193
Scheiffarth, F. 94, 108
Scheithauer, E. 11, 33
Schelling, E. L. 47, 144, 160
Schiff, F. 170, 194
Schiffer, D. 249, 250, 251, 263, 268
Schilp, A. O. 240, 267
Schilling, E. L. 47, 50, 70
Schlenk, H. 29, 36
Schmid, H. H. O. 7, 8, 29, 36, 39
Schmid, K. 86, 89, 100, 108
Schogt, J. C. M. 8, 36
Scholefield, P. G. 50, 59, 73
Schotz, M. C. 279, 305
Schrappe, O. 245, 266
Schremmer, J. M. 27, 40
Schroepfer, G. J. Jr. 238, 266
Schultz, A. M. 168, 185, 190
Schuster, M. W. 283, 305
Schutta, H. S. 238, 266
Schweet, R. S. 202, 225
Scott, J. F. 200, 223
Scow, R. O. 63, 73
Segi, M. 331, 337, 339, 340, 351
Seidman, I. 285, 298
Seifert, H. 128, 133, 243, 244, 266
Sela, B.-A. 168, 193

Selverstone, B. 241, 266
Serrone, D. 4, 39
Shabad, L. M. 198, 225
Shah, E. 80, 108
Shah, E. B. 63, 69, 77, 79, 81, 86, 89, 104, 108
Shah, S. N. 236, 263, 266
Shamberger, R. J. 329, 351
Shapiro, D. J. 278, 279, 281, 305
Sharon, N. 168, 193
Shaw, J. 171, 176, 193
Shediac, C. C. 279, 304
Shefer, S. 288, 305
Shein, H. M. 146, 160, 252, 253, 254, 255, 257, 262, 263
Sheinin, R. 128, 134, 154, 166
Shen, L. 169, 194
Sherman, C. D. Jr. 51, 72
Shigematsu, T. 309, 330, 341, 352, 353
Shils, M. E. 308, 351
Shipp, J. C. 59, 73
Shoham, J. 151, 164
Shohet, S. B. 146, 162
Shonk, C. E. 28, 32, 60, 70
Shubik, P. 322, 325, 346, 351
Shuck, A. E. 4, 5, 13, 36
Siddiqui, B. 128, 129, 132, 133, 168, 170, 192, 195, 243, 266
Silberberg, M. 319, 351
Silberberg, R. 319, 351
Siler, J. 212, 225
Silides, D. J. 228, 264
Silverstein, A. M. 170, 194
Silverstone, H. 308, 309, 313, 314, 316, 318, 329, 351, 352
Silvestri, L. 208, 225
Simer, F. 64, 70
Simmons, D. A. R. 188, 195
Simons, K. 139, 143, 152, 163, 164
Sinha, D. 325, 351
Siperstein, M. D. 148, 165, 259, 266, 270, 271, 272, 273, 275, 276, 277, 278, 279, 280, 287, 288, 289, 291, 292, 295, 297, 298, 304, 305, 306
Sirtori, C. R. 253, 254, 259, 263, 266
Skipski, V. P. 52, 63, 69, 77, 79, 80, 81, 86, 87, 89, 91, 94, 95, 96, 101, 102, 104, 108, 111, 113, 115, 116, 117, 118, 119, 120, 121, 122, 123, 124, 125, 127, 128, 131, 133, 134, 153, 163, 219, 222, 243, 265
Skogerson, L. 212, 225
Slagel, D. E. 4, 11, 36, 228, 241, 242, 266
Slakey, L. L. 278, 279, 282, 299, 305
Slomiany, A. 171, 177, 179, 195
Slomiany, B. L. 177, 179, 195
Slot, E. 13, 37
Slotboom, A. J. 27, 37
Sly, W. S. 144, 145, 151, 162
Smith, E. L. 171, 175, 176, 193, 195
Smith, J. K. 11, 33
Smith, J. L. 148, 151, 162
Smith, M. B. 46, 48, 71, 151, 162, 273, 274, 284, 286, 289, 302
Smith, M. E. 231, 234, 238, 262, 265
Smith, R. R. 4, 9, 37, 239, 240, 261, 267, 268
Smith, S. 285, 305, 326, 349
Smith, S. W. 27, 37
Smolowe, A. F. 117, 119, 134
Snyder, C. 16, 37
Snyder, F. 2, 3, 4, 6, 7, 8, 10, 11, 14, 16, 17, 18, 19, 20, 23, 25, 27, 28, 29, 30, 31, 32, 35, 37, 38, 39, 40, 60, 68, 75, 144, 158, 165, 168, 190, 195, 217, 225, 239, 267
Soboroff, J. M. 57, 58, 62, 68, 74
Soderberg, J. 12, 39, 126, 134
Soeldner, J. S. 285, 298
Sokoloff, L. 280, 306
Soloff, B. L. 57, 65, 70
Someda, K. 252, 254, 261
Sommerau, J. 198, 206, 224
Soodsma, J. F. 25, 29, 39
Sorokina, I. B. 14, 32, 33
Spanner, S. 25, 26, 31
Spector, A. A. 44, 45, 49, 50, 52, 53, 54, 55, 56, 57, 58, 59, 60, 61, 62, 63, 64, 65, 66, 67, 68, 70, 71, 73, 74, 138, 144, 145, 155, 156, 165, 254, 255, 267
Spikes, J. L. Jr. 88, 102, 111
Spohn, M. 251, 267
Srere, P. A. 283, 284, 299, 306
Ställberg-Stenhagen, S. 206, 222

Stanier, R. Y. 290, 306
Starr, J. L. 216, 226
Staszewski, J. 343, 351
Steele, W. 7, 30, 39
Stein, A. A. 4, 9, 39, 240, 252, 253, 267
Stein, L. 67, 72
Stein, O. 22, 24, 39, 279, 306
Stein, Y. 22, 24, 39, 279, 306
Steinberg, D. 45, 49, 50, 54, 57, 58, 59, 60, 61, 64, 65, 66, 74, 145, 155, 156, 165
Steiner, S. 168, 169, 171, 187, 195
Steinke, J. 285, 298
Stellner, K. 168, 170, 171, 173, 175, 178, 179, 180, 183, 192, 195
Stener, B. 5, 14, 31
Stenhagen, E. 206, 222
Stepanov, A. V. 170, 171, 195
Stephens, N. 6, 7, 8, 35, 38
Stephenson, M. L. 200, 223
Stevenson, G. T. 314, 347
Stewart, A. G. 51, 74, 100, 111
Stewart, L. 309, 350
Stock, C. C. 77, 80, 89, 94, 95, 96, 108, 113, 115, 118, 119, 121, 122, 123, 124, 125, 131, 133
Stoffel, W. 29, 39, 116, 124, 134
Stoffyn, P. 113, 134
Stoffyn, P. J. 113, 131
Stoll, E. 283, 302
Stonehill, E. H. 293, 304, 306
Stoker, M. 136, 165
Stoker, M. G. P. 138, 165
Stokes, W. M. 232, 233, 262
Strehler, B. L. 136, 161
Strobel, G. 168, 191
Stroufová, A. 198, 217, 224
Strycharz, G. D. 170, 171, 172, 192
Strychmans, P. A. 168, 193
Sturtevant, J. M. 281, 300
Subba Rao, G. 278, 283, 306
Subbarayan, C. 278, 282, 305
Suda, M. 284, 307
Sugiyama, T. 281, 282, 306
Sullivan, R. C. 117, 134
Sulya, L. L. 239, 261
Suovaniemi, O. 228, 263

Surgenor, D. M. 86, 89, 100, 108
Suzuki, C. 171, 175, 176, 195
Suzuki, K. 237, 238, 262, 267
Suzuki, S. 170, 171, 196
Svennerholm, L. 119, 134, 153, 165, 228, 230, 242, 243, 244, 263, 267
Swahn, B. 245, 262, 267
Swan, D. 208, 225
Swann, A. 275, 291, 306
Swanson, R. F. 284, 306
Sweeley, C. C. 116, 124, 128, 133, 134
Sweeney, M. J. 285, 306
Sweet, W. H. 252, 254, 261
Swell, L. 201, 225
Swim, H. E. 136, 145, 158, 165
Swyryd, E. A. 278, 279, 282, 299, 300
Szepsenwol, J. 229, 267, 313, 319, 351

Tabakoff, B. 29, 39
Taguchi, T. 52, 68, 71
Takahashi, T. 29, 39
Takano, A. 335, 341, 352
Takaoka, T. 50, 74, 143, 165
Takasugi, Y. 279, 280, 306
Taketa, K. 284, 306
Tan, C. S. 287, 304
Tanaka, A. 57, 74, 156, 165
Tannenbaum, A. 308, 309, 310, 311, 313, 314, 315, 316, 318, 319, 320, 324, 325, 327, 328, 329, 351, 352
Tarnowski, G. S. 113, 123, 124, 133
Tavaststjerna, M. G. 11, 33
Teather, C. 168, 192
Teebor, G. 285, 298
Tellem, M. 182, 195
Temin, H. M. 136, 138, 165
Tepper, S. A. 233, 234, 236, 237, 264
Terebus-Kekish, O. 52, 63, 69, 77, 79, 80, 81, 86, 87, 89, 91, 94, 95, 96, 101, 102, 104, 108, 124, 131
Tetsuro, M. 239, 240, 241, 242, 264
Tevethia, S. S. 136, 137, 151, 159, 165
Theise, H. 217, 226
Thomas, A. J. 44, 50, 54, 57, 59, 72
Thompson, E. B. 295, 306
Thompson, G. A. Jr. 10, 17, 39
Thompson, W. R. 170, 194

Thorne, R. F. W. 283, 306
Tichy, J. 246, 267
Tietz, A. 25, 39
Tijo, J. H. 136, 166
Tilley, C. A. 183, 194
Tillman, S. F. 62, 63, 68, 69, 155, 156, 158
Todaro, G. 137, 166
Toman, R. 343, 348
Tomita, M. 168, 195
Tomkins, G. M. 295, 300, 306
Torkhovskaya, T. I. 14, 32, 33, 35, 127, 131
Tot, P. D. 182, 194
Toyne, P. H. 325, 349
Toyoda, N. 94, 111
Trautman, R. 86, 89, 109
Troitskaya, L. P. 14, 35
Trouillas, P. 229, 267
Trucco, R. E. 281, 297
Trueman, L. S. 137, 159
Tsao, S. S. 140, 166
Tschudy, D. P. 292, 297
Tsintsadze, T. M. 88, 110
Tsuji, T. 4, 9, 11, 35
Tubbs, P. K. 284, 306
Tuna, N. 7, 8, 36
Turinsky, J. 285, 306
Turpeinen, O. 338, 347
Tyson, S. 342, 350
Tytell, A. E. 256, 264

Ueta, N. 5, 12, 40, 139, 143, 166
Uezumi, N. 4, 7, 14, 34, 39
Uhlenbruck, G. 128, 133, 134, 243, 244, 266, 267
Uhr, J. W. 99, 111
Ukita, T. 168, 195
Unger, R. H. 285, 307
Urna, G. 234, 238, 262
Uroma, E. 86, 89, 100, 108
Usardi, M. 4. 11, 34, 239, 240, 253, 263
Uzman, B. G. 238, 265

Vagelos, P. R. 283, 307
Vance, D. E. 116, 124, 134
Vance, J. E. 285, 297

Vandenheuvel, F. A. 246, 265
Vanhoude, J. 168, 193
Vasquez, D. 214, 226
Veerkamp, J. H. 11, 12, 39
Vicar, G. 188, 191
Vitetta, E. S. 99, 111
Vlahakis, G. 277, 300
Vogt, P. K. 153, 161, 168, 192
Vost, A. 63, 74

Waard, F. de 330, 337, 352
Wada, E. 45, 46, 50, 71
Wagener, H. 13, 35
Wagner, H. 119, 134
Walker, A. R. P. 341, 346, 352
Wallace, B. H. 182, 193
Wallace, J. C. 292, 307
Wallach, D. F. H. 12, 39, 126, 134, 228, 267
Wallis, K. 290, 299
Walton, M. 7, 17, 30, 31, 40, 139, 143, 144, 158
Warner, E. D. 58, 74
Warner, G. A. 168, 180, 183, 195
Warner, H. R. 25, 40
Warner, R. C. 285, 300
Warren, L. 154, 166, 168, 190, 195
Watabe, H. 294, 307
Watanabe, K. 170, 171, 173, 175, 178, 179, 187, 188, 192, 195
Waterhouse, J. 337, 340, 347
Waters, M. 285, 303
Watkin, D. M. 52, 72, 97, 110
Wattenberg, L. W. 329, 352
Watson, A. F. 309, 311, 319, 323, 352
Watson, J. A. 49, 74, 145, 148, 166, 280, 294, 307
Watson, J. D. 216, 226
Weber, G. 11, 40, 45, 75, 295, 296
Weber, M. J. 142, 166, 168, 196
Webster, G. C. 203, 226
Wechsler, W. 238, 239, 267
Weinhouse, S. 44, 50, 52, 54, 57, 58, 59, 64, 65, 70, 72, 73, 75
Weinstein, D. B. 139, 143, 152, 166
Weis, H. J. 278, 307
Weisburger, J. H. 294, 302, 343, 352
Weiss, C. 228, 264

Weiss, E. de C. 234, 248, 249, 251, 252, 254, 256, 258, 259, 261, 267, 268
Weiss, H. 50, 71
Weiss, J. F. 231, 232, 233, 234, 235, 241, 246, 247, 248, 249, 250, 251, 252, 253, 256, 258, 259, 261, 262, 263, 267, 268
Weiss, S. B. 19, 23, 27, 35, 37, 40
Welsch, C. W. 327, 346, 352
Wherrett, J. R. 118, 128, 132, 134, 154, 166, 170, 171, 176, 177, 178, 185, 191, 195
White, F. R. 329, 352
White, H. B. Jr. 2, 4, 9, 37, 142, 166, 229, 231, 239, 240, 241, 242, 261, 262, 267, 268
White, L. W. 278, 282, 307
Whitmore, W. F. 309, 340, 353
Wicks, M. B. 277, 304
Wiegandt, H. 244, 268
Wieland, O. 50, 71, 278, 300
Williams, C. D. 145, 150, 151, 158, 166, 280, 294, 296
Williams, D. J. 291, 307
Williams, G. M. 294, 302
Williams, R. E. O. 309, 341, 342, 348
Williams, R. H. 285, 297
Williamson, W. P. 241, 261
Willis, D. B. 216, 226
Wilson, C. B. 4, 11, 36, 228, 241, 242, 246, 252, 254, 256, 261, 264, 266
Wilson, D. E. 43, 71
Wilson, L. 44, 55, 56, 69
Windeler, A. S. 116, 124, 134
Windisch, F. 217, 226
Winkley, J. 342, 350
Wintrobe, M. M. 123, 134
Witebsky, E. 170, 195
Wolff, P. 119, 134
Wolffram, E. von 94, 111
Wollemann, M. 229, 268
Woo, N.-C. 342, 350
Wood, R. 3, 7, 8, 11, 14, 16, 17, 38, 40, 60, 68, 75, 144, 165, 168, 195, 230, 239, 267, 268
Wool, I. G. 216, 225

Wooten, W. B. 243, 244, 263
Wray, V. L. 152, 163
Wright, J. D. 57, 75
Wu, H. 168, 195
Wykle, R. L. 10, 16, 17, 18, 19, 20, 23, 27, 28, 29, 32, 37, 38, 40
Wynder, E. L. 308, 309, 330, 335, 340, 341, 342, 343, 350, 352, 353

Yamada, T. 14, 39
Yamagishi, M. 283, 284, 303
Yamaguchi, Y. 181, 185, 194
Yamakawa, T. 5, 12, 40, 139, 143, 166, 170, 171, 172, 196
Yamamoto, A. 116, 124, 134, 230, 231, 266
Yamashita, S. 17, 18, 20, 41
Yanagihara, T. 228, 268
Yang, H.-J. 128, 134, 170, 171, 178, 185, 186, 196
Yasuda, M. 52, 75
Yau, T. M. 142, 166, 168, 196
Yavin, E. 25, 26, 41
Yogeeswaran, G. 128, 134, 154, 166, 168, 190, 255, 268
Yokono, O. 279, 299
Yoshida, T. 292, 302
Yoshikawa-Fukada, M. 151, 166
Yoshimura, S. 338, 349
Yosizawa, Z. 171, 175, 176, 180, 194, 195
Youlos, J. 169, 191
Yu, R. K. 116, 134
Yugari, Y. 284, 307
Yukio, Y. 239, 240, 241, 242, 264

Zajdela, F. 341, 349
Zamecnik, P. C. 200, 220, 223, 226
Zamfirescu-Gheorghiu, M. 5. 13, 35
Zanetti, M. E. 291, 301
Zbarskii, I. B. 14, 35
Zimmerman, H. M. 238, 248, 268
Zuckerman, N. J. 273, 274, 307
Zülch, K. J. 249, 250, 262

Subject Index

A,B,H, *see* Blood groups, Fucolipids, Glycolipids, Glycoproteins
Acatalasemic mice 281
Acetate
   effect on fatty acid oxidation 59, 64
   incorporation into cholesterol 148–150, 232, 241, 254, 255, 280, 282
   incorporation into fatty acids 44, 45, 48, 50, 146, 254, 255, 274
   incorporation into lipids 17, 51, 146
   oxidation 45, 59, 274
   supply of 283
   uptake by Ehrlich cells 58
Acetoacetate 59
2-Acetylaminofluorene 314, 315, 321, 322
Acetyl CoA 18, 46, 281–283, 289, 290
Acetyl CoA carboxylase 46, 48, 151, 274, 283–286, 289
Acetyl CoA carboxylase kinase 284, 285
Acetyl CoA synthetase 50
N-Acetylgalactosamide 176
N-Acetylgalactosamine 115, 116, 118, 130, 176, 180, 181, 183, 184, 188
N-Acetylglucosamine 176, 177, 181, 182, 185, 186
N-Acetylneuraminic acid 124, 130
   see also Sialic acid
Acids, *see* Fatty acids
Acylation of lysophospholipids 21, 22
Acyl CoA 273, 284, 285, 289

Acyl CoA reductase 29
Acyl dihydroxyacetone,
   see Dihydroxyacetone
Acyl transferases, *see* Transferases
Adenocarcinoma,
   see also individual tissues
   effect of diet on incidence of 313, 319
   fatty acid synthesis by 44, 45
   growth retardation on fat-free diet 52
   human
      blocked synthesis of glycolipids in 183, 184
      deletion of blood group antigens 182
      fucolipids in 128, 170–172, 178, 183–187, 189
      neoproteolipid in 114
Adenoma, pulmonary
   disaturated phosphatidylcholine in 16
Adenosine triphosphate 19, 22, 283, 289
Adenyl cyclase 295
Adhesiveness, cellular 151, 168
Adipocytes 43
Adipose tissue 5, 8, 52, 66, 324, 327, 341
Adrenalectomy
   influence on cholesterol synthesis in hepatomas 271, 272
Adrenalin, *see also* Epinephrine
   effect on HMG CoA reductase 278
Adrenals 326

Subject Index

Adrenocorticotropin 66
Aflatoxin 276, 291, 293
Age
  effect on serum lipoproteins 95, 101, 104
  in relation to cancer incidence 338
  in relation to desmosterol levels 234
Agglutination of human erythrocytes 169
Agglutination reaction, mixed cell 182
Agglutinin, wheat germ 185–187
Albumin
  bound to free fatty acids 55–59, 64, 66, 96, 97, 100, 154
  bound to fatty acid methyl ester 62
  in ascites plasma 53
  in culture medium 54, 58, 66
  production of, by rat liver slices 199
Albumin-fatty acid ratio 53, 61, 62
Alcohols, see Fatty alcohols
Aleukemia 335, 338
Alk-1-enylacylglycerolipids,
  see also Plasmalogens
  biosynthesis 19
  fatty acid composition 8, 16
  hydrolysis by pancreatic lipase 27
  occurrence 3, 7–11, 14–16, 144, 168, 239
  thin-layer chromatography 3
O-Alkyl cleavage enzyme 30
Alkylacylglycerolipids
  conversion to alk-1-enylglycerolipids 19
  fatty acid composition 7, 16
  hydrolysis by pancreatic lipase 27
  occurrence 3, 5–8, 10, 11, 14–16, 30, 168, 239
  thin-layer chromatography 3, 6
1-Alkyl-2-acyl-sn-glycerol-3-phosphate 22
Alkylacylphosphorylcholine 11, 14–16, 19
Alkylacylphosphorylethanolamine 11, 14–16, 19
Alkyldihydroxyacetone,
  see Dihydroxyacetone

Alkyl ethyleneglycols 29
1-Alkyl-sn-glycerol-3-phosphate 22
Alkylglycerols 3, 8, 10
  biosynthesis 23
  cleavage 24–26, 29
  conversion to alk-1-enylglycerols 17
Alloxan diabetes 292
Allylisopropylacetamide 292
Americans
  cancer mortality relative to Japanese 339
  fecal steroid excretion 341
  linoleic acid intake 338
Amino acids
  activation 199
  incorporation into protein 199, 201
  in proteolipids 126
Amino acid-tRNA ligase 200, 202–208, 211
Amino acid-tRNA synthetase 206, 211
Aminoacyl-tRNA 199–207, 209, 211, 212, 214, 216, 220
Aminoazo dyes 318, 324, 344
Aminoglutethimide phosphate 291
Aminolevulinic acid synthetase 292
Ammonium sulphate fractionation 202, 203, 209
Animal protein, see Protein
Antibodies against glycolipids 169, 185, 186
Antigenic activity, coincident with ethanolamine phospholipids 13
Antigens
  allogeneic 168
  blocked synthesis 182–185
  blood group 169–189
  carcinoembryonic 180, 187, 188, 229
  deletion of 180–182
  embryonal 293
  fetal 294
  Forssman 170, 186, 187
  fucolipids as 169
  Lewis 170, 178, 180, 181, 185, 186
  syngeneic 168
  transplantation 168, 178
  tumor 151, 186–188
  Wasserman's syphilis 170

Antioxidants, lipid 250
Antisera 53, 168, 169, 186
Aorta 63, 156
Apoproteins of serum lipoproteins 99
Arachidonic acid
  biosynthesis in cultured cells 143, 255
  in brain tumors 240, 242
  in cultured cells 47, 142, 143, 168
  in 2-position of glycerides 7, 16
Ascites plasma
  as source of cellular lipids 44, 54–56, 62–64
  fatty acids 52–54
  lipoproteins 53, 54, 62–64
  triglycerides 53, 62–64
Ascites tumors, see individual types
Ascorbic acid 25, 26
Aspartate transcarbamylase 292
Astrocytes 146, 252, 254, 255
Astrocytoma cells 254–258
Astrocytomas 146, 238–240, 243, 246, 251–258
  see also Brain tumors
Atherosclerosis, dietary treatment of 338
ATP, see Adenosine triphosphate
Autoradiography 170
AY-9944 238, 254, 259

Baboon cell lines, fucolipids in 169, 188
Bacteria
  blood group B producing 187
  intestinal, role in carcinogenesis 341, 342, 345
  role of lipids in aminoacyl-tRNA synthesis 205
Barrett mammary adenocarcinoma, see Mammary tumors
Base exchange reactions in phospholipid biosynthesis 19
Beef liver peroxidase 281
Benz(α)pyrene 218, 310, 311, 320
BHK21 cells 137, 139, 143, 145
Bicarbonate, effect on fatty acid biosynthesis 45, 50

Bile
  carcinogens in 342
  protein inhibiting cholesterogenesis 283
Bile acids
  carcinogenic properties 341
  fecal 341
  metabolism 345
Bile salts, effect on cholesterol biosynthesis 272, 287
Biliary passages, cancer of 334, 335
Biotin 289
Bladder
  carcinoma cells, absence of A and B antigens 182
  neoplasms in relation to dietary fat 330, 335
Blood, radioactivity after injection of tritiated estradiol 326
Blood-brain barrier 228, 245
Blood cells, see also Erythrocytes, Leukemic cells
  neoproteolipids in 120–123
Blood group
  antigens 169–189
    blocked synthesis 182–185
    deletion of 180–182
    of erythrocytes 170–175
    of gastrointestinal and glandular tissues 176, 177
    of tumor tissue 177–188
    relation to carcinoembryonic antigen 188
  glycolipids 167–189
  glycoproteins 180–186
  haptens 170–175, 180, 187
  polysaccharide 180
Blood lipids,
  see also Blood plasma/serum
  comparison with cerebrospinal fluid lipids 245
Blood plasma/serum, see also Serum
  cholesterol 14-methylhexadecanoate in 218, 219, 222
  lipids in cancer 51, 52
  lipoproteins 77–107

Subject Index

see Lipoproteins
neoproteolipids 119–130
    see Neoproteolipids
Blood supply of hepatomas 294
Body weight and cancer incidence 329, 330
Bone cells, normal and transformed
    phospholipid biosynthesis 145
Bovine lymphosarcoma cells
    fatty acid composition 54
BP8/C3H ascites carcinoma 16
Brain
    chick embryo, sterol synthesis in 231, 232, 236
    developing 228, 230–238, 240, 243
    enzymes cleaving plasmalogens 25, 26
    fetal 242
    lipids 11, 28, 227–260
    microsomes 27, 242
    sterols 227–260
Brain tumors 4, 9, 11, 128, 227–260
    chemotherapy 229
    cholesterol 232, 239–241, 248, 250–252, 255, 259, 260
    cholesterol esters 239, 240, 255
    classification 238, 239, 243
    cultures of 145, 254–260
    desmosterol 240, 241, 248, 253, 255–260
    diagnosis 243, 246, 259
    effect on cerebrospinal fluid lipids 245, 246
    effect on hepatic lipids 253
    effect of dietary fat on 229, 319
    fatty acid composition 239, 240, 242, 253, 255
    glycolipids 128, 239, 242–244, 246, 252, 255, 260
    growth of 241, 253, 254, 259
    induction of 248–250
    lipid composition 4, 9, 11, 128, 239–244, 252–260
    models for research 248–260
    phospholipids 11, 146, 229, 239, 241, 242, 252, 255, 260
    reviews on 229

sterols 229, 232, 239–241, 248, 250–260
transplanted 252–254, 259
triglycerides 4, 239, 252, 254
Branched-chain fatty acids, see Fatty acids
Breast cancer, see also Mammary tumors
    effects on serum lipoproteins 80, 82, 83, 86, 88, 91, 95, 96
    epidemiology of 309, 331, 335–342, 345
    in migrant populations 342, 343
    in relation to
        body weight 330
        caloric intake 336, 338
        dietary fat 331, 335–342, 345
        dietary fibre 341
        dietary protein 334, 336–338, 341
        intestinal bacteria 342
        hormonal influences 342
    neoproteolipids in 99, 114, 125, 126
BRL-62 cells, regulation of lipid biosynthesis in 150
Bronchiogenic cancer 125, 186, 333, 335
Brown fat, phospholipids of 14
Brush border membranes
    glycolipids and fucolipids in 170, 171, 177
Buccal cavity, neoplasms
    in relation to dietary fat 335
Butter, in diet 310, 311, 316
Butyrate, oxidation in hepatomas 59

Calcium
    requirement by base exchange enzymes 20
    uptake and release by mitochondria 283
Calf liver, cholesteryl 14-methylhexadecanoate in 201
Caloric intake, in relation to tumorigenesis 308, 309, 322, 327–329, 330, 336, 338, 339, 343–345
Caloric restriction 328, 329
Cancer, see also Tumors, Tumorigenesis
    blood test for 129

Subject Index

defective control of lipid biosynthesis in 270–278, 286–296
definition of 269, 270
environmental influences 342, 343
epidemiology of 329–345
etiology 340, 342, 343
family history of 80–85, 89–92, 98, 107
genetic factors in 342
geographical distribution 320–343
glycerolipids and 1–32
incidence, in relation to mortality 340
lipoproteins in relation to 76–107
mortality 309, 329–339, 345
neoproteolipids in 113, 114, 118–126, 129–131
role of cholesteryl 14-methylhexadecanoate in 217–222
role of nutrition, reviews on 308
Cancer cells, see Cell lines, Cells in culture, Neoplastic cells
Capillary, triglyceride hydrolysis in 63
Carbon dioxide, effect on fatty acid biosynthesis 45, 46
Carbohydrate, dietary
effect on fatty acid biosynthesis 48
relation to other dietary constituents 338, 339
Carbon tetrachloride, impairment of liver function 219
Carcass, loss of lipid in tumor-bearing animals 51, 52
Carcinoembryonic antigen, see Antigens
Carcinogenesis, see also Tumors, Tumorigenesis
chemical 218, 294
mode of action of dietary fat 322–329
transplacental 249
Carcinogens
absorption from skin 323, 324, 344
chemical
2-acetylaminofluorene 314, 315, 321, 322
aflatoxin 276, 291, 293
aminoazo compounds 276, 293, 318, 324, 344

coal tar 310, 311, 323, 340
ethionine 276, 292, 293
N-2-fluorenylacetamide 101, 105, 276, 293
3-hydroxyxanthine 114
nitrosoureas 248–252, 254, 257, 259
polycyclic hydrocarbons 101–104, 218, 248, 252, 309–329, 344
stilbestrol 314, 315, 325
in bile 342
in intestine 341, 345
Carcinolipin, see Cholesteryl 14-methylhexadecanoate
purification 198
role in protein synthesis 199–217
Carcinoma, see also individual tissues
BP8/C3H ascites 16
775 13
Ehrlich ascites, see Ehrlich carcinoma
epidermoid bronchogenic 87, 92, 125
KB 27, 140, 147
Krebs 51, 217
Landschutz ascites 5, 9, 12, 16, 22, 24
Shionogi 115 4, 9
squamous cell 182
TA3 64
Carcinosarcoma, see Walker 256
Cardiolipin 153, 170
Carnitine 59
Cartilage, fibroblasts derived from 138
Cecal tumor, fucolipids in 186
Cell
adhesiveness 151
cultures, see Cells in culture
differentiation 157, 220, 242
division 147, 152, 293
fractions, see Microsomes, Mitochondria, Ribosomes
neuronal and glial, isolation of 228
genetics 157
lines 7, 28, 31, 47, 50, 54, 99, 137–156, 168, 169, 254–259
membranes, see Membranes
sap 199, 201, 207, 220, 221
surface 24, 57, 189, 228, 242
antigenic specificities 168

marker 169, 170
membranes 151, 152, 167, 169, 231
transport 151
Cells in culture
characteristics 136
ether-linked lipids 7, 28, 31, 144, 147
fatty acids 47, 54, 141–143, 255
growth requirements 47, 136, 138, 156, 259
lipid biosynthesis 138, 144–147, 254, 255, 280
composition 7, 139–144, 152, 156, 187–189, 248, 255
metabolism 31, 144–154, 254–260
transport 63, 154–156
lipids, reviews of 138
membranes 151, 152, 157
nutrition 138
surface properties 151–157
transformed 28, 128, 136–157, 168, 171, 185, 187–189, 256
Ceramides
in brain tumors 242–244
in cultured cells 153, 154
in fucolipids 170, 172–179, 185, 189
in hepatomas 128
in neoproteolipids 118, 130
Cerebrosides
immunological activity 229
in brain 229, 230, 243, 245
in brain tumors 242, 243
in cerebrospinal fluid 245, 246
in cultured cells 255
in plasma 245
Cerebrospinal fluid, lipids of 229, 243, 245–248, 260
Cerebrum, sterols in 234, 235, 250–252
Cervical cancer
antigens 182
effect on serum lipoproteins 94
lipid composition 9, 13
Chang liver cells, lipid content 139, 140
CHD3 Chinese hamster cells, essential fatty acid requirement 47
Chick embryo
fibroblasts, see Fibroblasts
sterols 231, 232, 236, 237

Children
cancer in 83–85, 87, 93, 96, 125
serum lipoproteins 77, 83–85, 87, 93, 96, 125
CHL-1 cells, arachidonate synthesis by 255
Chlorophenoxyisobutyrate, effect on fatty acid uptake by Ehrlich cells 58
Cholestanol 233
Cholesterol 200, 290, 294
biosynthesis 290
diurnal rhythm 271, 272, 276, 278, 279, 282
effect of drugs 238, 240, 241, 253, 254, 258, 260, 272
effect of fasting 270, 271, 276–278, 282, 283, 296
feedback control 49, 148–150, 257, 259, 270–272, 275–283, 286–288, 291, 293–295
in brain 231–234, 236, 238
in brain tumors 240, 241, 253, 254, 259, 275
in chick embryo tissues 231, 232
in cultured cells 49, 144, 145, 148–150, 157, 254, 255, 257, 259, 280, 294
in hepatomas 49, 270–272, 286–288, 294, 295
in liver 276, 278–283, 293, 294
catabolism 240, 291
dietary
effect on tumorigenesis 229, 313, 319
effect on sterol biosynthesis 49, 259, 271, 279, 282, 294, 295
effect on protein synthesis 200, 201, 206
esterification 200, 221, 237
fecal 341
in brain 228, 230–238, 251
in cerebrospinal fluid 245–247
in cultured cells 138–142, 156, 248
in lipoproteins 86
in neoproteolipids 115–117, 130
in nerve 228, 234–236

Subject Index

in serum 51, 94, 95, 245
in tumors 217, 239–241, 250–252
metabolism, reviews of 270
precursors 233–236, 248, 259
uptake 138, 148, 251–253, 294
Cholesteryl esters 60, 87, 200, 205
effect on cholesterol synthesis 280
fatty acid composition 237, 240
in ascites plasma 53
in brain 231, 236, 237
in cerebrospinal fluid 245, 246
in cultured cells 139, 142, 155, 156
in liver 279
in plasma 240
in tumors 217, 239, 240
radioactive 221
role in gene expression, see
    Cholesteryl 14-methylhexadecanoate
synthesis 221, 237, 255
Cholesteryl 14-methylhexadecanoate
and malignant growth 217–222
distribution 198, 217–222
concentration 217–219
effects on protein synthesis 199–217
effect on transcription 208
quantitative determination 199
synthesis 219, 221, 222
transport 219
Cholestyramine 271, 276
Choline, incorporation into lipids 146
Choline phosphate 147
Chorioallantoic membranes,
see Membranes
Choriocarcinoma 86, 92
Chromatography
clathrate 198
column
    cholesteryl 14-methylhexadecanoate
        198, 202, 203, 209
    glycolipids 171, 173–176
    neoproteolipids 113, 115, 121–123, 125–127
gas-liquid
    amino acids 115
    fatty acids 46, 116, 124, 198, 199
    monosaccharides 116, 124

sphingosine bases 116, 124
sterols 232, 234, 236, 240, 246
gas-liquid, mass spectra
    sterols 232, 240
gel filtration 188, 242
thin-layer
    cholesteryl esters 199
    glycerolipids 3, 5–7
    glycolipids 172, 174–176, 185, 243
    neoproteolipids 112, 113, 115, 116, 118–126
Chylomicra 53, 81, 100, 101
Citrate 283, 284, 289
Citrate cleavage enzyme 282
CMH, see Cholesteryl 14-methylhexadecanoate
Coal tar 310, 311, 323, 340
Cocarcinogenesis 309, 322, 323, 325
Coconut oil 310–312, 314–316, 327
hydrogenated 318
Cod liver oil 311
Colon, cancer
and dietary fat 331, 334, 341, 345
effect on serum lipoproteins 86, 92, 96
incidence in emigrants 343
lipids in 125, 128
Column chromatography,
see Chromatography
Conjunctiva cells, triglyceride uptake 155
Contact inhibition 151, 154, 167, 168
Corn oil 310, 312, 314–318, 320, 324–328
Coronary disease, correlation with cancer 330
Correlation coefficients
between dietary components 339
cancer mortality and dietary fat 335, 337, 338
Cottonseed oil 316
partially hydrogenated 310–315, 318, 320, 321, 328
Croton oil, promoting action 325
Culture medium
cholesterol free 259

Subject Index

cholesterol in 149
fatty acids 50, 54, 65, 66, 144
glucose effects on lipid metabolism 58, 60, 64–66
lipid content 30, 49
lipid deficient 47, 256–259, 294
lipid free 50, 68, 138, 143, 144, 150
lipoproteins 53, 54
Neumann-Tytell 256
serum free 54, 143, 255, 259
serum in 54, 138, 143–145, 148, 150, 255, 256, 258, 259
serum requirement 138
triglycerides 62
vitamin B_{12} 255
Cyclic adenosine monophosphate (cyclic AMP)
binding protein 295
dibutyryl derivative 273, 285
relation to lipid biosynthesis 272, 273, 285, 286, 288, 295
Cytidine diphosphate choline (CDP-choline) 19
Cytolipin-H 243, 244

Decanoic acid 58
7-Dehydrocholesterol 238, 254, 258, 259
reductase 254
7-Dehydrodesmosterol 238, 254, 257
Dehydrogenase shuttle enzymes 147
Demyelination 228, 236
Desmosterol 233, 259, 272
in brain tumors 240, 241, 253, 255, 260
in cerebrospinal fluid 246–248, 259
in cultured cells 248, 255–260
in culture medium 259
in developing nervous system 234–238
in plasma 254
in skin 236
in tumors 250, 251, 253, 254, 259
synthesis 231, 240, 241, 246
Desmosterol/cholesterol ratio 235, 236, 246, 250, 251, 256, 258
Desmosterol reductase
inhibitors of 254, 257–259

Determinants, blood group
blocked synthesis, in tumors 182–185
deletion 180–182
Diabetes 216, 285, 292
Diacylglycerols 4, 8–10, 16, 20, 24
20,25-Diazacholesterol 237, 248, 254, 257, 259
Dibenzanthracene 310, 320, 321
Dietary carbohydrate
correlations with other dietary variables 338, 339
effect on fatty acid biosynthesis 48
Dietary cholesterol
effect on cholesterol biosynthesis 49, 271, 272, 276–283, 286–288, 294, 295
effect on tumor incidence 229, 313, 319
Dietary fat
and tumorigenesis 308–345
correlation with cancer mortality 330–345
correlation with other dietary variables 338, 339, 345
effect on fatty acid biosynthesis 48, 272–274, 276, 277, 283–286, 289, 294–296
effect on fecal steroid excretion 341
effect on growth rate of tumors 52
effect on serum lipids 51
effect on serum lipoproteins 87, 93, 101, 105
effect on tumor incidence 229, 309–322
mode of action 322–329
effect on tumor lipids 52, 68
Dietary fiber 341
Dietary protein
correlation with cancer mortality 336–338, 341, 343
correlation with other dietary variables 338, 339, 345
effect on tumor incidence 329
Dihydroxyacetone, alkyl 22
Dihydroxyacetone phosphate 17, 18, 28, 45, 60, 144

Subject Index

acyl- 18, 19, 23, 27, 28, 144, 147
alkyl- 22, 23, 27
   biosynthesis 19, 28, 29
p-Dimethylaminoazobenzene 318
Dimethylbenz(α)anthracene 101–104, 314–317, 324–326, 328, 337, 344
  distribution and metabolism 324
N,N'-Diphenyl-p-phenylenediamine 249, 250
Diploid cell lines,
  see also Cells in culture
  characteristics 136, 137
  compared to established cell lines 138–157
Diurnal rhythm, of cholesterol synthesis 271, 272, 276, 278, 279, 282
DMBA, see Dimethylbenz(α)anthracene
DNA 146, 208, 324
Docosadienoic acid 47
Docosahexenoic acid 7, 16
Docosatrienoic acid 47
Dog
  heart, lipids of 6
  intestine, fucolipids 171, 175–177
  prostatic cancer 343
DPPD, see N,N'-Diphenyl-p-phenylenediamine
Drugs
  acting on brain tumor sterols 229, 240, 241, 253–260
  affecting cholesterol biosynthesis 238, 240, 241, 253–260, 272
  affecting fatty acid metabolism 43
  effect on lipoproteins 43, 96
  hypocholesterolemic 52, 237, 238, 257, 259, 291
  hypolipidemic 58
  means of getting into tumor cells 68
Dunning hepatoma, see Hepatomas
Dura mater
  desmosterol in 240
  fatty acid composition 240
Dutch ovarian tumor,
  see Ovarian tumors
Dysplasia, diminished blood group reactivities 182
Dysplastic epithelium, isoantigens in 182

Ear duct tumors 321, 322
Eel serum, determination of blood group activity 180, 181
Egg
  consumption, correlation with cancer mortality 330
  lipids, effects on tumor incidence 313, 314, 319
  yolks, source of carcinolipin 198
Ehrlich ascites carcinoma
  effect on blood lipids 51, 105
Ehrlich ascites cells
  ether-linked lipids 5, 6, 14, 27, 147
  fatty acids 5, 12, 13
    biosynthesis 44–46, 50, 67, 145
    desaturation and chain elongation 46, 47, 67
    release 65, 66
    sources 44
    turnover 55, 64–67
    utilization 54–62, 66, 67
  glycerol-3-P dehydrogenase activity 28
  lipid biosynthesis 17, 28, 45, 49, 50, 145, 147
  neutral lipids 5–7, 9, 62, 139
  triglyceride uptake 62, 63, 155, 156
Ehrlich ascites plasma
  lipids and lipoproteins 52–56, 62–64, 68
Ehrlich carcinoma
  cholesteryl 14-methylhexadecanoate 217
  fatty acids 13
  neoproteolipid 114
  phospholipids 13, 23
Eicosadienoic acid 47
Eicosatrienoic acid 47
Eicosenoic acid 47, 329
Electrophoresis
  carcinoembryonic antigen 188
  glycoproteins 180
  lipoproteins 51, 53, 77–79, 83, 87–89, 97, 101, 102, 105, 106
Elipin, see Aminoglutethimide phosphate
Embryo
  chick, see Chick embryo

neoproteolipid 119, 120, 130
rat, see Rat embryo
Endometrium, cancer of, association with obesity 330
Environmental influences on tumor formation 322
Environmental temperature, and cancer incidence 330
Enzymes, see specific enzymes
  membrane-bound 151, 274, 279
E0771 adenocarcinoma 114
Ependymoblastoma, Zimmerman 253
Ependymomas 238, 252, 253
  see also Brain tumors
Epidemiology of human cancer 309, 329–334
Epidermis, malignant,
  see also Skin cancer
  phospholipid and fatty acid composition 13
Epidermoid bronchogenic carcinoma 87, 92, 125
Epinephrine 66
  see also Adrenalin
Episome 137
Epithelial mucosa
  enzymes converting glycolipids 183, 184
Epitheliomas, see Skin tumors
Epithelium, metaplastic, dysplastic 182
Erbland-Marinetti pathway 22
Erucic acid 329
Erythrocyte
  blood group
    antigens 169
    haptens 170–175
  fucolipids 169–175, 177, 179, 189
  membrane, see Membranes
  neoproteolipid 123, 129, 130
  uptake of lipoprotein phospholipids 62, 155
Esophagus, neoplasms
  mortality in relation to fat consumption 335
Essential fatty acids
  biosynthesis 46

deficiency 46, 47, 68, 101, 105, 231
requirement by cultured cells 47, 138
role in tumor cells 46–48
Established cell lines,
  see also Cells in culture
  characteristics 136, 137
  compared to diploid cell lines 138–157
Esterified fatty acids, see Fatty acids
Estradiol
  inhibition of mammary tumor induction 326, 327
  tritiated, distribution in tissues 325, 326
  urinary, in relation to breast cancer 342
Estriol 342
Estrogens, effect on serum lipoproteins 96, 99
Estrone 342
Ethanolamine plasmalogen 11, 14–16
  biosynthesis 19
  cleavage 26, 27
  fatty acid composition 16
  in brain 230
Ether glycerolipids
  biosynthesis 17–23, 30, 31, 147
  catabolism 25–27, 29
  occurrence in brain 230, 231
  occurrence in neoplastic tissues 3–16, 144, 157, 168, 239, 241
  regulation in cancer cells 28–31
Ethionine 276, 292, 293
Ethyleneglycols, alkyl 30
Ethylnitrosourea, see Nitrosourea
Ewing's sarcoma, see Sarcoma

Fabry's disease, accumulation of fucolipids in 171, 177, 178
Family history of cancer, serum lipoproteins and 80–85, 89–92, 98, 107
Farnesyl pyrophosphate 282
Fasting
  effect on cholesterol biosynthesis 270, 271, 276–278, 282, 283, 296
  effect on fatty acid biosynthesis 48, 272–274, 276, 282–286, 289, 294

Subject Index

effect on lipoproteins of ascites
    plasma 53
serum lipoproteins in 77, 78, 80, 101
Fat, *see also* individual fats and oils
    animal 338, 339, 343
    brown 14
    dietary, *see* Dietary fat
    heated 309, 312
    intestinal 341, 345
    mobilization in tumor-bearing
        animals 100
    saturated versus unsaturated 314–316,
        329, 344
    vegetable 338, 339, 343
Fatty acid binding protein, intracellular
    67
Fatty acid synthetase 46, 48, 274, 286
Fatty acids, *see also* individual acids
    biosynthesis
        effect of drugs 43, 273
        control of 48, 49, 151, 272–277,
            282–286, 288, 289, 294–296
        in adipose tissue 327
        in brain tumors 253
        in Ehrlich ascites cells 44–46, 50
        in cultured cells 50, 143–145, 151
        in hepatomas 44–46, 48, 50,
            272–275, 283, 288, 289, 294, 295
        in other tumors 44, 45
    branched-chain 198, 205, 206
    chain elongation and desaturation 46,
        47, 60, 67, 143, 255, 274
    dietary 52, 231
    essential, *see* Essential fatty acids
    esterified 52, 53, 56, 62–67
    formation from fatty aldehydes 25
    free, *see* Free fatty acids
    hydroxy 176
    in ascites cells 7, 12, 13, 16, 47, 60
    in ascites tumor plasma lipids 52–54
    in brain 230, 231, 237
    in carcinolipin 198
    in cholesteryl esters 198, 237, 240
    in cultured cells 47, 54, 141–143, 255
    in fucolipids 176

    in glycerolipids 4, 5, 7, 9, 11–13, 16,
        60, 242
    positional distribution 7, 60
    in mammary gland 327
    in mammary tumors 5, 13
    in neoproteolipids 115–117, 124, 127,
        130
    in tumors 4, 7, 12, 13, 17, 239, 240,
        242, 253
    metabolism 43–69
    incorporation into lipids 13, 55,
        60–62, 65
    interconversion with fatty alcohols 29
    methyl esters 62, 63
    oxidation 44, 55, 58–60, 62–65, 67,
        274
    polyunsaturated 12, 47
    requirement by tumors 43, 44, 47, 50,
        56, 138
    sources of 44–54, 56
    turnover 51, 64–67
    uptake 57, 58, 138, 155, 156
    utilization 49, 50, 54–67
Fatty acyl CoA synthetase 59
Fatty alcohols
    biosynthesis 25, 29
    in ether glycerolipids 4, 7, 8, 16–19,
        28, 30
Fatty aldehydes 24–26, 29
Feedback inhibition,
    see Cholesterol biosynthesis
Feline intestine, fucolipid 176
    sarcoma-leukemic virus, *see* Virus
Fetal antigen, *see* Antigens
α-Feto-protein 106
Fiber, dietary 341
Fibroblasts 68, 294
    see also Cells in culture
    avian 136
    chick 189
    chick embryo 62, 141, 142, 145, 146,
        152, 155, 156
    cornea 156
    diploid 136–157, 259
    established, *see* Transformed
    hamster embryo 141, 142

Subject Index

human 136, 139, 168, 280
  aorta 156
  connective tissue 139, 140
  skin 138, 150, 259
  L 27, 54, 63, 144, 148, 149, 155, 255–257, 280
  L-929 139, 143, 145, 156, 256
  L-M 7, 30, 139, 143, 144
  MAF 139, 140
  mouse 63, 169
  transformed 28, 136–157, 168, 189
  WI-38 27, 138, 140–143, 145, 148–150, 156
  WI-38VA13A 27, 140–143, 148–150
Fibroma 114
Fibrosarcoma 114
Fish, formation of fatty alcohols in 29
N-2-Fluorenylacetamide 101, 105, 276, 293
Food consumption, estimation of 339, 340
Forssman antigen, see Antigens
Free fatty acids
  binding to albumin, see Albumin
  esterification 58, 60–62
  in ascites tumor cells 5
  in ascites tumor plasma 52–56, 64
  in brain 231
  in plasma/serum 52, 55, 56, 96, 97, 100, 154
  mobilization 52
  modulators of fatty acid synthesis 49
  pools 57, 58
  release
    from adipocytes 43
    from tumor cells 65–67
  turnover 51, 55, 66
  uptake 57, 58
  utilization, by tumors 54–62
Fructose
  effect on fatty acid biosynthesis 45
  effect on fatty acid oxidation 64
Fucolipid research, development of 171
Fucolipids 167–189
  see also Glycolipids
  A and B active 171–175, 178–185, 189
  antigenic properties 169, 170
  as blood group hapten 170–175
  as cell surface marker 169, 170
  as Lewis antigens 170, 171, 175, 185, 186, 189
  changes in transformed cells 187, 188
  definition 169
  fatty acid profile 176
  occurrence 169–171, 188, 189
  of erythrocytes 170–175
  of gastrointestinal tract and glandular tissue 175–177
  of tumor tissue 177–188
  structures 178, 179
Fucose
  incorporation into glycolipids 170, 188, 189
  in glycolipids 128, 167–189
  in glycoproteins 169, 181
  in neoproteolipids 115, 116, 118, 124, 130
Fucosidosis 170, 171
Fucosphingolipids, see Fucolipids

Galactosamine 124, 172, 176
Galactose
  effect on fatty acid oxidation 59
  in glycolipids 170, 172–180, 183, 185, 186, 189, 242
  in glycoproteins 180, 181
  in neoproteolipids 115, 116, 118, 124, 130
Gall bladder, cancer of 330
Gangliosides 169
  biosynthesis 154
  changes in cultured and transformed cells 154, 168, 189, 255
  immunological activity 229
  in brain 229, 230, 243
  in brain tumors 128, 242, 243, 252
  in cerebrospinal fluid 245
  in myelin 230
  in neonatal and adult liver 128
  in neoproteolipids 129
Gardner lymphosarcoma, see Lymphosarcoma

Gas-liquid chromatography,
  see Chromatography
Gastric cancer, see Stomach cancer
Gastric juice, sulfated glycoprotein in 187
Gastric mucosa
  adenocarcinoma 183, 184, 186
  blood group glycoproteins 181
  blood group polysaccharides 180
  fucolipids 169, 171, 177, 179
  glycolipid synthesis 183, 184, 186
Gastrointestinal cancer,
  see Colon, Intestinal, Stomach cancer
Gastrointestinal tract
  adenocarcinoma 182–184, 186, 189
  embryonic, antigen 188
  ether-linked lipids 4
  experimental production of tumors in 343
  fucolipids 169, 171, 175–177, 189
  tumor antigens 187, 188
Gel filtration, see Chromatography
Gene expression
  role of cholesteryl 14-methylhexadecanoate 197–222
Genetic disorders affecting myelination 236
Genetic factors in cancer 342
Geographical distribution of cancer 330–342
Geranyl pyrophosphate 282
Glial cells
  desmosterol in 250, 256–259
  fatty acid composition 255
  phospholipids 242
Glial cell lines
  C_6 254–257, 259
  EA-285 259
Glial cell membranes 228, 231
Glioblastomas 4, 232, 238, 240, 242, 243, 246
  see also Brain tumors
Gliomas 4, 9, 11, 231, 239–259
  see also Brain tumors
Globosides 169, 171, 173, 174
Glucagon 278, 283, 285, 286

Glucokinase 274
Gluconeogenesis 292
Glucosamine 172, 175, 181, 189
Glucose
  effect on fatty acid
    biosynthesis 45
    incorporation 60, 65
    mobilization 52
    oxidation 59, 64
    release 66
    utilization 58
  in glycolipids 170, 172–179, 185
  in neoproteolipids 115, 116, 118, 124, 130
  incorporation into
    cholesterol 148
    fatty acids 44–46, 50, 254, 255
    glycerol 49, 60
    lipids 30, 49, 50, 60, 61, 145, 254
Glucosyl transferase 154
Glutamic acid 127
Glutathione 25
sn-Glycero-3-phosphorylcholine 26
Glycerol 17, 25, 49, 60, 310, 312
Glycerol-3-phosphate 17, 18, 28, 45, 60
  acyl and alkyl derivatives 20, 22, 23
Glycerol-3-phosphate dehydrogenase
  in cultured cells 144, 147
  in tumors 27, 60, 144
Glycerolipids, see also individual classes
  and cancer, reviews on 3
  biosynthesis 17–23, 28–31, 49, 144–147
  catabolism 23–27
  chromatography, see Chromatography
  occurrence
    in brain 230, 231
    in cultured cells 140–144
    in neoplastic cells 3–16
    in tumors of the nervous system 239, 241, 242, 252, 253
  regulation, in cancer cells 27–31
Glyceryl ethers, see Ether glycerolipids
Glycocalyx 170
Glycolipids
  blood group 167–189
  fucose-containing, see Fucolipids

in cultured cells 128, 129, 153, 154, 157, 187, 188, 255
in liver 128
in membranes 168–177, 187–189
in neoproteolipids 117, 118, 124, 127, 129
in nervous system 228, 230
in tumors 128, 129, 180–186, 239, 242–244, 246, 252, 260
synthesis 49, 168, 173, 182–186, 189
Glycolysis 45, 59, 274
Glycoproteins 168, 169, 172, 179, 180–187
Glycosphingolipids, see Glycolipids
Glycosyltransferases 168, 184
Gonadotrophin 327
Growth
and neoplasia 230, 344
autonomous, of cell populations 322, 323
malignant 217–222
of cells in culture 47, 136–138, 151, 255, 257–259
of nervous tissue 234, 236
of tumors 52, 144, 217–222, 241, 253, 254, 259, 275, 309
Growth factors 136, 138
Growth hormone 285
Growth inhibitors 99
Growth medium, see Culture medium
Guanosine triphosphate 211, 212, 214, 216
Guinea pig
feedback control of cholesterol synthesis 275, 291
leukemia 275, 291
Gynecological cancer
and serum lipoproteins 94, 96, 97

Hamster
astrocytes and astrocytoma cell cultures 255
cell lines,
see BHK21 cells, Fibroblasts
glycolipids of 153
tumors grown in 252, 253

Hapten
amino sugar 181
blood group 170–175, 180, 187
carcinoembryonic antigen 187
fucolipid 170–175, 178
glycolipid, X-hapten 128, 180
lipid, see Cytolipin-H
Harderian gland tumors
ether-linked lipids 4
Heart
fatty acid metabolism 59
phospholipid composition 140
HEK cells, lipid synthesis by 145
HeLa cells
fatty acids 47, 54, 143, 255
fucolipid 169
growth of 47, 50, 99
in lipid-deficient medium 47, 50
lipid synthesis 145, 147, 150
lipid uptake 155, 156
Hemagglutination 180–182, 185, 187
Hematosides
in brain tumors 128, 252
in liver and hepatomas 128
Heme formation 292
Heparin 53, 63
Hepatectomy, partial 106, 219
Hepatoma
98/10 44
98/15 58, 64
ascites 4, 12, 14, 50, 51, 58, 218, 295
BW 7756 48, 271
cholesterol biosynthesis 49, 148, 150, 270–272, 277, 278, 280, 286–288, 290, 291, 294, 295
cholesteryl esters 217, 218
diacylglycerols 9
Dunning 23
effects on plasma lipoproteins 57, 67, 102, 105, 106
ether-linked lipids 6, 7, 15, 16, 28, 30, 31, 144
fatty acid biosynthesis 44, 45, 48–50, 67, 151, 272–275, 277, 283, 288, 289, 295

fatty acid composition 4, 7, 11, 12, 273, 274
fatty acid oxidation 58, 59, 64
glycerol-3-P dehydrogenase 28
glycolipids 128
growth of 52
H-4 and H-35 271
heme synthesis 292
HTC cells 49, 50, 145, 148, 280, 294, 295
incidence, effects of dietary fat 318, 319, 324, 344
JTC-16 rat ascites 50
Morris 6, 7, 14, 16, 23, 27, 29, 45, 46, 48, 49, 51, 59, 67, 102, 105, 106, 113, 114, 123, 124, 128, 130, 271, 273, 294
neoproteolipid 113, 114, 123, 124, 130
Novikoff 7, 11, 23, 26, 30, 45, 52, 59, 114
phospholipase, A_2 27
phospholipids 11, 12, 14–16, 23, 145
primary 253, 271, 277, 295
Reuber H35 59
spontaneous 271, 295, 318
sterol metabolism 253
transplanted 253, 277, 294, 295
triacylglycerols 4, 9
Yoshida ascites 4, 12, 218
Zajdela ascites 14
Heteroploid cell lines, see Established cell lines
Hexadecanol uptake by Ehrlich cells 57
Hexokinase in hepatoma 274
Hibernomas 5, 14
High density lipoproteins
in ascites plasma 53
in cancer patients 82–85, 87, 89–96, 98–100
in cerebrospinal fluid 245
in human serum 77–79
in tumor-bearing animals 51, 67, 100, 102–107
metabolism of 63, 98, 99
neoproteolipids in 112, 124–126
High fat diets, see Dietary fat

HMG-CoA, see 3-Hydroxy-3-methylglutaryl coenzyme A
^3H_2O, measurement of fatty acid biosynthesis 50
Hodgkin's disease
correlation with fat consumption 330
lipoproteins in 87, 93
Hog stomach mucosa, fucolipids 177, 179
Hormones, see also individual hormones
binding sites, destruction by aflatoxin 291
effects on cholesterol synthesis 278
effects on ether-linked lipid levels 31
fat-mobilizing 66
influence on tumorigenesis 323, 325–327, 340–344
Host lipids 50–56, 67, 217, 253, 273
HTC cells, see Hepatoma
Human
blood group activities 171
brain 232–234
cancer
mortality in relation to diet 309, 329–345
role of nutrition, reviews on 308
serum lipids and lipoproteins in 52, 79–100, 106, 107, 124–126
cultured cells 136, 154
see also Cells in culture, Fibroblasts
erythrocytes 57, 169–175, 177, 179
platelets, fatty acid uptake 58
serum lipoprotein spectrum 77–79
tumor lipids 3–9, 13, 16, 114, 170–172, 176, 178, 180–187, 239–243, 253
lipid synthesis 49, 271, 295
morphology 249, 250
tumors of the nervous system 238–248
Hydrocarbons
carcinogenic, see Carcinogens
in brain 231
in meningioma 239
Hydrolases 62, 168, 183–185, 237
Hydroxylase, tetrahydropteridine-dependent 24

Subject Index

3-Hydroxy-3-methyl-glutaryl
  coenzyme A
    reductase 148, 278–282, 286, 287
    synthetase 282
4-Hydroxyoctadecasphinganine 185
3-Hydroxyxanthine 114
Hyperlipemia in cancer 100
Hyperlipidemia during tumor growth 51, 52
Hyperlipoproteinemia 88, 97, 98, 102
Hypocholesterolemic agents,
  see also Drugs
  effect on developing nervous system 237, 238
Hypophyseal tumors, effect of dietary fat 319

^{131}I, induction of hypophyseal tumors 319
Immune competence, relation to high density lipoproteins 99
Immunity reaction, protein synthesis in spleen 216
Immunodiffusion, radial, of lipoproteins 95
Immunoelectrophoresis
  of ascites plasma 53
  of serum lipoproteins 89
Immunofluorescence 182
Immunoglobulins of lymphocyte 99
Immunological activity
  of blood group substances 181
  of brain lipids 229
Immunological resistance in cocarcinogenesis 323
Inflammatory reaction 218, 219
Incubation medium, see Culture medium
Initiation of tumorigenesis 322–324, 343, 344
Insulin and lipid biosynthesis 278, 283, 285, 286, 289
Insurance statistics, human cancer 329, 330
Intestinal bacteria, role in carcinogenesis 341, 342, 345

Intestinal cancer
  and dietary fat 309, 330, 332, 335, 337, 339, 343
  correlation with animal protein intake 343
  effect on serum lipoproteins 87, 92, 95, 96
Intestinal fat 341, 345
Intestinal mucosa 19, 170, 171
Intestine
  cholesterogenesis 288
  cholesterol uptake 253
  fucolipids 171, 175–177
  neoproteolipid 119, 120
Intracellular membranes, see Membranes
Irradiation
  cell transformation by 136
  of dietary fat 310, 312
Isoantigens 182, 189

Japanese
  cancer mortality relative to Americans 339
  fecal sterol excretion 341
  linoleic acid intake 338
  serum cholesterol and lipoproteins 95
Jensen sarcoma, see Sarcoma
JTC-16 rat ascites hepatoma,
  see Hepatoma

Karyotype of cells in culture 136, 137
KB carcinoma cells, see Carcinoma
Ketone bodies 281, 291, 329
Kidney 216, 217
  tumors 292
Krebs carcinoma, see Carcinoma
Krebs cycle enzymes 274

L cells, see Fibroblasts
L-5178 cells 145, 156
Lactate 59
  conversion to fatty acids 45
  effect on fatty acid oxidation 64
β-Lactoglobulin 57
Lacto-N-fucopentaose I 177

Lacto-N-fucopentaosyl-III-ceramide 185
Landschutz ascites carcinoma cells,
   see Carcinoma
Lanosterol 232
Lard 310, 312–315, 318–321
Larynx
   neoplasms, correlation with dietary
      fat 335
   squamous cell carcinoma 182
Latent period 314, 324
Lauric acid 58, 60, 61
   ethyl ester 310, 312
Lecithin 60, 65, 67, 147, 155, 255, 280
   see also Phosphatidyl choline
Lecithin-lysolecithin cycle 57
Lectins 151, 187
Leiomyosarcoma 86, 92, 125, 126
Leptomeninges, fatty acid composition
   240
Lettre-Ehrlich tumor 54
Leucine 106, 206
Leukemia
   and dietary fat 319–322, 330, 335,
      338, 344
   effect on serum free fatty acids 97
   effect on serum lipoproteins 81, 87,
      89, 92, 93, 96
   serum neoproteolipid in 125, 126
Leukemic cells
   lipids 4, 13, 140, 152
   lipid synthesis 23, 49, 60, 145, 275, 291
   lipid uptake 156
Lewis antigens, see Antigens
Ligase, see Amino acid-tRNA ligase
Linoleic acid
   conversion to longer chain fatty
      acids 47, 143, 255
   effect on fatty acid synthesis 285
   in adipose tissue 338
   in ascites plasma 53
   in brain tumors 240, 242, 253
   in cerebrospinal fluid 245
   in cultured cells 143
   in dietary fats 329, 338
   in phospholipids 12, 242
   in serum free fatty acids 97

oxidation 60, 65
synthesis 46
uptake 55, 58
Lipase 26, 65, 155
   pancreatic 10, 23, 26
Lipemia 76, 77
Lipid, see also Fat
   antioxidants 250
   biosynthesis 17–23, 28–31, 49, 50,
      60–62, 144–147, 241–243, 254, 255
      defective control of 48, 49, 148–151,
         269–296
   catabolism 24–28, 30, 58–60, 64, 65
   composition
      of blood and cerebrospinal fluid
         245, 246
      of cultured cells, see Cells in culture
      of normal and neoplastic tissues
         3–16, 170–180, 230–243, 252, 253
   metabolism
      in cultured cells 144–151, 254, 255
      in developing nervous system 229
   peroxidation 310, 329
   transport 154–156
Lipids, see individual lipid classes
   in cultured cells 135–157
   in tumors of the nervous system
      227–260
   supplied by the host 50–54
Lipogenesis, see Lipid biosynthesis
Lipoidoses 97
Lipolytic enzymes 8, 10
Lipomas 5, 14
Lipopeptide 127
Lipoprotein lipase 63, 68
Lipoproteins, see also individual classes
   effect of drugs on 43
   effects on cholesterol biosynthesis
      279–281
   in ascites plasma 53, 54, 56, 62–64
   in cerebrospinal fluid 245
   in children 77, 81, 84, 85, 87, 93, 96
   in culture medium 49, 57, 68, 148, 155
   in normal and tumor-bearing animals
      100–106, 219
   in relation to age 95, 101, 104

in relation to cancer 76–107
in subjects with family history of
  cancer 80–85, 89–92, 98, 107
neoproteolipids in 112, 124–126
part of glial carcinoembryonic
  antigen 229
spectrum of, in serum 77–79
synthesis 105, 106
α-Lipoproteins,
  see High density lipoproteins
β-Lipoproteins,
  see Low density lipoproteins
Liver
  cancer 98, 106, 291, 330, 334, 335,
    337, 338
  chick embryo 231, 232, 237
  cholesterol
    biosynthesis 233, 270–272, 276,
      278–283, 295
    esterification 221
    uptake 253
  enzymes 23–28, 46, 200, 207, 208,
    210, 211, 220, 280, 292
  fatty acid
    binding protein 67
    biosynthesis 45, 48, 50, 272–276,
      283–286, 294
  function, impairment 219
  lipid biosynthesis 269–296
  lipids 4, 6, 8, 9, 12, 14–16, 140, 171,
    217–222, 253
  lipoproteins 281
  peptidolipids 127
  peroxidase 281
  precancerous 273, 276, 293, 295
  regenerating 11, 14, 105, 277
  subcellular preparations 27, 58, 140,
    197, 199, 208, 209
  tumors, see Hepatomas
L-M fibroblasts, see Fibroblasts
Low density lipoproteins
  in ascites plasma 53
  in cancer patients 82, 84, 86–89, 91
  in cerebrospinal fluid 245
  in human serum 77–79
  in tumor-bearing animals 51, 100–105

  metabolism 99
  neoproteolipids 125
Low fat diets, see Dietary fat
Lung
  cancer 86, 92, 114, 186, 333, 335, 337,
    342
  cholesterol 14-methylhexadecanoate
    in 217, 220
  neoproteolipid in 119, 120
  nodules 310
  tumors 16, 186, 239, 319–321, 344
Lymph nodes
  cholesterol synthesis in 275
  lipids of 5, 13
Lymphangiomatosis 93
Lymphatic leukemia, see Leukemia
Lymphoblastoid cell culture 137
Lymphoblasts (MBIII) 155, 156
Lymphocytes 99, 146
Lymphoma 81, 93, 97
Lymphosarcoma 23, 54
  in relation to dietary fat 319, 330
  lipid composition 7, 13, 16
Lysolecithin 139, 142
Lyso-lipids 18, 28
Lyso-phosphatidyl choline 13, 20–22,
  24, 26
Lyso-phosphatidyl ethanolamine 24, 26
Lyso-phospholipase 24, 27
Lyso-plasmalogens 26, 30

Macrophages 47
MAF human fibroblasts, see Fibroblasts
Magnesium 19, 23, 25, 26, 284
Malignancy 323
  and cell surface changes 153, 167
  and lipid composition 13, 128, 129,
    168, 189, 236, 243, 246
  change of blood group antigen 180,
    182
Malignant cells 256, 257, 290
Malignant growth 217–222
Malignant transformation,
  see Transformation
Malignolipin 127
Malonyl coenzyme A 46, 48, 285, 289

Mammary gland 4, 5, 13, 28, 45,
    324–327, 344
Mammary tumors, *see also* Breast cancer
    adenocarcinoma 52, 114, 313
        Barrett 45
        TA3 44
    blood group antigens 182
    carcinoma 4, 5, 13
    effects of dietary fat on 309, 313–319,
        324, 328, 337, 340, 344
    growth of 52
    lipids in 4, 5, 13, 28, 44, 45, 114
Manganese 53
Margaric acid, methyl ester 199
Mass spectrometry
    of fatty acids 199
    of sterols 232, 240
Mast cells 147
Mastectomy 80, 125, 126
    effect on plasma lipoproteins 91, 95
MBIII cells, *see* Lymphoblasts
Melanoma 92
    malignant 342
    neoproteolipid in 114, 125
Membrane-bound enzymes 62, 151, 279
Membranes
    alterations in transformed cells
        151–156, 189
    brush border 170, 171, 177
    chorioallantoic 45, 141, 145
    erythrocyte 13, 57, 170–175, 177, 189
    formation 61, 238
    fucose in 169
    glial cell 228, 231
    in cultured cells 151, 152, 157
    intracellular 170, 188, 290, 291
    isoantigens 189
    lipids of 58, 63, 68, 152, 157, 228
    microsomal 8, 280, 281, 291
    microvillus 170, 171, 177
    plasma 8, 123, 152, 168, 170, 176,
        177, 187–189, 228, 291
    receptors in 57, 58
    ribosomal 228, 291
    Schwann cell 231
    tumor 12

tumor cell 167, 168
yolk sac 231, 232, 237
Meninges, lipids of 239
Meningiomas 4, 11, 238–241, 243, 244,
    246
    see also Brain tumors
MER-29, *see* Triparanol
Messenger RNA 208, 209
Metastatic cancer 80, 96, 114, 125, 126,
    128, 168, 180, 182, 239, 319
Methionine incorporation into protein
    197
Methylation pathway of phospholipid
    biosynthesis 23, 147
Methylcholanthrene 252, 310–312,
    320–322, 325, 326
3′-Methyl-4-dimethylaminoazobenzene
    276, 293
14-Methylhexadecanoic acid 199, 200,
    205, 206
    methyl ester 199, 206
Methylnitrosourea, *see* Nitrosourea
Methyl pentose 172
    see also Fucose
Mevalonate kinase 282
Mevalonic acid 149, 150, 231, 232, 240,
    241, 281, 282
Microsomal
    cholesterol synthesis 281, 291
    membranes, *see* Membranes
    protein 289
    system for fatty acid synthesis 46
Microsomes
    in protein synthesis 199
    enzymes 19, 23–29, 288
    lipids 5, 9, 12, 14, 217, 242
Microvillus 170, 171, 177
Migrant populations, cancer incidence
    342, 343
Milk 281, 330
Minimal deviation hepatoma,
    see Hepatoma, Morris
Mitochondria
    calcium uptake and release 283
    enzymes 47, 48, 281
    fatty acid oxidation 59

fatty acid uptake 58
lipids 4, 5, 9, 12–14
Mitochondrial
  function 47, 138
  system for fatty acid synthesis 46
Mitosis and lipid biosynthesis 146
Mixoploid cell line 136
  see Established cell lines
Models for brain tumor research 248–260
Molecular species of glycerolipids 3, 10
Monoacylglycerols 8–10
Monoglyceride uptake 155
Morris hepatoma, see Hepatoma
Mortality, cancer, see Cancer
Mouse
  acatalasemic 281
  blood plasma antisera 53
  cells in culture 30, 137, 147, 154
  embryo cells 148
  epidermis, phospholipids of 13
  essential fatty acid deficient 46, 47
  fasted 53
  fibroblasts 63, 169, 256, 257
  genetic disorders affecting myelination 236
  hepatoma-prone 276, 277, 293
  leukemia cells 145, 156
  liver 9, 217, 276, 280
  mammary glands, lactating 28
  mammary tissue 45
  preputial gland, ether lipids in 31
  tissues, cholesteryl 14-methylhexadecanoate in 217
  toxic effects of AY-9944 254
  tumors 198
    as models for brain tumor research 248, 253, 254, 259
    ascites forms 23, 49, 52–55, 60, 64, 67, 291
    defective control of cholesterol synthesis 271, 295
    effect on body weight 51
    embryonal antigens in 293
    fatty acid synthesis in 44, 45, 48, 272, 273
    incidence, effects of dietary fat 229, 309–322, 327, 328, 344
    lipids in 4, 6, 9, 14, 114, 123, 130, 217
Murine virus, see Virus
Muscle cells 216
Myelin 228, 230, 231, 238, 241, 242, 260
Myelination
  genetic disorders affecting 236
  lipid changes during 230, 234, 237, 242
  retardation of 238
Myeloma 86–88, 97, 102
Myristic acid 58

NAD, cofactor in oxidation of fatty alcohols 29
NADH, cofactor in glycerolipid biosynthesis 19, 22, 23, 28
NADPH
  generating system 45, 46
  cofactor in glycerolipid biosynthesis 18, 19, 22, 23, 28
  cofactor in reduction of fatty acids to fatty alcohols 29
Nakahara-Kukuoka sarcoma, see Sarcoma
Neoplasms, see specific tissues and specific types
Neoplastic cells, see also Tumor cells
  ester and ether glycerolipids in 3–32
Neoproteolipids
  composition 113, 115–118, 124, 127, 129, 130
  definition 113
  in erythrocytes 123, 129
  in serum 99, 112, 119, 120, 123–126, 129, 130
  in tumors 113, 114, 118–124, 129–131
  thin-layer chromatography 118–121
Neoproteolipid-S 113, 114, 123–127, 129, 130
Neoproteolipid-W 113–124, 127, 129, 130
Nephrotic rats 216

Subject Index

Nerve
  lipids 228
    during development 230–238
    neoplastic 251
  trigeminal, see Trigeminal nerve
Nervous system
  developing 229–238, 250, 260
  lipids of 227–260
    reviews on 231
  tumors of 228, 229, 238–260
Neural lipids
  developmental changes in 230–238
Neuraminic acid, N-acetyl,
  see Sialic acid
Neurinoma 238
  glycerolipids in 4, 11
Neuroblastoma 238
  see also Brain tumors
  cell lines 254, 255
  effect on serum lipoproteins 86, 92
  neoproteolipid in 125
Neutral lipids,
  see also specific lipid classes
  in cultured cells 139, 141, 142, 168
  in lipoproteins 100
  in neoproteolipid 130
  in nervous tissue 231
  in tumors of the nervous system 239
Nitrosourea
  1,3-bis(2-chloroethyl)-1- 248
  ethyl 249, 250
  methyl 249, 252
Nitrosourea-induced tumors 250–252, 254, 257, 259
Noradrenalin, effect on HMG-CoA reductase 278
Norepinephrine 66
Novikoff hepatoma, see Hepatoma
Nuclei, phospholipid composition 14
Nutrition
  in relation to cancer 308, 335
  of cultured cells 138

Obesity and cancer incidence 330, 344
Octadecenoic acids 30
Octanoic acid 58, 59, 61

Ocular orbit, tumors 322
Oils, dietary, see Dietary fat
Oleic acid
  biosynthesis 46
  conversion to longer chain fatty acids 47, 60
  in ascites plasma 53
  in cultured cells 142, 168
  in phospholipids 12
  in tumors 12, 47
  incorporation into lipids 60–62, 146
  oxidation 60, 65
  uptake by cells 55, 58, 65
Oligodendroglial cells 231, 238
Oligodendrogliomas 238
  see also Brain tumors
  glycolipids 242, 243
  phospholipids 241
  sterols 240, 250, 251, 256
Olive oil 314–316, 329
Oncogenic virus, see Virus
Oncornavirus 187, 189
Oophorectomy 80
Optical rotation of blood group substances 180
Oral epithelium, neoplastic 182
Osmotic shock 48
Ovarian hormone production 327
Ovarian tumors
  adenocarcinoma 182
  carcinoma 86, 92, 94, 96, 125
  cholesteryl 14-methylhexadecanoate in 218
  correlation with dietary fat 330, 335
  Dutch 218
  effect on serum lipoproteins 86, 92, 94, 96
  fatty acid biosynthesis in 44, 45
  isoantigens in 182
  neoproteolipid in 125
  phospholipid in 13
Ovaries 326, 327
Oxidase, mixed function 19
Oxidative phosphorylation 47, 59
Oxidoreductases 18, 22, 24

Oxygen consumption by tumor cells 59, 64, 65

^{32}P incorporation 17, 146, 152, 153, 241, 254, 255
Paired feeding 322, 327, 344
Palmitic acid
  biosynthesis 46
  conversion to other fatty acids 61
  in ascites plasma 53, 55
  in glycerolipids 4, 9, 12
  in serum free fatty acids 97
  incorporation into lipids 60–62, 64–66
  inhibition of fatty acid synthesis by 48
  oxidation 44, 59, 60, 65, 274
  uptake by cells 55–58
Palmitoleic acid 46, 47
Palmityl coenzyme A 48
Pancreas
  adenocarcinoma 128, 186
  fucolipids in 128, 177, 178, 186
  neoplasms and dietary fat 309, 330, 335, 342
Parotid cancer 87, 92
Peanut meal 322
Pentose phosphate pathway 45
Peptide chain
  initiation 208, 209
  elongation 208, 209, 212, 214, 216, 220
  termination 208
Peptidolipids 127
Peptidyl transferase 214–216
Peptidyl-tRNA translocase 212
Perese tumor, triglyceride 252
Peripheral nerve, tumors 238–239
Peroxidase from beef liver 281
Peroxidation of lipid, see Lipid
pH
  effects on cell metabolism 30, 57
  effects on enzymes 284, 289
pH 5 enzymes 200, 201, 203, 207, 220
Pharynx, neoplasms, and dietary fat 335
Phenylalanine 210, 213
Phenylalanine-tRNA 214
Phosphatases 27, 152, 284

Phosphate transport 152
Phosphatidic acid 27, 230
  biosynthesis 18, 20, 153
Phosphatidyl choline, see also Lecithin
  biosynthesis 20–23, 146, 152, 153
  occurrence 11–16, 139, 140, 142, 230, 241, 242
Phosphatidyl ethanolamine
  biosynthesis 20, 146, 153
  catabolism 24
  conversion to phosphatidyl choline 23
  occurrence 11–16, 139, 142, 230, 241, 242, 245
  turnover 152
Phosphatidyl glycerol 11, 230
Phosphatidyl glycerophosphate 11
Phosphatidyl inositol
  biosynthesis 146, 153
  occurrence 12, 139, 142, 230, 241
Phosphatidyl serine
  biosynthesis 146, 153
  occurrence 139, 142, 230
  turnover 152
Phosphoenolpyruvate carboxykinase 292
Phosphohydrolases 22, 27
Phospholipase A 10, 21, 24, 26
Phospholipase C 10, 16, 19, 24, 27
Phospholipase D 24, 27
Phospholipids, see also individual classes
  biosynthesis 19–23, 144–147, 152, 153, 241, 255
  catabolism 23–28, 30
  fatty acid composition 11–13, 16, 242
  in ascites cells 12, 13, 16, 61, 62, 66, 126
  in ascites tumor plasma 53
  in brain 228, 230, 241, 242
  in cerebrospinal fluid 245
  in cultured cells 140–142, 144–147, 152, 153, 155–157, 168
  in lipoproteins 62, 86, 87, 94, 100
  in liver 11
  in neoproteolipids 116, 117, 130
  in serum 51, 245
  in tumors 11–16, 229, 239, 241, 242, 252

in Rauscher leukemia virus 5
  metabolism 152, 153, 260
  molecular species 10
  turnover 152, 157
  uptake by cells 62, 155, 156
Phosphorus in neoproteolipid 115, 116
Phosphotransferase, see Transferase
Phytoagglutinins 168
  see also Wheat germ agglutinin
Pineal gland, tumors of 238
Pituitary gland
  tumors of 238, 319
  uptake of estradiol 326
Plant growth retardants
  effect on cholesterol synthesis 272, 287
  effect on fatty acid synthesis 273
Plant lectins 151, 187
Plant sterols 280
Plasma
  desmosterol 240, 254
  lipoproteins 77–79
    see also Lipoproteins
  membranes, see Membranes
  phospholipids 13
Plasmalogenic acid 11
Plasmalogens,
  see also Alk-1-enylacylglycerolipids
  biosynthesis 19
  cleavage 25, 26
  in brain 230
  in neoplastic cells 11
Platelets 58
Plating efficiency 136
Polymerase, see RNA polymerase
Polyoma virus, see Virus
Polypeptides in neoproteolipid 115, 124, 126, 130
Polyphenylalanine, synthesis 213
Polyphosphoinositides 255
Polysomes 215, 279
Porcine intestine, fucolipids 171, 175, 176
Portacaval shunt 294
Precancerous changes 293, 294
Pregnancy 94, 95, 249, 326

Premalignant lesions 187
Pretumorous tissue, lipid metabolism in 275, 276, 295, 296
Preputial glands
  ether lipids 30
  tumors, glycerolipids in 6, 7, 14, 16, 28
Primary culture, definition 136
Progesterone, effect on mammary tumors 325–327
Prolactin, effect on mammary tumors 325
Promoting action 323
  of dietary fat 325–329
Prostate gland, cancer
  and dietary fat 309, 330, 331, 335, 337, 339, 340, 343, 345
  effect on serum lipoproteins 95
Protein
  animal 330, 338, 339, 341, 343
  dietary, see Dietary protein
  ribosomal 216
  sterol carrier 283
  synthesis 106, 197–217, 220–222
  vegetable 338, 339
Proteolipids 113
  see Neoproteolipids
Pulmonary adenoma
  disaturated phosphatidylcholine in 16
Pyrimidine synthesis 292
Pyruvate incorporation into fatty acids 45
Pyruvate carboxylase 292

Rabbit
  anti-human serum 89
  perfused aorta 63
Rapeseed oil 316, 329
Rat
  age and sterols in developing nervous system 234–238
  tissues, distribution of tritiated estradiol 325, 326
  tumor-bearing
    carcass lipids 51, 52

cholesteryl 14-methylhexadecanoate in 217, 218
neoproteolipid in 118–123, 129, 130
serum lipoproteins in 100–106
tumors
  as models for brain tumor research 249–252
  defective control of lipid biosynthesis 271–277, 295
  incidence, effects of dietary fat 309, 313–322, 344
  neoproteolipid in 113, 114, 123, 124
Rauscher leukemia virus, see Virus
Rectal cancer 95
  and dietary fat 330, 332, 335, 337, 341
  incidence in emigrants 343
Red blood cells, see Erythrocyte
Reductases 22, 23, 29, 233, 254, 257–259
Reduction procedures, chemical 10
Refeeding
  effect on cholesterol synthesis 271
  effect on fatty acid synthesis 48, 272–274, 276, 284–286, 289
Regenerating liver, see Liver
Regression of tumors, effect on cholesteryl 14-methylhexadecanoate content 218
Reticuloendotheliosis, effect on serum lipoproteins 81, 87, 93
Reticulum cell sarcoma, see Sarcoma
Reuber hepatoma H35, see Hepatoma
Rhabdomyosarcoma MC-1A 44, 64
Riboflavin and tumor incidence 319, 324, 329
Ribosomal membrane 288, 291
Ribosomal proteins 216
Ribosomes 201, 208–220, 279, 291
Ridgway osteogenic sarcoma 114
RNA polymerase, DNA dependent 208
Rodent, see also Mouse, Rat
  carcass lipids 51
  tumors, growth rate 52
Rous sarcoma virus, see Virus

Sarcoma
  37 44, 64, 217
  180 12, 13, 23, 46, 52, 113, 114, 123, 124, 130, 148, 217
  BP8/C3H ascites 16
  derived transplantable 13
  effect of dietary fat on 319, 320, 344
  Ewing's 89
  Fischer rat 2
  induced by bile acids 341
  Jensen 14, 114, 127
  Nakahara-Fukuoka 4, 9, 128
  neoproteolipid in 113, 114, 123, 124, 130
  of meninges 238
  reticulum cell 6
Sarcomatosis 88
Schwann cell 231, 250
Schwannomas 231, 238, 249, 251
Serum
  albumin, see Albumin
  cholesteryl 14-methylhexadecanoate in 218, 219, 222
  enzymes converting glycolipids 183
  factors affecting cell growth 99, 136–138
  for determination of antigenic activity 180, 181
  in culture media, see Culture medium
  lipids 170, 171, 245
  lipoproteins, see Lipoproteins
  neoproteolipids, see Neoproteolipids
Shionogi carcinoma 115
  see Carcinoma
Sialic acid 152, 169, 172, 180, 181
  see also N-acetylneuraminic acid
Skin
  absorption of carcinogens 323, 324
  desmosterol in 236
  fibroblasts derived from 138, 259
Skin cancer
  and dietary fat 333–335, 337, 340
  geographical distribution 340
  isoantigens in 182
Skin tumors, effects of dietary fat 309–313, 314, 319, 322–325, 327–329, 344
Somantin, see Growth hormone

Soybean oil 316
Sphingolipids, see Glycolipids, Sphingomyelin
Sphingomyelin 13, 155
  biosynthesis 147
  in cultured cells 139, 140, 142, 146, 147
  in developing brain 130
Sphingosine bases in neoproteolipids 112, 115–118, 124, 130
Spinal cord tumors 239, 246
Spleen 216, 217, 220
  neoproteolipid in 119, 120, 123, 130
Squalene 232, 233, 239, 283
Squamous cell carcinoma, see Carcinoma
Stearic acid
  biosynthesis 46
  in ascites plasma 53
  in glycerolipids 4, 9, 12
  in serum free fatty acids 97
  oxidation 60, 65
  uptake by Ehrlich cells 58
Stereospecific analysis 7, 10, 16, 60
Steroids, fecal excretion 341
Sterol carrier protein 283
Sterol esters, see Cholesteryl esters
Sterols, see also Cholesterol, Desmosterol
  biosynthesis 148–151, 231–234, 240, 241, 270–272, 275–283, 286–288
  in cerebrospinal fluid 245–248
  in cultured cells 139–142, 148–151, 155–157, 255–260
  in nervous system 231–238
  in tumors of nervous system 240, 241, 250–254
Stilbestrol 314, 315, 325
Stomach cancer
  blood group fucolipids and glycoproteins in 180–185
  effect on serum lipoproteins 95
  in relation to body weight 330
  in relation to dietary fat 334, 335, 338
  incidence in emigrants 343
Stomach mucosa, see Gastric mucosa

Submaxillary gland tumors 321, 322
Succinate dehydrogenase 47
Suckling rats 281
Sugar consumption and cancer mortality 330
Sulfatides 230, 242, 255
Sulfomucopolysaccharide 187
Sunflowerseed oil 316
Surgery, effect on serum lipoproteins 96
SV_{40} virus, see Virus
Synthetases 183, 206, 211

3T3 cells 30, 137, 152, 153, 168
TA3 carcinoma, see Carcinoma
Tallow 316
Tangier disease 98
Taper liver tumors 7, 16, 114
Temperature, environmental, and breast cancer mortality 330
Teratoma 238
Testes 16
Testicular carcinoma 114
Testosterone 30
Thin-layer chromatography, see Chromatography
Thromboplastic activity 13
Thymidine incorporation 251
Thymus 6, 99
Thyroid
  activity 236
  neoplasms, correlation with fat consumption 335
Thyroxine 278
Tissue culture 30, 135–157, 254–260
  see also Cells in culture
dl-α-Tocopheryl acetate 249
Tongue, squamous cell carcinoma 182
Tonsils 209, 210, 213
Trachea, neoplasms, and dietary fat 333, 335
Transcription of genetic message 208
Transfer RNA 200–207, 220
Transferases
  acyl 10, 17, 18, 20–23, 59
  glycosyl 154, 168, 185

nucleotidyl 208
peptidyl 214–216
phospho- 22, 23, 59, 274
phosphorylbase 10, 20, 23
Transferase I 209–213, 216, 217, 221
Transferase II 212–214
Transformation 322
  effect on antigenic components 182, 183
  in relation to cell surface changes 178
  malignant 136, 156, 157, 168, 188, 189
  spontaneous 137, 141, 142
  viral 136, 137, 140–142, 152, 153, 168, 188
Transformed cells, see Cells in culture
Translation of genetic message 208–217, 219
Translocase, peptidyl-tRNA,
  see Transferase II
Transport
  in relation to cell transformation 152–156, 168
Triacylglycerols, see also Triglycerides
  deacylation 23, 27
  fatty acid composition 4, 5
  occurrence 4, 5, 9
Trigeminal nerve
  sterols in 234–236, 250–252, 256
  tumors of 249–252, 256
Triglycerides, see also Triacylglycerols
  biosynthesis 60–62, 146, 254, 255
  fatty acid composition 253
  in chylomicrons 100, 101
  in cultured cells 139, 142
  in Perese tumor 252
  metabolism in liver 253
  reconstituted 310, 312
  uptake and utilization 44, 62, 63, 68, 155
Triparanol 237, 240, 246–248, 254, 256–258, 272, 287
Triton WR 1339 272
Trout hepatoma 271, 276, 295
Trypsin 169, 284
Tumor
  antigens, see Antigens

cells in culture 135–157, 187, 188, 254–260
  see also Cells in culture, Ehrlich ascites cells
development 308, 322, 344
growth 43, 52, 68, 217–222, 241, 253, 254, 259, 275, 309
incidence, effects of dietary fat 310–329
induction 248–251
  two-stage theory of 322, 323
lipids, see individual lipid classes
promotion 323, 325–329
regression 218
transplantation 218
Tumorigenesis, effects of dietary fat 308–345
Tumors,
  see also specific tissues and types
  benign, lipids of 11, 242
  classification 238, 239
  effects on serum lipoproteins 100–106
  fatty acid metabolism in 42–69
  malignant 11, 114, 239, 242, 243, 247, 249, 250
  mouse, see Mouse
  primary 80, 96, 114, 238, 239, 248–253, 259, 270, 277, 287, 319, 321
  rat, see Rat
  spontaneous 6, 114, 313–315, 318, 319, 344
  transplanted 114, 221, 252–254, 259, 270
Turpentine 218

Ubiquinone 282
Ultracentrifugation of lipoproteins 77–79
Ultraviolet light, induction of skin tumors by 310, 311
Unsaponifiable lipid 310, 312, 329
Urethan 16, 88
Urinary tract cells, A and B antigens in 182
Urine 182, 285

Subject Index

Uterus
  carcinoma 125
  malignant, glycerolipids of 5, 9
  neoplasms in relation to fat consumption 330, 335
  uptake of tritiated estradiol 326

Vanadium, inhibition of cholesterol synthesis 241
Vegetarians, fecal steroid excretion 341
Very high density lipoproteins 96, 97
Very low density lipoproteins
  in ascites plasma 44, 53, 62–64
  in cancer patients 81–86, 88, 92, 93
  in human serum 77–79
  in subjects with family history of cancer 91, 98
  in tumor-bearing animals 51, 67, 100, 101, 103, 105
  metabolism 63, 98, 99
  neoproteolipids 125
Viral action in carcinogenesis 323
Viral conversion 152
  reviews on 136
Viral transformation 136, 137, 141, 142, 153, 168, 187–189
Virus
  feline sarcoma-leukemia 188
  murine leukemia 187, 188
  murine sarcoma 187
  oncogenic 136, 137, 140, 152, 248
  oncorna- 187, 189
  polyoma 141, 142, 152, 153
  Rauscher leukemia 5

Rous sarcoma 45, 141, 142, 146, 189
  simian 256
  SV_{40} 140–142, 147, 168
Vitamin B_{12} 255
Vitamin E 249

Walker 256 carcinosarcoma
  cholesterol and cholesteryl esters in 217, 218, 221
  effect on liver enzymes 219
  effect on plasma lipids 51, 52, 100–102, 104, 105
  ether lipids in 16
  lipid complex in 127
  lipoprotein lipase in 63
  neoproteolipid 112–115, 119–123, 129, 130
Wallerian degeneration, triacylglycerols in 4
Water, tritiated, incorporation into cholesterol 150
Wheat germ agglutinin 185–187
Wheat germ oil 310
WI-38 cells, see Fibroblasts
WI-38VA13A cells, see Fibroblasts

Xanthomatosis 87, 88
X-ray 96

Yolk sac membrane, see Membranes
Yoshida ascites hepatoma, see Hepatoma

Zajdela ascites hepatoma, see Hepatoma
Zimmerman ependymoblastoma 253

144253
v.10-